PUBLIC HEALTH

Health and social welfare are the cornerstones to the wellbeing of the community as a whole. They are underpinned by those networks of relationships among persons, firms, and institutions in a society, together with associated norms of behaviour, trust, cooperation, etc., that enable a society to function effectively. The **SAGE Library of Health and Social Welfare** focuses on the critical and reflective nature of theory, policy and practice in health and social welfare. By bringing together seminal works from across the health and social welfare spectrum, from social work and nursing to social care and community work, these multi-volume collections encourage dialogue and debate on those issues relating to public and government intervention into social life.

Chris Heginbotham is Emeritus Professor of Mental Health Policy and Management at the University of Central Lancashire, a Fellow of the Royal Society for Public Health and a Board member of the Global Health Equity Foundation. In a career spanning over 45 years he was, for most of the 1980s, Chief Executive of Mind, the National Association for Mental Health, then Fellow in Health Services Management at the Kings Fund (1989–2003), and subsequently Chief Executive of two NHS Trusts and two health authorities. He has published widely, including five books variously on mental health and wellbeing, most recently with Dr. Karen Newbigging, *Commissioning Health and Wellbeing* (SAGE, 2013).

Karen Newbigging is a Senior Lecturer in Health Care Policy and Management at the Health Services Management Centre at Birmingham University and a Fellow of the Royal Society for Public Health. Originally qualifying as a clinical psychologist, Karen has over thirty years' experience in the health and social care sector, including direct service provision and commissioning. Since 1999, Karen has been involved in research, consultancy, teaching and system development for a broad range of organisations including government. She has a particular interest in participatory methods, addressing inequalities, prevention and mental health and has published widely.

SAGE LIBRARY OF HEALTH AND SOCIAL WELFARE

PUBLIC HEALTH

VOLUME I

*Global Public Health and Theoretical Foundations
of Public Health and Health Inequalities*

Edited by

Chris Heginbotham
and Karen Newbigging

Los Angeles I London I New Delhi I Singapore I Washington DC I Melbourne

Los Angeles I London I New Delhi
Singapore I Washington DC I Melbourne

SAGE Publications Ltd
1 Oliver's Yard
55 City Road
London EC1Y 1SP

SAGE Publications Inc.
2455 Teller Road
Thousand Oaks, California 91320

SAGE Publications India Pvt Ltd
B 1/I 1, Mohan Cooperative Industrial Area
Mathura Road
New Delhi 110 044

SAGE Publications Asia-Pacific Pte Ltd
3 Church Street
#10-04 Samsung Hub
Singapore 049483

Editor: Becky Taylor
Assistant editor: Colette Wilson
Permissions: Enid Andrews
Production controller: Prasanta Barik
Proofreader: Asish Sahoo
Marketing manager: Kay Stefanski
Cover design: Wendy Scott
Typeset by Chennai Publishing Services, Chennai
Printed in the UK

Library of Congress Control Number: 2016938004

British Library Cataloguing in Publication Data

A catalogue record for this book is available from the British Library

ISBN: 978-1-4739-2504-5 (set of four volumes)

Contents

Equity and Inequality

Volume II: Epidemiology of Infectious and Noncommunicable Diseases

Epidemiology of Non-Communicable and Communicable Diseases: History and Future Developments

Volume III: Ecological Public Health, Case Studies and Evaluation

Ecological Public Health

Evaluation and Health Systems

Country-based Case Studies

Volume IV: Public Health, Climate Change and Conflict

Conflict, War and Fragile States

Appendix of Sources

All articles and chapters have been reproduced exactly as they were first published, including textual cross-references to material in the original source.

Grateful acknowledgement is made to the following sources for permission to reproduce material in this book.

1. 'Declaration of Alma-Ata', *World Health Organization (1978)*
   International Conference on Primary Health Care, Alma-Ata, USSR,
   6–12 September 1978.
   Reprinted with permission of World Health Organization (WHO).

2. 'The Ottawa Charter for Health Promotion', *World Health Organization (1986)*
   First International Conference on Health Promotion, Ottawa,
   21 November 1986.
   Reprinted with permission of World Health Organization (WHO).

3. 'Beijing Declaration and Platform for Action at the Fourth World Conference on Women, Beijing 4–15th September', *The United Nations (1995)*
   Report of the Fourth World Conference on Women, Beijing,
   4–15 September 1995.
   © United Nations. Reprinted with the permission of the United Nations.

4. 'Transforming Our World: The 2030 Agenda for Sustainable Development',
   The United Nations (2016)
   Transforming Our World: The 2030 Agenda for Sustainable Development,
   A/RES/70/1 (2016).
   © United Nations. Reprinted with the permission of the United Nations.

5. 'Action on Global Health: Addressing Global Health Governance
   Challenges', *Ilona Kickbusch*
   Public Health: Journal of The Royal Institute of Public Health, 119(11)
   (2005): 969–973.
   © 2005 Published by Elsevier Ltd on behalf of The Royal Institute of Public Health. Reprinted with permission from Elsevier via Copyright Clearance Center's RightsLink service.

6. 'Global Health Inequalities: An International Comparison', *J.P. Ruger and H.-J. Kim*
   Journal of Epidemiology & Community Health, 60(11) (2006): 928–936.
   Reproduced with permission from BMJ Publishing Group Ltd via Copyright Clearance Center's RightsLink service.

7. 'Stigmatized Ethnicity, Public Health, and Globalization', *S. Harris Ali*
Canadian Ethnic Studies/Études ethniques au Canada, 40(3) (2008): 43–64.
Published by Canadian Ethnic Studies Association. Reprinted with
permission.

8. 'Mental Disorders, Health Inequalities and Ethics: A Global Perspective',
Emmanuel M. Ngui, Lincoln Khasakhala, David Ndetei and Laura Weiss Roberts
International Review of Psychiatry, 22(3) (2010): 235–244.
Copyright © Institute of Psychiatry. Reprinted by permission of Taylor &
Francis Ltd, www.tandfonline.com on behalf of Institute of Psychiatry via
Copyright Clearance Center's RightsLink service.

9. 'The 2006 Hugh Rodman Leavell Lecture "Globalization, Poverty, and
Health"', *Paulo Marchiori Buss*
Journal of Public Health Policy, 28(1) (2007): 2–25.
Copyright (2007) published by Palgrave Macmillan. Reprinted by
permission from Macmillan Publishers Ltd via Copyright Clearance Center's
RightsLink service.

10. 'Placing Gender at the Centre of Health Programming: Challenges and
Limitations', *Carol Vlassoff and Claudia Garcia Moreno*
Social Science & Medicine, 54(11), International Health in the 21st Century:
Trends and Challenges (2002): 1713–1723.
© 2002 Elsevier Science Ltd. All rights reserved. Reprinted with permission
from Elsevier via Copyright Clearance Center's RightsLink service.

11. 'Deprivation and Health', *Douglas Black*
British Medical Journal, 307(6919) (1993): 1630–1631.
Reproduced with permission from BMJ Publishing Group Ltd via Copyright
Clearance Center's RightsLink service.

12. 'Development as Capability Expansion', *Amartya Sen*
Journal of Development Planning, 19 (1989): 41–58.
© United Nations. Reprinted with the permission of the United Nations.

13. 'Capitals and Capabilities: Linking Structure and Agency to Reduce Health
Inequalities', *Thomas Abel and Katherine L. Frohlich*
Social Science & Medicine, 74(2) (2012): 236–244.
© 2011 Elsevier Ltd. All rights reserved. Reprinted with permission from
Elsevier via Copyright Clearance Center's RightsLink service.

14. 'Social Determinants of Health Inequalities', *Michael Marmot*
The Lancet, 365(9464) (2005): 1099–1104.
Reprinted from *The Lancet* with permission from Elsevier via Copyright
Clearance Center's RightsLink service.

15. 'Income Inequality and Population Health: A Review and Explanation of the Evidence', *Richard G. Wilkinson and Kate E. Pickett*
    Social Science & Medicine, 62(7) (2006): 1768–1784.
    © 2005 Elsevier Ltd. All rights reserved. Reprinted with permission from Elsevier via Copyright Clearance Center's RightsLink service.

16. 'Inequalities in Health: Some International Comparisons', *Julian Le Grand*
    European Economic Review, 31(1–2) (1987): 182–191.
    © 1987, Elsevier Science Publishers B.V. (North-Holland). Reprinted with permission from Elsevier via Copyright Clearance Center's RightsLink service.

17. 'Health Inequalities and Welfare State Regimes: Theoretical Insights on a Public Health "Puzzle"', *Clare Bambra*
    Journal of Epidemiology & Community Health, 65(9) (2011): 740–745.
    Reproduced with permission from BMJ Publishing Group Ltd via Copyright Clearance Center's RightsLink service.

18. 'Is Wealthier Always Healthier in Poor Countries? The Health Implications of Income, Inequality, Poverty, and Literacy in India', *Keertichandra Rajan, Jonathan Kennedy and Lawrence King*
    Social Science & Medicine, 88 (2013): 98–107.
    © 2013 Elsevier Ltd. All rights reserved. Reprinted with permission from Elsevier via Copyright Clearance Center's RightsLink service.

19. 'Sutherland, Snow and Water: The Transmission of Cholera in the Nineteenth Century', *Stephanie J. Snow*
    International Journal of Epidemiology, 31(5) (2002): 908–911.
    © International Epidemiological Association 2002. Reprinted by permission of Oxford University Press.

20. 'Smoking and Carcinoma of the Lung: Preliminary Report', *Richard Doll and A. Bradford Hill*
    British Medical Journal, 2(4682) (1950): 739–748.
    Reproduced with permission from BMJ Publishing Group Ltd via Copyright Clearance Center's RightsLink service.

21. 'Sick Individuals and Sick Populations', *Geoffrey Rose*
    International Journal of Epidemiology, 14(1) (1985): 32–38 (reprinted 30(3) (2001): 427–432).
    © International Epidemiological Association 1985. Reprinted by permission of Oxford University Press.

22. 'Economic and Social Determinants of Disease', *Michael Marmot*
    Bulletin of the World Health Organization, 79(10) (2001): 988–989.
    Copyright (2001). Reprinted with permission from World Health Organization (WHO).

23. 'Rose's Population Strategy of Prevention Need Not Increase Social Inequalities in Health', *Lindsay McLaren, Lynn McIntyre and Sharon Kirkpatrick*
    International Journal of Epidemiology, 39(2) (2010): 372–377.
    © The Author 2009; all rights reserved. Published by Oxford University Press on behalf of the International Epidemiological Association. Reprinted by permission of Oxford University Press.

24. 'Measuring the Global Burden of Disease', *Christopher J.L. Murray and Alan D. Lopez*
    The New England Journal of Medicine, 369(5) (2013): 448–457.
    Copyright © (2013) Massachusetts Medical Society. Reprinted with permission from Massachusetts Medical Society via Copyright Clearance Center's RightsLink service.

25. 'Epidemiology and Quantitative Risk Assessment: A Bridge from Science to Policy', *Irva Hertz-Picciotto*
    American Journal of Public Health, 85(4) (1995): 484–491.
    Published by American Public Health Association. Reprinted with permission.

26. 'Theories for Social Epidemiology in the 21st Century: An Ecosocial Perspective', *Nancy Krieger*
    International Journal of Epidemiology, 30(4) (2001): 668–677.
    © International Epidemiological Association 2001. Reprinted by permission of Oxford University Press.

27. 'Modelling the Epidemiology of Infectious Diseases for Decision Analysis: A Primer', *Mark Jit and Marc Brisson*
    PharmacoEconomics, 29(5) (2011): 371–386.
    © 2011 Adis Data Information BV. All rights reserved. Reprinted with permission of Springer via Copyright Clearance Center's RightsLink service.

28. 'Genetic Epidemiology and Public Health: Hope, Hype, and Future Prospects', *George Davey Smith, Shah Ebrahim, Sarah Lewis, Anna L. Hansell, Lyle J. Palmer and Paul R. Burton*
    The Lancet, 366(9495) (2005): 1484–1498.
    Reprinted from *The Lancet* with permission from Elsevier via Copyright Clearance Center's RightsLink service.

29. 'Non-communicable Diseases in the Arab World', *Hanan F. Abdul Rahim, Abla Sibai, Yousef Khader, Nahla Hwalla, Ibtihal Fadhil, Huda Alsiyabi, Awad Mataria, Shanthi Mendis, Ali H. Mokdad and Abdullatif Husseini*
    The Lancet, 383(9914) (2014): 356–367.
    Reprinted from *The Lancet* with permission from Elsevier via Copyright Clearance Center's RightsLink service.

30. 'Non-communicable Diseases in Sub-Saharan Africa: What We Know Now',
*Shona Dalal, Juan Jose Beunza, Jimmy Volmink, Clement Adebamowo,
Francis Bajunirwe, Marina Njelekela, Dariush Mozaffarian, Wafaie Fawzi,
Walter Willett, Hans-Olov Adami and Michelle D. Holmes*
International Journal of Epidemiology, 40(4) (2011): 885–901.
© The Author 2011; all rights reserved. Published by Oxford University
Press on behalf of the International Epidemiological Association. Reprinted
by permission of Oxford University Press.

31. 'Burden of Non-communicable Disease: Global Overview',
Samira Humaira Habib and Soma Saha
Diabetes & Metabolic Syndrome: Clinical Research & Reviews, 4(1) (2010):
41–47.
© 2008 Diabetes India. Published by Elsevier Ltd. All rights reserved.
Reprinted with permission from Elsevier via Copyright Clearance Center's
RightsLink service.

32. 'Ecological Public Health: The 21st Century's Big Idea? An Essay by
Tim Lang and Geof Rayner', *Tim Lang and Geof Rayner*
British Medical Journal, 345: e5466 (2012).
doi: http://dx.doi.org/10.1136/bmj.e5466
© BMJ Publishing Group Ltd 2012. Reproduced with permission from BMJ
Publishing Group Ltd via Copyright Clearance Center's RightsLink service.

33. 'A Perspective on the Future Public Health: An Integrative and Ecological
Framework', *Phil Hanlon, Sandra Carlisle, Margaret Hannah, Andrew Lyon
and David Reilly*
Perspectives in Public Health, 132(6) (2012): 313–319.
Published by SAGE Publications Ltd. Reprinted with permission.

34. 'An Ecological Public Health Approach to Understanding the Relationships
between Sustainable Urban Environments, Public Health and Social
Equity', *Michael Bentley*
Health Promotion International, 29(3) (2014): 528–537.
© The Author (2013). Reprinted by permission of Oxford University Press.

35. 'Building the Field of Health Policy and Systems Research: Social Science
Matters', *Lucy Gilson, Kara Hanson, Kabir Sheikh, Irene Akua Agyepong,
Freddie Ssengooba and Sara Bennett*
PLOS Medicine, 8(8): e1001079 (2011).
doi:10.1371/journal.pmed.1001079
© 2011 Gilson et al. This is an open-access article distributed under
the terms of the Creative Commons Attribution License, which permits
unrestricted use, distribution, and reproduction in any medium, provided
the original author and source are credited.

44. 'Bangladesh – BRAC Community Health Workers Program', *Global Health Workforce Alliance*
Global Experience of Community Health Workers for Delivery of Health Related Millennium Development Goals: A Systematic Review, Country Case Studies, and Recommendations for Integration into National Health Systems, (2010): 251–264.
Reprinted with permission from World Health Organization (WHO).

45. 'The Paradoxes of Gender Mainstreaming in Developing Countries: The Case of Health Care in Papua New Guinea', *Gina Lamprell, David Greenfield and Jeffrey Braithwaite*
Global Public Health: An International Journal for Research, Policy and Practice, 10(1) (2015): 41–54.
© 2014 Taylor & Francis. Reprinted by permission of Taylor & Francis Ltd, http://www.tandfonline.com via Copyright Clearance Center's RightsLink service.

46. 'The Governance of AIDS in Chile: Power/Knowledge, Patient-User Organisation, and the Formation of the Biological Citizen',
Hernán Cuevas Valenzuela and Isabel Pérez Zamora
International Social Science Journal, 62(205–206) (2011): 377–389.
© UNESCO 2013. Published by Blackwell Publishing Ltd. Reprinted with permission from John Wiley and Sons Ltd.

47. 'Below the Poverty Line and Non-communicable Diseases in Kerala: The Epidemiology of Non-communicable Diseases in Rural Areas (ENDIRA) Study', *Jaideep Menon, N. Vijayakumar, Joseph K. Joseph, P.C. David, M.N. Menon, Shyam Mukundan, P.D. Dorphy and Amitava Banerjee*
International Journal of Cardiology, 187 (2015): 519–524.
© 2015 Elsevier Ireland Ltd. All rights reserved. Reprinted with permission from Elsevier via Copyright Clearance Center's RightsLink service.

48. 'Determinants of Basic Public Health Services Provision by Village Doctors in China: Using Non-communicable Diseases Management as an Example', *Tongtong Li, Trudy Lei, Zheng Xie and Tuohong Zhang*
BMC Health Services Research, 16: 42 (2016).
doi: 10.1186/s12913-016-1276-y
© 2016 Li et al. This article is distributed under the terms of the Creative Commons Attribution 4.0 International License (http://creativecommons.org/licenses/by/4.0/), which permits unrestricted use, distribution, and reproduction in any medium, provided you give appropriate credit to the original author(s) and the source, provide a link to the Creative Commons license, and indicate if changes were made. The Creative Commons Public Domain Dedication waiver (http://creativecommons.org/publicdomain/zero/1.0/) applies to the data made available in this article, unless otherwise stated.

49. 'Mental Health First Aid Is an Effective Public Health Intervention for Improving Knowledge, Attitudes, and Behaviour: A Meta-Analysis', *Gergö Hadlaczky, Sebastian Hökby, Anahit Mkrtchian, Vladimir Carli and Danuta Wasserman*
    International Review of Psychiatry, 26(4) (2014): 467–475.
    Copyright © Institute of Psychiatry. Reprinted by permission of Taylor & Francis Ltd, www.tandfonline.com on behalf of Institute of Psychiatry via Copyright Clearance Center's RightsLink service.

50. 'Lessons from Local Engagement in Latin American Health Systems', *Geoffrey D. Meads, Frances E. Griffiths, Sarah D. Goode and Michiyo Iwami*
    Health Expectations, 10(4) (2007): 407–418.
    © 2007 The Authors. Journal compilation © 2007 Blackwell Publishing Ltd. Reprinted with permission from John Wiley and Sons Ltd.

51. 'Culture and Health', *A. David Napier, Clyde Ancarno, Beverley Butler, Joseph Calabrese, Angel Chater, Helen Chatterjee, François Guesnet, Robert Horne, Stephen Jacyna, Sushrut Jadhav, Alison Macdonald, Ulrike Neuendorf, Aaron Parkhurst, Rodney Reynolds, Graham Scambler, Sonu Shamdasani, Sonia Zafer Smith, Jakob Stougaard-Nielsen, Linda Thomson, Nick Tyler, Anna-Maria Volkmann, Trinley Walker, Jessica Watson, Amanda C. de C. Williams, Chris Willott, James Wilson and Katherine Woolf*
    The Lancet, 384(9954) (2014): 1607–1639.
    Reprinted from *The Lancet* with permission from Elsevier via Copyright Clearance Center's RightsLink service.

52. 'Hidden Heroes of the Health Revolution: Sanitation and Personal Hygiene', *Allison E. Aiello, Elaine L. Larson and Richard Sedlak*
    American Journal of Infection Control, 36(10), Supplement: Against Disease – The Impact of Hygiene and Cleanliness on Health (2008): S128–S151.
    Copyright 2007–2008 SDA. All Rights Reserved. Reprinted with permission from Elsevier via Copyright Clearance Center's RightsLink service.

53. 'Ecological Public Health and Climate Change Policy', *George P. Morris*
    Perspectives in Public Health, 130(1) (2010): 34–40.
    Published by SAGE Publications Ltd. Reprinted with permission.

54. 'Climate Change: The Public Health Response', *Howard Frumkin, Jeremy Hess, George Luber, Josephine Malilay and Michael McGeehin*
    American Journal of Public Health, 98(3) (2008): 435–445.
    Published by American Public Health Association. Reprinted with permission.

55. 'Global Climate Change: Implications for International Public Health Policy', *Diarmid Campbell-Lendrum, Carlos Corvalán and Maria Neira*
    Bulletin of the World Health Organization, 85(3) (2007): 235–237.
    Reprinted with permission from World Health Organization (WHO).

56. 'The Effectiveness of Public Health Interventions to Reduce the Health
Impact of Climate Change: A Systematic Review of Systematic Reviews',
Maha Bouzid, Lee Hooper and Paul R. Hunter
PLOS One, 8(4) (2013): e62041.
doi:10.1371/journal.pone.0062041
© 2013 Bouzid et al. This is an open-access article distributed under
the terms of the Creative Commons Attribution License, which permits
unrestricted use, distribution, and reproduction in any medium, provided
the original author and source are credited.

57. 'How Can a Climate Change Perspective Be Integrated into Public Health
Surveillance?', *M. Pascal, A.C. Viso, S. Medina, M.C. Delmas and P. Beaudeau*
Public Health: Journal of The Royal Institute of Public Health, 126(8)
(2012): 660–667.
© 2012 The Royal Society for Public Health. Published by Elsevier Ltd.
All rights reserved. Reprinted with permission from Elsevier via Copyright
Clearance Center's RightsLink service.

58. 'A Public Health Framework to Translate Risk Factors Related to Political
Violence and War into Multi-Level Preventive Interventions',
Joop T.V.M. De Jong
Social Science & Medicine, 70(1), Conflict, Violence, and Health, Edited by
Catherine Panter-Brick (2010): 71–79.
© 2009 Published by Elsevier Ltd. Reprinted with permission from Elsevier
via Copyright Clearance Center's RightsLink service.

59. 'Promoting Health Equity in Conflict-affected Fragile States',
Olga Bornemisza, M. Kent Ranson, Timothy M. Poletti and Egbert Sondorp
Social Science & Medicine, 70(1), Conflict, Violence, and Health, Edited by
Catherine Panter-Brick (2010): 80–88.
© 2009 Elsevier Ltd. All rights reserved. Reprinted with permission from
Elsevier via Copyright Clearance Center's RightsLink service.

60. 'Health in Fragile States', *Brooke Benton, Jibril Handuleh,*
Katharine Harris, Mahiben Maruthappu, Preeti Patel, Brian Godman
and Alexander E. Finlayson
Medicine, Conflict and Survival, 30(1) (2014): 19–27.
© 2013 Taylor & Francis. Reprinted by permission of Taylor & Francis Ltd,
http://www.tandfonline.com via Copyright Clearance Center's RightsLink
service.

61. 'Comparative Public Health: The Political Economy of Human Misery
and Well-Being', *Hazem Adam Ghobarah, Paul Huth and Bruce Russett*
International Studies Quarterly, 48(1) (2004): 73–94.
© 2004 International Studies Association. Published by Blackwell
Publishing. Reprinted by permission of Oxford University Press.

62. 'Gender and War: The Effects of Armed Conflict on Women's Health
and Mental Health', *Golie G. Jansen*
Affilia: Journal of Women and Social Work, 21(2) (2006): 134–145.
Published by SAGE Publications, Inc. Reprinted with permission.

63. 'Public Health Equity in Refugee Situations', *Jennifer Leaning, Paul Spiegel and Jeff Crisp*
Conflict and Health, 5(6) (2011).
doi: 10.1186/1752-1505-5-6
© 2011 Leaning et al; licensee BioMed Central Ltd. This is an Open Access article distributed under the terms of the Creative Commons Attribution License (http://creativecommons.org/licenses/by/2.0), which permits unrestricted use, distribution, and reproduction in any medium, provided the original work is properly cited.

64. 'Violence as a Public Health Problem: An Ecological Study of 169 Countries', *Achim Wolf, Ron Gray and Seena Fazel*
Social Science & Medicine, 104 (2014): 220–227.
© 2013 The Authors. Published by Elsevier Ltd. The article is published under CC-BY terms of Open Access License: https://creativecommons.org/licenses/by/3.0/.

Introduction: Global Public Health, Health Equity and Health Inequalities

Chris Heginbotham and Karen Newbigging

1. Introduction

The term public health originated in the 19th century with the realisation that tackling the causes of disease was as important as treating them. In the last 20 years, the interest and efforts of tackling upstream determinants has grown at a rapid pace (Gehlert et al., 2008; Scutchfield and Howard, 2011; Whitehead and Popay, 2010). The importance of concerted action for health improvement across the globe remains. Impacts on health result from a number of factors: the return of antibiotic resistant communicable diseases, such as tuberculosis (TB; Gilles and Traidl-Hoffmann, 2014; Spellberg et al., 2008); increased mortality and morbidity from chronic non-communicable diseases with a disproportionate burden falling on low- and middle-income countries (Habib and Saha, 2010); the problems of air pollution and climate change; the impact of fragile states and localised warfare as in Syria and the Middle East and elsewhere; and the increasing prevalence of mental health and neurological conditions including dementia (Wittchen et al., 2011).

All of this demands a new paradigm, one that recognises the multifactorial, complex and inter-related environment in which people live whilst understanding the role of both structural and individual determinants of health (Davison, Frankel and Smith, 1992). Any compendium of articles must distinguish between the historical value of academic research of the preceding 150 years and the imperative for action to tackle the present impasse in global health governance (Benatar, Gill and Bakker, 2011). We believe only a *critical* reading of past and current articles will suffice (Nixon, 2006): a fundamental reappraisal of what is required to tackle public health in the 2020s and 2030s and to recognise the importance of a *global* public health focus, especially in the global south (Colvin, 2011). The moment is ripe for a clear statement about the way that thinking about public health has changed in the past 20 years and to offer a direction for action over the next 20.

Global public health describes the health of populations in an international context (Brown, Cueto, and Fee, 2006), and requires health research and interventions as well as legal and policy frameworks that place a priority on improving health and achieving healthcare equity across national borders, even if that means states place national interests secondary to international concerns (Fox and Thomson, 2013). Global public health demands world-wide health improvement, a reduction of inequalities and protection from global threats that disregard national borders.[1] Public health is concerned with equity and equality for all, often framed as tackling inequity and health inequalities. Global public health recognises the need for high-quality national population heath measures but also underscores human rights and health equity within and across populations. Similarly global public health requires multi-sectorial and multi-disciplinary action as well as emphasising both interdisciplinary and transdisciplinary approaches that achieve a new synthesis of health with other disciplines, in which knowledge, methods and solutions are developed holistically in order to address the complexity and diversity of the issues presenting (Matlin et al., 2015). We will see in Volume III how ecological public health both demands and assists in achieving transdisciplinary mechanisms.

The terms inequity and inequalities appear in the various articles in these volumes, reflecting the authors' preferences and some of the theoretical underpinnings described here. Throughout the articles these terms are used to reflect inequalities that could be avoided (i.e. inequities) and addressed with effective political action and interventions at macro-, meso- and micro-levels. Understanding the lack of equity in health and healthcare requires careful analysis of economic, educational, political and social systems (Braveman et al., 2011; Whitehead, 1992). Health inequities are often hidden: equity gap analysis demands rigorous assessment of population level outcomes and a number of factors many of which many not be measured or publicised (Stegeman and Costongs, 2012). We see a focus on equity and equality as taking a critical approach to public health, recognising that whilst the history of public health is essentially a western discipline and draws heavily on British, European and American research, the future of public health *must* include and be relevant to all countries. The implications for public health of economic disadvantage, poverty, clean water, climate change, and other factors that may be obscured by the immediate issues facing health service delivery, are crucial to effective interventions. It is also clear that the discipline of public health is being recast with an emerging 'new public health' paradigm concerned with ecology, system leadership, evaluation, collaboration across different disciplines and sectors and engagement with communities and individual lifestyles (Beaglehole and Bonita, 2012; Beaglehole et al., 2011).

Early achievements in public health are noteworthy, particularly interventions at a population level and especially in those areas amenable to relatively straightforward solutions, such as centralised sewerage systems or provision of clean water. Over the last 50 years, similar advances have supported a global approach through the creation of the World Health Organisation (WHO) and other international bodies. Although they initially set ambitious targets and produced policy in working towards health for all they are not without their critics (e.g. a critique of the World Health Assembly in Kickbusch, 2014). Alongside this, there has been a developing theoretical and empirical contribution from the social sciences that has enabled new methodological insights about the meaning of health and illness, the challenges of implementing public health measures, the relationship between social determinants (i.e., structure) and individual behaviour (i.e., agency) and the value of coproducing solutions with the communities directly affected (Cepiku and Giordano, 2014; Fledderus, Brandsen and Honingh, 2015; Van Eijk and Steen, 2016).

Equity and equality are fundamental to the implementation of effective public health measures. Although a great deal of public health research in the developed world focuses on the effects of long-term conditions (LTCs) and noncommunicable diseases (NCDs), in the developing world, especially in the 'global South'[2] – sub-Saharan Africa, the countries of East and South East Asia, and countries of South America – the implications of infectious diseases are more prevalent (such as the impact of malaria or the zika virus) as are the implications of the lack of clean water and the effects of climate change. Water borne parasites, food borne infections, the potential for viral pandemics, and environmental pollution affect the developing more than the developed world. Nonetheless, as these countries are also subject to the forces of rapid globalisation, new patterns of disease are emerging and attention is being drawn to health conditions that have been subsumed beneath the focus on increased mortality: examples include mental illnesses or HIV/AIDS and non-communicable diseases, such as diabetes and coronary vascular diseases.

The public health implications are as important in those states with universal health coverage as in those without the safety net that states provide, and health inequalities are as evident in welfare states as anywhere else. Because of the support provided to populations in high income welfare states it might be expected that public health would be enhanced. One group of countries that appear to flout this generalisation are those in Scandinavian that counterintuitively do not have the smallest health inequalities (Bambra, 2011, this volume; Igene, 2008). Biggs and colleagues (2010) argued that 'wealthier is not always healthier'. Both gross domestic product (GDP) per capita (in purchasing power parity (PPP)) and income inequality are determinants of public health, but the way these interact with existing inequalities, and the health of the population is affected by the levels of all three factors. Data for 22 Latin American countries was analysed from 1960 to 2007. Three common measures of public health (life expectancy, infant mortality rates, and TB mortality rates) varied as the relationship between GDP and health varied during times of increasing, decreasing or constant poverty and inequality.

Although their results are complex, overall what Biggs and colleagues found was that increases in GDP per capita had little or no effect on the three measures when poverty or inequality were rising. But when poverty and inequality were constant or decreasing increasing GDP led to improvements in all three measures. In other words, wealthier societies are generally healthier but the way the resources are distributed is crucial (Wilkinson and Pickett, 2009): inequality and poverty 'exert independent, substantial effects on the relationship between national income level and health' (Biggs et al., 2010).

2. Selected Papers

These four volumes provide an overview of global public health from an international perspective while drawing heavily on European and US journals for the majority of the articles. We believe the compendium brings together key articles to form a reference work for students and practitioners of global public health. It will have international relevance and includes articles from Asia, Africa, the US, Australasia and Europe. The compendium is divided into four volumes to form a coherent, albeit wide ranging, compilation. In choosing these articles, we have focused on health equity and inequality and an ecological public health approach. This spans communicable diseases (Ebola virus, polio, AIDS, TB, etc.), noncommunicable diseases, such as diabetes, mental health and neurological conditions; using articles chosen as case studies we highlight those geographical regions where inequities are most apparent, although health inequities are evident in the global north as well as the global south.

The articles are organised thematically. The first volume provides a broad overview of the theoretical foundations of global public health and health inequalities; Volume II discusses epidemiological approaches to public health; Volume III covers ecological public health and important aspects of programme evaluation, providing some case studies as an illustration; and Volume IV deals with articles on public health, climate change and conflict. The foundation on which these are built is a view of public health that engages notions of disadvantage and social exclusion and ways in which the public health paradigm can be used to tackle deprivation and disadvantage, improve equity and reduce inequality and build more resilient and healthy communities through engaging those for who public health policy is directly targeted.

The theme that emerged most strongly from our consideration of the literature is one that deals with the implications of public health for disadvantaged communities, wherever they are located. Deprivation and disadvantage are relative and the country context must be taken into consideration: a person in the UK earning $80 per day may perceive themselves to be as socially deprived as a person in rural India who receives only $2 per day: a theory of deprivation must recognise the costs of engagement in society and those of accessing safe and effective social services. A focus on globalised disadvantage demands we consider PPPs in addition to GDP measures of national income and take account of their implications for public health interventions and outcomes.

There are two other dominant themes evident in the literature: the need, on the one hand, to take a more political approach to public health, recognising the importance of global governance and democratic processes; and on the other hand, the importance of ecological public health that engages systems theory and complexity, and recognises the way that most public health problems require a complex and multilayered understanding of the global influences and local context.

The health of a community depends on many other factors. For example, a discussion of public health in India requires a wide ranging debate (Chauhan, 2011) in which we must consider the implications of caste, the availability of sanitation and clean water (Kumar, Kar and Jain, 2011) accessibility of sufficient nutritious food (Das and Bose, 2015), gender discrimination and intimate partner abuse (Spiwak et al., 2013), communicable and noncommunicable diseases (Shetty, 2012), HIV and AIDS (Sheikh and Porter, 2010), and the implications of enormous discrepancies of wealth and power. In other countries with similar features we need to consider the nature of governmental fragility, the direct effects and long-term implications of conflict and war, refugee status and anxieties created by terrorist groups on minority populations. The effects of direct discrimination on minority groups is pervasive, often serving to exclude them from mainstream society and access to health care with flagrant human rights abuses including inhumane treatment. Examples include the attitude of the Bulgarian health service to Roma communities (Rechel et al., 2009), or the attitudes of the Hindu majority in India towards the Moslem population (Tausch, Hewstone and Roy, 2009) and health care for lesbian, gay, bisexual and trans (LGBT) people in countries where being a sexual minority is illegal or punishable by death (Meyer and Northridge, 2007). Similarly the implications of colonialism and imperialism and their aftermath continue to feed myths and structures that can foster social disadvantage and, therefore, have public health implications (e.g., Estrada, 2008).

A naive view of inequality takes the stance that inequality matters because it creates poverty or income differences that seem at best unfair and are probably seriously discriminatory. However, as Wilkinson and Pickett (2009) have demonstrated the relative gap between richest and poorest in different countries are more important than absolute wealth. They argue that greater income inequalities increase social distance, accentuating social class or status differences (Pickett and Wilkinson, 2015). Thus, addressing health inequities requires reducing material differences, 'which are so often constitutive of the cultural markers of social differentiation' (Pickett and Wilkinson, 2015: 324).

We are mindful that this discussion must also take account of ways in which health and wellbeing can be strengthened, how illness prevention can be created and sustained, and ways in which social capital can be grown and become part of a social fabric that supports public health activities. Awareness of the influence of social determinants of health has come of age in the last 10 years with significant areas of research and action, and, within the Global North especially, tackling the public health time-bomb of NCDs, notably diabetes, has gained increasing importance. As people live longer, the increasing numbers of people with dementia will become a challenge for developing countries as well, as it is currently for developed countries. Finally we want to consider ecological aspects as an innovative approach to public health action.

3. Health Equity and Health Inequalities

For most people their 'health', rather than specific illnesses or diseases, is the result of many differing ecological, economic, educational, political and social influences, and the social determinants paradigm encompass all these to a greater or lesser degree, as we shall see in this volume. The recent emphasis on social determinants of health and health equity identifies a new way to think about health and mental health. Research on social determinants of health highlights the need to focus on heath inequities wherever these arise (Östlin et al., 2011). Östlin and colleagues propose a new paradigm that focuses on four critical elements: global factors that affect health equity, structures and processes that differentially affect people's chance to be healthy within a given society, health system factors that affect health equity, and policies and interventions that may reduce health equity (Heginbotham and Newbigging, 2014, p. 36.). By bringing together health equity, ecology and social determinants it becomes possible to create a framework to drive significant improvements in public health.

3.1. Health Equity

Health 'inequity' refers to differences in health that are not only unnecessary and avoidable but are unfair and unjust. Margaret Whitehead (1992: 6) provided a helpful definition:

> *'Equity in health implies ideally that everyone should have a fair opportunity to attain their full health potential and, pragmatically, if it can be avoided, no one should be disadvantaged from achieving this potential.'*

On this definition, and whilst governments may not be able to eradicate all public health differences between communities or groups, we may not be able to expect them to aim to reduce or remove entirely those public health differences that are considered (by an unbiased observer) to be both avoidable (readily avoidable at reasonable cost?) and unfair (on some rational criterion?) and should outlaw adverse discrimination. As we see in Volumes II and III, each country will have different criteria. Individual, community and political values dictate the broad thrust of this policy, which is acceptable as long as that does not include direct or indirect discrimination against groups on the basis of inherent characteristics or central to their identity (ethnicity, tribal affiliation, language, religion or faith, gender, age, previous political association).

Equity is one of those 'essentially contested concepts' that occasionally spring up in health care. At least eight definitions are discussed in the literature and there may be more. The more technically rigorous do not readily lend themselves to practical application. Of the others, we prefer either the simple definition of equity as a basic minimum of health to all, or the more complex equity of ability to benefit (Heginbotham and Newbigging, 2014, p. 39). The EuroHealthNet, an EU wide organisation concerned with equity and equality in health care, has undertaken a lot of work on equity gaps and gradients, indicating the implications of government policies for equity (Stegeman and Costongs, 2012). Wilkinson and Pickett (2009) demonstrated clear evidence in 'The Sprit Level' that health inequity follows the gap between the bottom and top of the income distribution: income disparity is directly proportional to lack of health equity.

Taking action on health equity is both challenging and complex. Internationally the report of WHO, Closing the Gap in a Generation (2008), suggests areas for action. These include improving daily living conditions especially of poorer households, changing the inequitable distribution of power money and resources and researching the problem adequately. Many governments pay lip service to the health inequalities that exist because it is too difficult, usually for political reasons, to do anything serious about it. Inequity is systemic, produced by social norms and policy practices that not only tolerate inequity but actually promote unfair distribution of resources. Equity gaps show up when there is difference between population level indicator (e.g. at a national level) and the same statistic for a defined area or group (in the absence of some specific factor that explains the difference). Whilst equity gradients demand recognition of social stratification (based on age, class, education, gender, income or other relevant factor), differential vulnerability between social groups, variant consequences of disease or health events are dependent on particular socioeconomic circumstances. (Bornemisza et al., 2010, p. 81; Goldman et al., 2006).

3.2. Social and Political Determinants of Equity

Östlin and colleagues (2011) propose that tackling health equity will require a concerted and wide ranging programme to achieve significant improvements in public health. They suggest, *inter alia*, the importance of the following imperatives. We should:

- go beyond individual determinants of illness towards a holistic programme that incorporates cultural, economic, political and social influences;

- study the intersections between social hierarchies and power structures and their cumulative impact on public health;
- scrutinise the 'psychosocial pathways of the 'upstream' social determinants in addition to the usual risk factors in 'downstream' support;
- tackle patterns of health inequity and 'social reality';
- consider the dynamic as well as the static nature of equity including temporal and financial dimensions;
- recognise that some types of evidence although given predominance in the literature may not address the social determinants of health;
- involve the affected populations through carefully designed instruments.

Each of these elements adds to create a comprehensive programme that Östlin and colleagues claim genuinely tackles the health equity aspects of public health and does so in an ecologically sophisticated way (Ingram et al., 2012). However, when and how to make the changes to society that the research suggests are necessary is not always easy especially as there are political considerations (Braveman et al., 2011). Ecological determinants engage a wide range of influences on health and to a degree subsume other factors. Ecological public health has an encompassing literature of its own, not only on adult health but also including child and adolescent health (Karpati et al., 2002; Smith, 2005; Viner et al., 2012; Mignone and O'Neil, 2005). Economic factors play a major role in health over the life course, especially the part that deprivation, lack of employment (see, e.g., the implications for Italian workers in Minelli et al., 2014), poor housing (Bentley, Baker and Mason, 2012; Fertig and Reingold, 2007; Herrin, Amaral and Balihuta, 2013; Hood, 2005), food insecurity and poverty (e.g., Cook et al., 2004), and low income (e.g. Sepehri and Guliani, 2015) play in creating and sustaining poor quality lifestyles and life chances (Costa-Font, 2008; Masseria, Mladovsky and Hernández-Quevedo, 2010; Spencer, 2003; Tipper, 2010).

Educational disadvantage is also a major factor in generating health risks over the life course (Dupre, 2008; Stansfeld et al., 2011), and has its effects in unhealthy behaviours such as lack of exercise, obesity and poor nutrition (Lee, Harris and Lee, 2013). Political factors, such as democratic accountability (Powell-Jackson et al., 2011) matter in enabling public health but are especially important within fragile states and those affected by war and social displacement. Public health is always affected adversely by war and its consequences but public health can be a 'prime mover' in encouraging improved public policies in war zones (De Jong, 2010, Volume IV). Culture, a contested concept, has been identified as one of the key determinants of the population's reaction to the implications of war and fragility (Napier et al., 2014, Volume IV).

To be successful in tackling the social determinants of health all governments and states parties should create, or recreate, programmes that confront poverty, poor housing, lack of jobs, poor hygiene, lack of clean water and the implications of disease. Housing is one of the traditional areas of concern for public health, although it has been relatively neglected over recent decades. Housing is important for psychosocial reasons as well as its protection against the elements, but it can also be the source of many physical, chemical hazards. Problems associated with poor housing include cramped and crowded conditions that give rise to poor hygiene by providing places for vermin to breed and transmit diseases via fleas, ticks and other vectors; poor household hygiene leads to food and water contamination; poor indoor air quality leads to respiratory problems and inadequate lighting leads to eyesight problems and poor housing leads to higher stress levels. Poor housing leaves populations vulnerable to the effects of climate change and demands innovative technological solutions (Kern, Bolay and Thanh, 2012); but sustainable development is best carried out in conjunction with the communities effected (Kern and Bolay, 2014).

4. Global Public Health

In this volume we have included articles that offer two broadly related aspects of global public health: overarching international declarations and policy on public health and health inequalities, and the importance of good governance in promoting global public health. The first set of contributions reflects four of the most important international meetings of the last 40 or so years. These discussions came about as a result of international pressure about the need for imperatives to tackle long held concerns.

4.1. Declaration of Alma-Ata

The Alma-Ata Declaration of 1978 emerged as a major milestone for public health from a conference on primary care, identifying primary health care as the key to Health for All. The conference strongly reaffirmed that health, which it considered 'a state of complete physical, mental, and social wellbeing, and not merely the absence of disease or infirmity', to be a fundamental human right. The attainment of the highest possible level of health was stated to be one of the most important world-wide social goals. Existing health inequalities, especially those between high income and low and middle income countries, as well as those within countries, are politically, socially and economically unacceptable.

Primary health care was considered by the Alma-Ata conference as essential health and health care 'based on practical, scientifically sound, and socially acceptable methods and technology made universally accessible to individuals and families in the community through their full participation and at a cost that the community and country can afford to maintain at every stage of their development in the spirit of self-reliance and self-determination'. This should form an integral part of every country's health system and of the overall social and economic development of the community.

Alma Ata proposed that it should be possible to obtain an acceptable level of health for all the people of the world by the year 2000 through better use of the world's resources. Although this was a tough challenge, and ultimately unachievable, it was hoped that a genuine policy of independence, peace, détente and disarmament could and should release additional resources to be devoted to peaceful aims including health and health care. The Alma-Ata Declaration led to many criticisms, in particular that the slogan 'Health for All by 2000' was impracticable and that the declaration did not have clear targets. The lessons from this period are twofold: firstly, resources are highly unlikely to become available on the optimistic timescales suggested; secondly, if so it is important to persuade governments to tackle public health as an essential and separate priority and not to wait for a promise of future funding that may never materialise.

In response to widespread criticism, the idea of Selective Primary Health Care (PHC) was introduced as opposed to the PHC of the Alma-Ata Declaration. Selective PHC had the objective of obtaining low-cost solutions to very specific, common causes of death with targets that were clear, concise, measurable and easy to observe. They were known as GOBI (growth monitoring, oral rehydration treatment, breast-feeding and immunisation) and later GOBI-FFF (adding food supplementation, female literacy and family planning). Within GOBI-FFF there were the beginnings of a global campaign to tackle gender discrimination and to place a higher value on the roles of women in society.

4.2. Ottawa Charter

The Ottawa Charter for Health Promotion was the next major milestone agreed at the First International Conference on Health Promotion, Ottawa, in 1986. Primarily a response to growing expectations for a new public health movement around the world, this conference in practice came to represent the first step towards the ecological and all-encompassing trends we observe today. Discussions at Ottawa focused on the needs of industrialised countries, but took into account similar concerns from other regions, and built on a recent debate at the World Health Assembly on intersectoral action for health.

Perhaps the main outcome of Ottawa was to give legitimacy and purpose to health promotion as an important and valuable discipline. Until then health promotion had been seen as the poor relation within many health systems but Ottawa gave new hope to public health practitioners. Health was framed as a positive concept emphasising social and personal resources as well as physical capacities and highlighting healthy life-styles and wellbeing.

This was the first time the importance of social determinants of health within an ecological programme was articulated. Initially the debate did not have international prominence but in the longer-term social determinants of health and ecology have become the major planks of public health policy. The fundamental prerequisites for health were listed as peace, shelter, education, food, income, sustainable resources, social justice and equity within a stable ecosystem. Good health is a major resource for social, economic and personal development and an important dimension of quality of life. Political, economic, social, cultural, environmental, behavioural and biological factors can all work towards health improvement or be harmful to it. Health promotion includes advocacy for health and focuses on achieving equity, in reducing differences in current health status and ensuring equal opportunities and resources to enable everyone to achieve their fullest potential.

Health promotion, as Ottawa affirmed, demands coordinated action by all concerned: by governments, by health and other social and economic sectors, by nongovernmental and voluntary organisation, by local authorities, by industry and by the media. People in all walks of life are involved as individuals, families and communities. Professional and social groups and health personnel have a major responsibility to mediate between differing interests in society for the pursuit of health. Health promotion strategies and programmes can be adapted to local needs and all countries are encouraged to work towards these goals taking into account differing social, cultural and economic systems.

4.3. Beijing Platform: Gender Equality

The next significant international declaration was on gender equality. The founding United Nations charter in 1945 included a provision for equality between men and women. From 1945 to 1975 various political leaders, mainly women, attempted to turn these principles into action. Twenty-five years after the first session of the Commission on the Status of Women was held at Lake Success, New York, in 1947, the UN General Assembly in 1972 passed a resolution (resolution 3010) that 1975 should be International Women's Year. In December 1975, the UN General Assembly passed a further resolution (resolution 31/136) that 1976–1985 should be made the 'Decade of Women'. Subsequently the first world conference on women was held in Mexico City in 1975 and the second world conference in Copenhagen in 1980. The third world conference took place in Nairobi in 1985 and set out areas by which progress in women's equality could be measured: constitutional and legal measures; equality in social participation; equality in political participation and decision making. The conference also acknowledged that women should be able to participate in all areas of human activity, not just those areas that relate to gender.

The fourth world conference took place in Beijing in 1995 with a powerful plan for gender equality and the empowerment of all women, everywhere. A preconference Declaration and Platform for Action aimed at achieving greater equality had been prepared, but in fallout from preconference meetings, the Catholic Church publicly disagreed with positions outlined by the

United States and other nations concerning abortion, reproductive rights, and other issues that dismayed the Church. From a public health perspective, religious moral codes are often backward looking and interrupt sensible attempts to improve the overall health of the community. The impact of the Catholic Church on women's health is a good example but it is not alone. Other fundamentalist religious teachings and cultural practices that seek to restrict women's behaviour also undermine women's rights to health.

A remarkably diverse 17,000 participants and 30,000 activists travelled to Beijing for the opening of the Fourth World Conference. The agreed Platform for Action made comprehensive commitments under 12 critical areas of concern incorporating inter alia health and health-related matters, including women and the environment, women and poverty, violence against women, women and health and women and armed conflict. The Platform for Action imagined that every girl and woman should be able to exercise freedom of choices, realise all her rights, go to school, participate in decisions and earn equal pay for equal work. Since then, governments, civil society and the public have signed up to the Platform for Action's objectives and worked on implementing these, although there remains a great deal to do, especially within the area of public health (Ruger and Kim, 2006). The Beijing Platform for Action promoted the idea of mainstreaming, encouraging governments and organisations to assess the implications for gender equality of legislation, policies or programmes, in all areas and this clearly included health. This does not negate the need for action that is targeted at women or positive legislation and affirmative action (Stratigaki, 2005). The experience of gender mainstreaming has attracted criticism and perhaps the most powerful of these is that it has deflected attention away from the structural forces that shape gender inequality (Subrahmanian, 2004).

We will see in Volume III the extent to which women are denied their rights in many countries, which are particularly apparent during times of war and conflict but also evident in the wider lack of equality, participation and representation across the globe.

To be effective, however, it is essential that all states parties include women fully at the core of programmes (Vlassoff and Moreno, 2002, this volume), especially as women bear the brunt of so many of the world's ills. However, it is not only gender, albeit extremely important, that requires anti-discriminatory programmes. Discrimination on the grounds of ethnicity (Ali, 2008), physical disability and especially mental illness and learning disability (Ngui et al., 2010), and age – children and older people – must be tackled as they have direct and indirect effects on health. Religion and faith is a further category that leads to direct and indirect discrimination.

4.4. Addis Ababa: Sustainable Development

In 1977, the World Health Assembly decided the main social target of governments and of WHO should be the attainment by all the people of the world by the year 2000 of a level of health that would permit them to lead a socially and economically productive life. In other words, as a minimum, all people in all countries should have at least a level of health that ensured they were capable of working productively and of participating actively in the social life of the community in which they live. The third evaluation of progress in implementing the Global Strategy for Health for All by the year 2000 (carried out in 1997) showed significant but patchy improvements worldwide both in health status and in access to health care. Increasing numbers of countries undertook monitoring and evaluation of their health-for-all strategies at specified intervals. Although there was general acceptance of the programme, which had effects in some relatively safe conflict free areas, it did not fundamentally shift overall public health or the attitude of the Global North towards the health of the majority of the world's people.

Following the disappointing results from the Millennium goals in many areas of growth and development and increasing concern about sustainable progress amongst developing countries, a burgeoning debate in the early part of the 21st century considered the need to restate the international development goals on a sustainable platform. In 2014 the UN set out in the Addis Ababa Declaration an 'Agenda for Sustainable Development 2030'. Seventeen sustainable goals were articulated of which Goal 3 covered health in some detail. Goal 3 incorporated agreements to ensure healthy lives and promote wellbeing for all at all ages.

The specific goals included, by 2030, to: reduce the global maternal mortality ratio to less than 70 per 100,000 live births; end preventable deaths of babies and children younger than 5 years of age; end the epidemics of AIDS, TB, malaria and neglected tropical diseases and combat hepatitis, water-borne diseases and other communicable diseases; reduce by one third premature mortality from noncommunicable diseases; strengthen the prevention and treatment of substance abuse, including narcotic drug abuse and harmful use of alcohol; halve the number of global deaths and injuries from road traffic accidents; ensure universal access to sexual and reproductive health-care services; achieve universal health coverage, access to quality essential health-care services and access to safe, effective, quality and affordable essential medicines and vaccines for all; substantially reduce the number of deaths and illnesses from hazardous chemicals and air, water and soil pollution and contamination; strengthen the implementation of the World Health Organisation Framework Convention on Tobacco Control in all countries; support the research and development of vaccines and medicines for the communicable and noncommunicable diseases that primarily affect developing countries; substantially increase health financing and the recruitment, development, training and retention of the health workforce in developing countries; strengthen the capacity of all countries, in particular developing countries, for early warning, risk reduction and management of national and global health risks.

This is a formidable but necessary agenda of public health action. But the world has just over a decade to achieve these goals at a time of increasing global terrorism, growing complications from climate change, significant shocks to the world economic order, difficulties caused by antibiotic resistance and increasing scepticism amongst the world's population about governments' willingness to take the necessary action. Good governance to ensure global health challenges are tackled will require regular and detailed evaluation and audit of progress, an essential process in the fight to improve lives and tackle discrimination, as well as seeking to prevent war and conflict and the harmful effects of man-made disasters, especially climate change (Kickbusch, 2005, this volume). This explains why we focus on evaluation in Volume III and consider mechanisms for improving the delivery of health care. Public health is not simply health promotion important as that is. Public health is attained and sustained by an amalgam of prevention, promotion and treatment.

5. Ecological Public Health

What is ecological public health? By ecology we mean more-or-less the same as the term coined by the biologist Ernst Haeckel when he derived it from the Greek oîkos, 'house'; and logia, 'study of'; Haeckel unequivocally located its meaning within Darwinian thought and Darwin's principle of natural selection:

> *By ecology we mean the body of knowledge concerning the economy of nature – the investigation of the total relations of the animal both to its inorganic and to its organic environment; including above all, its friendly and inimical relations with those animals and plants with which it comes directly or indirectly into contact – in a word ecology is the study of all those complex interrelations referred to by Darwin as the conditions of the struggle for existence.*

It has been argued that public health thinking needs to be refurbished using ecological principles. Ecological approaches integrate material, biological, social and cultural aspects of life. Instead of distancing the technologies of public health as distinct and uninvolved in the life of communities, public health becomes the task of transforming the relationship between people, their circumstances and the biological world of nature and bodies. This demands a refocusing of public health actions onto the conditions on which human and ecosystems health interact. Public health in the 21st century is 'unavoidably complex', and 'requires stronger and more daring combinations of interdisciplinary work, movements and professions locally, nationally and globally' (Lang and Rayner, 2012).

Against this backdrop we have taken health inequalities as the most significant symptoms of a lack of ecological thinking. We cannot continue with an epidemiologically driven public health agenda that essentially offers a western perspective on global public health issues. We give disadvantage and inequality an essential place in the overall structure of these volumes in an approach that recognises the complex interplay of human and environmental factors. But it must not be a simplistic 'environmental layer' 'tacked on', as Lang and Rayner describe it (see Volume III), to an old framework. Ecological public health demands a recognition of complexity and a determination to find acceptable, as minimally complex as possible but not unsophisticated, ways through that complexity.

Public health policy will need to change if it is to tackle the critical problems in the 21st century and beyond. Five specific approaches are needed. First we need a systems approach to health and wellbeing; second it is essential that we have good data, using sophisticated methods for data collection and approaches to evaluation and research that reflect this systems approach and bring together the tradition of quantitative health research with the understanding from social sciences; third we must focus on new technologies especially the application of telecommunications and especially the internet; fourth we need to strengthen participatory approaches in policy, practice and research – engaging the diversity of stakeholders to build their capabilities for health and finally, we should all nurture learning communities that seek to address both global and local challenges to health.

In 2001, the World Bank published a report known as 'Dying for Change' on the experience of health and ill-health amongst poor people throughout the globe. The report highlighted the stark relationship globally between poverty and poor health: in the least developed countries, life expectancy was just 49 years in 2000, and one in ten children do not reach their first birthday (a 10% mortality rate), whereas in high-income countries the average life span was 77 years and the infant mortality rate was six per 1000 live births (0.6%).

In a report in 2000 entitled 'Environmental Health and the Poor: Our Shared Responsibility,' Sandy Cairncross and Peter Kolsky list the implications of poor living environments, including the effects of indoor air pollution, poor sanitation and lack of clean water (Singh and Singh, 2008). Poverty leads to many dangers including contaminated water supplies, inadequate sanitation, poor nutrition and lack of proper health care, and leads to a greater likelihood of a wide range of diseases including trichomoniasis, malaria, intestinal parasites, schistosomiasis, TB, AIDS, asthma, cardiovascular disease as well as tooth decay and gum disease.

6. Health Technology

The availability and use of new health technology is often related inversely to health and social need. Although health care systems in high-income countries make extensive use of high technology systems and equipment, people in the low- and middle-income countries (LMICs) where need is most acute often lack the most fundamental drugs and devices. Technology for global public health incorporates a broad category of interventions that may reduce malnutrition, improve sanitation and increase personal safety – the social determinants of health that govern the majority of illness and disability. They are distinct from specific health technologies designed to diagnose and treat illness, from the highly specific (e.g., a vaccine for a particular disease) to the more widely applicable (e.g., a blood pressure monitor). Preventive systems cut across both: for example, providing clean water reduces and in many cases eliminates water borne parasites and monocellular organisms that cause diseases in poorer countries.

New technologies could have an important effect on health outcomes in LMICs, where the greatest burden of disease is found. Insufficient resources have been dedicated to the development of so-called 'frugal' technologies to meet the needs of the world's poorest people. Even when the necessary technology has been developed, it is often unavailable, either because of cost or because of limits on energy or human resources. Efforts should also be made to ensure that technology is acceptable to, and will be adopted by, users.

All investment decisions should be evidence-based including the introduction of new technology for public health and require related process innovations to enable effective use. Ensuring that any new technology improves rather than damages health is a key determinant of appropriate intervention, and the demands of equity and equality apply here as anywhere else. Novel technologies are being created in low-income and middle-income countries that might prevent the burgeoning health-care costs seen in high-income countries.

It is important to think 'out of the box' taking a multidisciplinary perspective on the development and introduction of new ways of working. One technology that is sometimes not recognised as such is genetics, genetic engineering and knowledge of the human genome. Most public health concerns are best addressed by a combination of approaches, some of which are specific to health, such as drugs and medical devices, whereas others have health benefits that arise from use outside health, such as the internet or irrigation. It goes without saying that whenever possible, technology already available in low-income country settings (such as mobile telephones) should be used as a platform for health interventions.

Judith Kurland (2000) in an article in Public Health Reports suggested that, whereas technology must advance, 'the discipline of public health needs to be more aggressive about relating technology and medicine to the larger questions of human existence'. The use of cheap technologies with the wider availability of knowledge can dramatically improve global health outcomes. Other factors have played a role, such as increased income that not only allows for improved nutrition, but also helps to improve access to more complex preventive processes (Casabonne and Kenny, 2012).

Public health services and systems research (PHSSR) have emerged over the past two decades largely to produce the evidence needed to address the organisation, financing and delivery of effective public health strategies. An

American national PHSSR research process (Consortium from Altarum Institute, 2012) was developed to identify information needs that public health stakeholders face in relation to public health workforce, system structure and performance, financing and information technology. We say more about this in the Introduction to Volume II.

7. Global Governance

7.1. Global Governance and Diversity

Following the four International Declarations on community health, gender and sustainability, the articles we have included in the first main section of this volume seek to reflect diversity, equity, health inequalities and health governance. Ilona Kickbusch's article (2005, this volume) is about action on global health addresses global health governance. We see this as a critical element of the international management of public health and a way to ensure effective international engagement with public health concerns. Ideally public health would be the same the world over and this is patently not true. But within separate jurisdictions, health and health care should not depend on characteristics over which individuals have no control or social identities – race or ethnicity, membership of a political or social group, religious affiliation, gender, disability or age.

Respect for difference and celebration of diversity not only ensures everyone is treated with equal concern and respect, but also that people are treated according to their needs. We are not all the same: as George Bernard Shaw observed, 'do not treat others as you want to be treated, their tastes may not be the same'. The only way through this dilemma is to make every effort to find out what individuals and countries require in support of maximising public health and then allocate resources as equally as possible across the globe, giving greater resources to those with the poorest health.

Murray and Lopez (2013, see Volume II) set out fully a comparative global perspective on burden of disease. Other articles that we might have included discuss life expectancy in a similar way. Our purpose in including this article was to demonstrate three points: first, that the burden of disease measured in disability adjusted life years (DALYs – see Volume II Introduction) shows clearly the extent to which developing countries carry the largest proportion of disease burden, and second that NCDs create a large part of that burden: life expectancy figures are much kinder to NCD statistics. Any public health collection such as this must describe at the outset the total burden of disease and deprivation that a public health programme should address. It should be evident that an ecological approach to public health, that posits an integrated or at least an interdependent group of relevant causative factors drawn appropriately from economic, educational, political, social spheres, identifies not one medical or social model but a complex nexus of factors that draw one upon the other, modifying yet others in turn. Finding our way through that nexus is challenging. But the work of Murray and Lopez and other authors, offers one perspective on the way in which these factors inter-relate.

Although we have included the article based on DALYs, it may be worth noting that this approach is not without its critics. Adams (2016) suggests that 'if the DALY was the gold standard for the era of health development, then today the gold standard is a new kind of metric: the statistically robust, randomised and controlled, cost-effectively constituted, experimentally designed, outcome measureable intervention'. Although the RCT is hardly new, Adams suggests that the methodology could be applied to global health as a way of dealing with some of the more intractable political debates and problems of evidence, so that global health today 'means doing *research as intervention* through RCTs'.

Age, disability, ethnicity, gender and sexuality are perhaps the five most evident forms of discrimination that occurs in public health as anywhere else in society. The article we have chosen on ethnicity by Ali (2008, this volume) takes a critical stance in describing ethnicity as 'stigmatised' reflecting a radical and demanding position that sees ethnicity as one of the reasons that public health is not as well developed in poorer countries as it might be. Similarly Vlassoff and Moreno (2002) place gender at the centre of health planning. In the introduction to Volume III, we describe in some detail the extent to which women are so often discriminated against either directly or indirectly as a result of their place in society. Women suffer poorer health and are affected more by environmental issues, nutritional problems, the burden of feeding and clothing a family and many other factors.

Finally in this section we turn to mental illness and distress. Mental health is one of the most important aspects of any public health programme but is rarely accorded sufficient importance (at least until now). Ngui and colleagues (2010) argue for a practical yet philosophical stance on funding and care for people with mental illnesses, suggesting powerfully that this is an ethical matter of concern to everyone. The Hugh Rodman Leavell Lecture sums up one of the most important matters for health – poverty. Buss (2007) describes the importance of equality within globalisation. International organisations must do more to tackle inequalities, because, as Wilkinson and Pickett (2006) show, inequalities will in themselves undermine and attempts at amelioration if they are not fixed.

7.2. Capabilities and Social Determinants of Health

In the second section of this volume we cover those articles, many as it happens from the United Kingdom, that have acted as particularly sharp reminders about what really matters in public health. One of the prime movers in this debate was Douglas Black whose report in 1980 led the way on tackling deprivation and its links to poor health. This was followed 30 years later by Michael Marmot's seminal report on social determinants of health. In the meantime a number of important articles had been written that offer a guide to the debates of the last 30 years or so. Amartya Sen's article (1989, this volume) describes one of the most important aspects of achieving equity in health care, his capabilities approach. We have chosen to include an article on development as capability expansion, but the improvement in health equity that comes from improving everyone's 'ability to benefit' is a crucial underpinning of the attack on inequalities.

In describing health inequalities across the globe we have included an article by Ruger and Kim (2006) that provides an international comparison of global health inequalities. This makes grim reading as it identifies the way that the international order has allowed western governments to ignore the harsh realities of health problems on the ground in LMIC countries. For instance, by analysing World Development indicators compiled by the World Bank they demonstrated the very high adult and child mortality in western and sub-Saharan Africa and Afghanistan. Compared with countries with low child mortality, those with high child mortality had 'significantly higher rates of extreme poverty, people living in rural areas and female illiteracy, significantly lower per capita expenditure on healthcare, hospital beds and doctors, and lower rates of access to improved water, sanitation and immunisations' (Ruger and Kim, 2006). Between 1960 and 2000, they calculated that adult male mortality in countries with high mortality was increasing at more than 4 times the rate in countries with low mortality. Their conclusion was that inequalities in child and adult mortality are large and growing, and global efforts to deal with this problem require attention from rich countries.

Rajan, Kennedy and King (2013) make similar points, although they suggest that money is not the only or indeed the main reason for poorer health. A rather different objective is taken in Bambra et al.'s article (2011) on health inequalities and welfare state regimes. Bambra is keen to identify the reasons why some welfare states do not have the smallest health inequalities. She uses a series of theoretical inequalities models – artefact, selection, cultural-behavioural, materialist, psychosocial and life course – to generate valuable theoretical insights. The main reason for including this article here is in identifying the differing models and their use, demonstrating a helpful methodology that might be applied in other circumstances.

The other articles in this section unashamedly focus on health inequalities and health equity. Abel and Frohlich (2012) provide a sociopolitical theoretical model that other countries may find valuable and they support Amartya Sen's capabilities approach. Culyer and Wagstaff (1993) take a health economics approach to health equity and inequality and explore four definitions of equity in health care: equality of utilisation, distribution according to need, equality of access, and equality of health. From this theoretical position they argue that equality of health should be the governing principle and that equity in health care 'should therefore entail distributing care in such a way as to get as close as is feasible to an equal distribution of health'. Although this proposal sounds attractive it is, of course extremely difficult to achieve or to measure, as it concerns outcomes of whatever processes are put in place. Nonetheless it sets a critical objective for public health interventions and is included here because of its importance in the burgeoning debate in the 1980s and beyond.

Social determinants of health and the implications of inequalities in income and other social factors are the substance of the articles by Le Grand (1987), Marmot (2005) and Wilkinson and Pickett (2006). Each of these articles reinforces the position of the others. They are included here for two main reasons: first, because they define and promote a social determinants approach to health inequalities, and second, because the way the theoretical underpinning is constructed provides a helpful and powerful approach to the problem of public health inequalities (which we will discuss further in Volume III when we consider ecological public health).

8. Conclusions

Le Grand's article shows what can be achieved by a relatively small international comparison in shedding light on the reasons for inequalities; Marmot describes the importance of social determinants of health in providing a powerful set of reasons for differences in health within and between countries; and Wilkinson and Pickett in the article included here and in their seminal book, *The Spirit Level*, demonstrate conclusively that more unequal societies are bad for almost everyone within them from those who are better-off to those living in poverty or deprivation. Their book shows that almost every modern social and environmental problem (the social determinants of health) such as violence, drugs, obesity, mental illness or large prison populations are more likely to occur in less equal societies. We recommend the book to anyone who has not yet read it: it is an eye-opener!

The selection we have included offers a wide-ranging view of the field and we hope the articles included in these volumes will stimulate and excite further curiosity about global public health. In 64 articles we could not possibly do justice to the hundreds of thousands of articles on public health, nor cover the ground in anything like the depth that we would have liked. However, we hope that the selection offered here is sufficiently broad for there to be something for (almost) everyone.

Notes

1. A distinction is made in the literature between international health and global health, in which international health is defined as the branch of public health focusing on developing nations and foreign aid efforts by industrialized countries (White & Nanan, 2008). From our perspective it is difficult to tease out where international health stops and global health starts, whilst recognising the simple truth about granting true sovereignty to national governments.

2. The term global South is used to refer to developing countries, which are located mainly in the Southern hemisphere and generally this term is preferred to 'developing' or 'third world' countries and contrasts them with the wealthier states of the Global North, which tend to be more stable and have a more developed public sector (Eriksen et al., 2015). However, this terminology is not without its problems and the limitations, and potentially negative implications, of capturing global diversity within a binary concept (global South-global North) is discussed in detail by Eriksen et al. (2015). Whilst recognising this, we have chosen to use the terms developing countries and Global South interchangeably, reflecting their common usage in the articles in this collection but would urge readers to consider the debate.

References and Bibliography

Adams, V. Metrics of the global sovereign. In Adams, V. (ed.), *Metrics, What counts in Global Health*. Durham & London: Duke University Press, 2016.

Beaglehole, R. & Bonita, R. Tackling NCDs: a different approach is needed. *The Lancet*, 2012; 379(9829): 1873.

Beaglehole, R., Bonita, R., Alleyne, G., Horton, R., Li, L., Lincoln, P., Mbanya, J. C., Mckee, M., Moodie, R., Nishtar, S., Piot, P., Reddy, K. S. & Stuckler, D. UN high-level meeting on non-communicable diseases: addressing four questions. *The Lancet* 2011; 378(9789): 449–455.

Bambra, C., Smith, K.E., Garthwaite, K., Joyce, K.E. & Hunter, D.J. A labour of Sisyphus? Public policy and health inequalities research from the Black and Acheson Reports to the Marmot Review. *Journal of Epidemiology and Community Health*, 2011: 65(5): 399.

Benatar, S., Gill, S. & Bakker, I. Global health and the global economic crisis. *American Journal of Public Health*, 2011; 101(4): 646–653.

Bentley, R., Baker, E. & Mason, K. Cumulative exposure to poor housing affordability and its association with mental health in men and women. *Journal of Epidemiology and Community Health*, 2012: 66(9): 761.

Biggs, B., King, L., Basu, S. & Stuckler, D. Is wealthier always healthier? The impact of national income level, inequality, and poverty on public health in Latin America. *Social Science & Medicine*, 2010; 71(2): 266–273.

Bornemisza, O., Ranson, M.K., Poletti, T.M. & Sondorp, E. Promoting health equity in conflict-affected fragile states. *Social Science & Medicine*, 2010; 70(1): 80–88.

Braveman, P.A., Egerter, S.A., Woolf, S.H. & Marks, J.S. When do we know enough to recommend action on the social determinants of health? *American Journal of Preventive Medicine*, 2011; 40(1 Suppl 1): S58–S66.

Brown, T. Cueto, M. & Fee, E. The World Health Organization and the transition from "international" to "global" public health. *American Journal of Public Health*, 2006; 96(1): 62–72.

Cairncross, S. & Kolsky, P. *Environmental Health and the Poor: Our Shared Responsibility*. Loughborough: University of Loughborough, 2000.

Casabonne, U. & Kenny, C. The best things in life are (nearly) free: Technology, knowledge, and global health. *World Development*, 2012; 40(1): 21–35.

Cepiku, D. & Giordano, F. Co-production in developing countries: Insights from the community health workers experience. *Public Management Review*, 2014; 16(3): 317–340.

Chauhan, L.S. Public health in India: Issues and challenges. *Indian Journal of Public Health*, 2011; 55(2): 88–91.

Colvin, C.J. Think locally, act globally: Developing a critical public health in the global South. *Critical Public Health*, 2011; 21(3): 253–256.

Consortium from Altarum Institute. A National Research Agenda for Public Health Services and Systems. *American Journal of Preventive Medicine*, 2012; 42(5): S72–S78.

Cook, J.T., Frank, D.A., Berkowitz, C., Black, M.M., Casey, P.H., Cutts, D.B., Meyers, A.F., Zaldivar, N., Skalicky, A., Levenson, S. & Heeren, T. Food insecurity is associated with adverse health outcomes among human infants and toddlers. *The Journal of Nutrition*, 2004; 134(6): 1432–1438.

Costa-Font, J. Housing assets and the socio-economic determinants of health and disability in old age. *Health and Place*, 2008; 14(3): 478–491.

Culyer, A.J., & Wagstaff, A. Equity and Equality in Health and Health Care. *Journal of Health Economics*, 1993; 12: 431–457.

Das, S. & Bose, K. Adult tribal malnutrition in India: An anthropometric and socio-demographic review. *Anthropological Review*, 2015; 78(1): 47–65.

Davison, C., Frankel, S. & Smith, G.D. The limits of lifestyle: re-assessing 'fatalism' in the popular culture of illness prevention. *Social Science & Medicine*, 1992; 34(6): 675–685.

De Jong, J.T.V.M. A public health framework to translate risk factors related to political violence and war into multi-level preventive interventions. *Social Science & Medicine*, 2010; 70(1): 71–79 (included in Volume III).

Dupre, M.E. Educational differences in health risks and illness over the life course: a test of cumulative disadvantage theory. *Social Science Research*, 2008; 37(4): 1253–1266.

Eriksen, T.H. What's wrong with the Global North and the Global South? Internet resource 2015. Available at: http://gssc.uni-koeln.de/node/454 [accessed 051016].

Estrada, J.H. Imperialist times and its implications for public health. *Revista Facultad Nacional de Salud Pública*, 2008; 26(2): 215–222.

Fertig, A. & Reingold, D. Public housing, health, and health behaviors: Is there a connection? *Journal of Policy Analysis and Management*, 2007; 26(4): 831.

Fledderus, J., Brandsen, T. & Honingh, M. User co-production of public service delivery: An uncertainty approach. *Public Policy and Administration*, 2015; 30(2): 145–164.

Fox, M. & Thomson, M. Realising social justice in public health law. *Medical Law Review*, 2013; 21(2): 278–309.

Gehlert, S., Sohmer, D., Sacks, T., Mininger, C., Mcclintock, M. & Olopade, O. Targeting health disparities: A model linking upstream determinants to downstream interventions. *Health Affairs* (Project Hope), 2008; 27(2): 339–349.

Gilles, S. & Traidl-Hoffmann, C. The environment-pathogen-host axis in communicable and non-communicable diseases: Recent advances in experimental and clinical research. *JDDG: Journal der Deutschen Dermatologischen Gesellschaft*, 2014; 12(5): 395–399.

Goldman, N., Kimbro, R.T., Turra, C.M. & Pebley, A.R. Socioeconomic gradients in health for white and Mexican-origin populations. *American Journal of Public Health*, 2006; 96(12): 2186–2193.

Herrin, W.E., Amaral, M.M. & Balihuta, A.M. The relationships between housing quality and occupant health in Uganda. *Social Science & Medicine*, 2013; 81: 115–122.

Habib, S.H. & Saha, S. Burden of non-communicable disease: Global overview. *Diabetes & Metabolic Syndrome: Clinical Research & Reviews*, 2010; 4(1): 41–47.

Heginbotham, C. & Newbigging, K., 2014. Commissioning Health and Wellbeing. Sage.

Hood, E. Dwelling disparities: How poor housing leads to poor health. *Environmental Health Perspectives*, 2005; 113(5): A311–A317.

Igene, H. Global health inequalities and breast cancer: An impending public health problem for developing countries. *The Breast Journal*, 2008; 14(5): 428–434.

Ingram, R.C., Douglas Scutchfield, G.P., Mays, M.W. & Bhandari, M.W. The economic, institutional, and political determinants of public health delivery system structures. *Public Health Reports*, 2012; 127(2): 208–215.

Karpati, A., Galea, S., Awerbuch, T. & Levins, R. Variability and vulnerability at the ecological level: Implications for understanding the social determinants of health. *American Journal of Public Health*, 2002; 92(11): 1768–1772.

Kern, A.L. & Bolay, J.C. Participatory Processes in Urban Planning Projects in China: The Example of Caoyang Village, Shanghai. In Technologies for Sustainable Development (pp. 209–223). Springer International Publishing, 2014.

Kern, A.L., Bolay, J.C. & Thanh, L.N. Peri-Urbanisation and the Vulnerability of Populations to the Effects of Climate Change in Southern Vietnam: Innovating Solutions in Research. In Technologies and Innovations for Development (pp. 155–167). Paris: Springer, 2012.

Kickbusch, I. Making the World Health Assembly fit for the 21st century. *BMJ*, 2014; 348.

Kumar, G.S., Kar, S.S. & Jain, A. Health and environmental sanitation in India: Issues for prioritizing control strategies. *Indian Journal of Occupational and Environmental Medicine*, 2011; 15(3): 93–96.

Kurland, J. Public health in the new millennium I: Technology. *Public Health Reports*, 1974; 115(1): 3–4.

Kurland, J. Public health in the new millennium II: Social exclusion. *Public Health Reports*, 2000; 115(4): 298.

Lee, H., Harris, K.M. & Lee, J. Multiple levels of social disadvantage and links to obesity in adolescence and young adulthood. *Journal of School Health*, 2013; 83(3): 139–149.

Masseria, C., Mladovsky, P. & Hernández-Quevedo, C. The socio-economic determinants of the health status of Roma in comparison with non-Roma in Bulgaria, Hungary and Romania. *European Journal of Public Health*, 2010; 20(5): 549–554.

Matlin, S., Kickbusch, I., Schäffner, E. & Stöckemann, S. *Global Health Challenges: How are France and Germany Responding? Report from 4 Workshops on Franco-German Cooperation.* Paris-Berlin Centre Virchow-Villermé for Public Health Paris-Berlin (CVV), Universitätsmedizin Berlin: Berlin, 2015.

Meyer, I.H. & Northridge, M. E. *The Health of Sexual Minorities*, 2007; pp. 242–267. New York: Springer.

Mignone, J. & O'Neil, J. Social capital and youth suicide risk factors in First Nations communities. *Canadian Journal of Public Health-Revue/Canadienne De Sante Publique*, 2005; 96(Suppl 1): S51–S54.

Minelli, L., Pigini, C., Chiavarini, M. & Bartolucci, F. Employment status and perceived health condition: longitudinal data from Italy. *BMC Public Health*, 2014; 14(1): 946–958.

Nixon, S.A. Critical public health ethics and Canada's role in global health. *Canadian Journal of Public Health/Revue Canadienne de Santé Publique*, 2006; 97(1): 32–34.

Östlin, P., Schrecker, T., Sadana, R., Bonnefoy, J., Gilson, L., Hertzman, C. Popay, J., Sen, G., Vaghri, Z. Priorities for research on equity and health: Towards an equity-focused health research agenda. *PLoS Med*, 2011; 8(11): e1001115.

Pickett, K.E. and Wilkinson, R.G. Income inequality and health: A causal review. *Social Science & Medicine*, 2015; 128, pp. 316–326.

Powell-Jackson, T., Basu, S., Balabanova, D., Mckee, M. & Stuckler, D. Democracy and growth in divided societies: A health-inequality trap? *Social Science & Medicine*, 2011; 73(1): 33–41.

Rechel, B., Blackburn, C.M., Spencer, N.J. & Rechel, B. Access to health care for Roma children in Central and Eastern Europe: Findings from a qualitative study in Bulgaria. *International Journal for Equity in Health*, 2009; 8(24): 1–10.

Scutchfield, F.D. & Howard, A.F. Moving on upstream: The role of health departments in addressing socio-ecologic determinants of disease. *American Journal of Preventive Medicine*, 2011; 40(1 Suppl 1): S80–S83.

Sepehri, A. & Guliani, H. Socioeconomic status and children's health: Evidence from a low-income country. *Social Science & Medicine*, 2015; 130: 23–31.

Sheikh, K. & Porter, J. Discursive gaps in the implementation of public health policy guidelines in India: The case of HIV testing. *Social Science & Medicine*, 2010; 71(11): 2005–2013.

Shetty, P., 2012. Public health: India's diabetes time bomb. *Nature*, 485(7398), pp. S14–S16.

Singh, A.R. & Singh, S.A. Diseases of poverty and lifestyle, well-being and human development. *Mens Sana Monographs*, 2008; 6(1): 187–225.

Smith, G.D. Equal, but different? Ecological, individual and instrumental approaches to understanding determinants of health. *International Journal of Epidemiology*, 2005; 34(6): 1179–1180.

Spellberg, B., Guidos, R., Gilbert, D., Bradley, J., Boucher, H.W., Scheld, W.M., Bartlett, J.G. & Edwards, J. The epidemic of antibiotic-resistant infections 2004: A call to action for the medical community from the Infectious Diseases Society of America. *Clinical Infectious Diseases*, 2008; 46(2): 155–164.

Spencer, N. Social, economic, and political determinants of child health. *Paediatrics*, 2003; 112(2): 704–706.

Spiwak, R., Afifi, T.O., Halli, S., Garcia-Moreno, C. & Sareen, J. The relationship between physical intimate partner violence and sexually transmitted infection among women in India and the United States. *Journal of Interpersonal Violence*, 2013; 28(13): 2770–2791.

Stansfeld, S.A., Clark, C., Rodgers, B., Caldwell, T. & Power, C. Repeated exposure to socioeconomic disadvantage and health selection as life course pathways to mid-life depressive and anxiety disorders. *Social Psychiatry and Psychiatric Epidemiology*, 2011; 46(7): 549–558.

Stegeman, I. & Costongs, C. *The Right Start to a Healthy Life*. Brussels: EuroHealthNet, 2012.

Stratigaki, M. Gender mainstreaming vs positive action. An ongoing conflict in EU gender equality policy. *European Journal of Women's Studies*, 2005; 12(2): 165–186.

Subrahmanian, R. Making sense of gender in shifting institutional contexts: some reflections on gender mainstreaming. *IDS Bulletin*, 2004; 35(4): 89–94.

Tausch, N., Hewstone, M. & Roy, R. The relationships between contact, status and prejudice: An integrated threat theory analysis of Hindu-Muslim relations in India. *Journal of Community & Applied Social Psychology*, 2009; 19(2): 83–94.

Tipper, A. Economic models of the family and the relationship between economic status and health. *Social Science & Medicine*, 2010; 70(10): 1567–1573.

Van Eijk, C. & Steen, T. Why engage in co-production of public services? Mixing theory and empirical evidence. *International Review of Administrative Sciences*, 2016; 82(1): 28–46.

Viner, R.M., Ozer, E.M., Denny, S., Marmot, M., Resnick, M., Fatusi, A. & Currie, C. Adolescence and the social determinants of health. *The Lancet*, 2012; 379(9826): 1641–1652.

Wilkinson, R. & Pickett, K. *The Spirit Level: Why More Equal Societies Almost Always Do Better*. London: Allen Lane, 2009.

Wittchen, H.U., Jacobi, F., Rehm, J., Gustavsson, A., Svensson, M., Jönsson, B., Olesen, J., Allgulander, C., Alonso, J., Faravelli, C., Fratiglioni, L., Jennum, P., Lieb, R., Maercker, A., van Os, J., Preisig, M., Salvador-Carulla, L., Simon, R. & Steinhausen, H.-C. The size and burden of mental disorders and other disorders of the brain in Europe 2010. *European Neuropsychopharmacology*, 2011; 21(9): 655–679.

White, F. & Nanan, D.J. International and global health. In Maxcy-Rosenau-Last (ed.), *Public Health and Preventive Medicine*. New York: McGraw Hill. 2008.

Whitehead, M. The concepts and principles of equity in health. International Journal of Health Services, 1992; 22: 429–445.

Whitehead, M. & Popay, J. Swimming upstream? Taking action on the social determinants of health inequalities. *Social Science & Medicine*, 2010; 71(7): 1234–1236.

WHO Commission on Social Determinants of Health. Closing the gap in a generation: health equity through action on the social determinants of health: Final report of the commission on social determinants of health. Geneva: World Health Commission, 2008.

Overarching International Declarations on Public Health

1

Declaration of Alma-Ata
World Health Organization (1978)

The International Conference on Primary Health Care, meeting in Alma-Ata this twelfth day of September in the year Nineteen hundred and seventy-eight, expressing the need for urgent action by all governments, all health and development workers, and the world community to protect and promote the health of all the people of the world, hereby makes the following.

Declaration

I

The Conference strongly reaffirms that health, which is a state of complete physical, mental and social wellbeing, and not merely the absence of disease or infirmity, is a fundamental human right and that the attainment of the highest possible level of health is a most important world-wide social goal whose realization requires the action of many other social and economic sectors in addition to the health sector.

II

The existing gross inequality in the health status of the people particularly between developed and developing countries as well as within countries is politically, socially and economically unacceptable and is, therefore, of common concern to all countries.

Source: International Conference on Primary Health Care, Alma-Ata, USSR, 6–12 September 1978.

III

Economic and social development, based on a New International Economic Order, is of basic importance to the fullest attainment of health for all and to the reduction of the gap between the health status of the developing and developed countries. The promotion and protection of the health of the people is essential to sustained economic and social development and contributes to a better quality of life and to world peace.

IV

The people have the right and duty to participate individually and collectively in the planning and implementation of their health care.

V

Governments have a responsibility for the health of their people which can be fulfilled only by the provision of adequate health and social measures. A main social target of governments, international organizations and the whole world community in the coming decades should be the attainment by all peoples of the world by the year 2000 of a level of health that will permit them to lead a socially and economically productive life. Primary health care is the key to attaining this target as part of development in the spirit of social justice.

VI

Primary health care is essential health care based on practical, scientifically sound and socially acceptable methods and technology made universally accessible to individuals and families in the community through their full participation and at a cost that the community and country can afford to maintain at every stage of their development in the spirit of selfreliance and self-determination. It forms an integral part both of the country's health system, of which it is the central function and main focus, and of the overall social and economic development of the community. It is the first level of contact of individuals, the family and community with the national health system bringing health care as close as possible to where people live and work, and constitutes the first element of a continuing health care process.

VII

Primary health care:

1. reflects and evolves from the economic conditions and sociocultural and political characteristics of the country and its communities and is based

on the application of the relevant results of social, biomedical and health services research and public health experience;

2. addresses the main health problems in the community, providing promotive, preventive, curative and rehabilitative services accordingly;
3. includes at least: education concerning prevailing health problems and the methods of preventing and controlling them; promotion of food supply and proper nutrition; an adequate supply of safe water and basic sanitation; maternal and child health care, including family planning; immunization against the major infectious diseases; prevention and control of locally endemic diseases; appropriate treatment of common diseases and injuries; and provision of essential drugs;
4. involves, in addition to the health sector, all related sectors and aspects of national and community development, in particular agriculture, animal husbandry, food, industry, education, housing, public works, communications and other sectors; and demands the coordinated efforts of all those sectors;
5. requires and promotes maximum community and individual self-reliance and participation in the planning, organization, operation and control of primary health care, making fullest use of local, national and other available resources; and to this end develops through appropriate education the ability of communities to participate;
6. should be sustained by integrated, functional and mutually supportive referral systems, leading to the progressive improvement of comprehensive health care for all, and giving priority to those most in need;
7. relies, at local and referral levels, on health workers, including physicians, nurses, midwives, auxiliaries and community workers as applicable, as well as traditional practitioners as needed, suitably trained socially and technically to work as a health team and to respond to the expressed health needs of the community.

VIII

All governments should formulate national policies, strategies and plans of action to launch and sustain primary health care as part of a comprehensive national health system and in coordination with other sectors. To this end, it will be necessary to exercise political will, to mobilize the country's resources and to use available external resources rationally.

IX

All countries should cooperate in a spirit of partnership and service to ensure primary health care for all people since the attainment of health by people in any one country directly concerns and benefits every other country. In this context the joint WHO/UNICEF report on primary health care constitutes a solid basis for the further development and operation of primary health care throughout the world.

X

An acceptable level of health for all the people of the world by the year 2000 can be attained through a fuller and better use of the world's resources, a considerable part of which is now spent on armaments and military conflicts. A genuine policy of independence, peace, détente and disarmament could and should release additional resources that could well be devoted to peaceful aims and in particular to the acceleration of social and economic development of which primary health care, as an essential part, should be allotted its proper share.

The International Conference on Primary Health Care calls for urgent and effective national and international action to develop and implement primary health care throughout the world and particularly in developing countries in a spirit of technical cooperation and in keeping with a New International Economic Order. It urges governments, WHO and UNICEF, and other international organizations, as well as multilateral and bilateral agencies, nongovernmental organizations, funding agencies, all health workers and the whole world community to support national and international commitment to primary health care and to channel increased technical and financial support to it, particularly in developing countries. The Conference calls on all the aforementioned to collaborate in introducing, developing and maintaining primary health care in accordance with the spirit and content of this Declaration.

2

The Ottawa Charter for Health Promotion
World Health Organization (1986)

The first International Conference on Health Promotion, meeting in Ottawa this 21st day of November 1986, hereby presents this CHARTER for action to achieve Health for All by the year 2000 and beyond.

This conference was primarily a response to growing expectations for a new public health movement around the world. Discussions focused on the needs in industrialized countries, but took into account similar concerns in all other regions. It built on the progress made through the Declaration on Primary Health Care at Alma-Ata, the World Health Organization's Targets for Health for All document, and the recent debate at the World Health Assembly on intersectoral action for health.

Health Promotion

Health promotion is the process of enabling people to increase control over, and to improve, their health. To reach a state of complete physical, mental and social well-being, an individual or group must be able to identify and to realize aspirations, to satisfy needs, and to change or cope with the environment. Health is, therefore, seen as a resource for everyday life, not the objective of living. Health is a positive concept emphasizing social and personal resources, as well as physical capacities. Therefore, health promotion is not just the responsibility of the health sector, but goes beyond healthy life-styles to well-being.

Source: First International Conference on Health Promotion, Ottawa, 21 November 1986.

Prerequisites for Health

The fundamental conditions and resources for health are:

- peace,
- shelter,
- education,
- food,
- income,
- a stable eco-system,
- sustainable resources,
- social justice, and equity.

Improvement in health requires a secure foundation in these basic prerequisites.

Advocate

Good health is a major resource for social, economic and personal development and an important dimension of quality of life. Political, economic, social, cultural, environmental, behavioural and biological factors can all favour health or be harmful to it. Health promotion action aims at making these conditions favourable through advocacy for health.

Enable

Health promotion focuses on achieving equity in health. Health promotion action aims at reducing differences in current health status and ensuring equal opportunities and resources to enable all people to achieve their fullest health potential. This includes a secure foundation in a supportive environment, access to information, life skills and opportunities for making healthy choices. People cannot achieve their fullest health potential unless they are able to take control of those things which determine their health. This must apply equally to women and men.

Mediate

The prerequisites and prospects for health cannot be ensured by the health sector alone. More importantly, health promotion demands coordinated action by all concerned: by governments, by health and other social and economic sectors, by nongovernmental and voluntary organization, by local authorities, by industry and by the media. People in all walks of life are involved as individuals, families and communities. Professional and social groups and health personnel have a major responsibility to mediate between differing interests in society for the pursuit of health.

Health promotion strategies and programmes should be adapted to the local needs and possibilities of individual countries and regions to take into account differing social, cultural and economic systems.

Health Promotion Action Means

Build Healthy Public Policy

Health promotion goes beyond health care. It puts health on the agenda of policy makers in all sectors and at all levels, directing them to be aware of the health consequences of their decisions and to accept their responsibilities for health.

Health promotion policy combines diverse but complementary approaches including legislation, fiscal measures, taxation and organizational change. It is coordinated action that leads to health, income and social policies that foster greater equity. Joint action contributes to ensuring safer and healthier goods and services, healthier public services, and cleaner, more enjoyable environments.

Health promotion policy requires the identification of obstacles to the adoption of healthy public policies in non-health sectors, and ways of removing them. The aim must be to make the healthier choice the easier choice for policy makers as well.

Create Supportive Environments

Our societies are complex and interrelated. Health cannot be separated from other goals. The inextricable links between people and their environment constitutes the basis for a socioecological approach to health. The overall guiding principle for the world, nations, regions and communities alike, is the need to encourage reciprocal maintenance – to take care of each other, our communities and our natural environment. The conservation of natural resources throughout the world should be emphasized as a global responsibility.

Changing patterns of life, work and leisure have a significant impact on health. Work and leisure should be a source of health for people. The way society organizes work should help create a healthy society. Health promotion generates living and working conditions that are safe, stimulating, satisfying and enjoyable.

Systematic assessment of the health impact of a rapidly changing environment – particularly in areas of technology, work, energy production and urbanization – is essential and must be followed by action to ensure positive benefit to the health of the public. The protection of the natural and built environments and the conservation of natural resources must be addressed in any health promotion strategy.

Strengthen Community Actions

Health promotion works through concrete and effective community action in setting priorities, making decisions, planning strategies and implementing them to achieve better health. At the heart of this process is the empowerment of communities – their ownership and control of their own endeavours and destinies.

Community development draws on existing human and material resources in the community to enhance self-help and social support, and to develop flexible systems for strengthening public participation in and direction of health matters. This requires full and continuous access to information, learning opportunities for health, as well as funding support.

Develop Personal Skills

Health promotion supports personal and social development through providing information, education for health, and enhancing life skills. By so doing, it increases the options available to people to exercise more control over their own health and over their environments, and to make choices conducive to health.

Enabling people to learn, throughout life, to prepare themselves for all of its stages and to cope with chronic illness and injuries is essential. This has to be facilitated in school, home, work and community settings. Action is required through educational, professional, commercial and voluntary bodies, and within the institutions themselves.

Reorient Health Services

The responsibility for health promotion in health services is shared among individuals, community groups, health professionals, health service institutions and governments.

They must work together towards a health care system which contributes to the pursuit of health. The role of the health sector must move increasingly in a health promotion direction, beyond its responsibility for providing clinical and curative services. Health services need to embrace an expanded mandate which is sensitive and respects cultural needs. This mandate should support the needs of individuals and communities for a healthier life, and open channels between the health sector and broader social, political, economic and physical environmental components.

Reorienting health services also requires stronger attention to health research as well as changes in professional education and training. This must lead to a change of attitude and organization of health services which refocuses on the total needs of the individual as a whole person.

Moving into the Future

Health is created and lived by people within the settings of their everyday life; where they learn, work, play and love. Health is created by caring for oneself and others, by being able to take decisions and have control over one's life circumstances, and by ensuring that the society one lives in creates conditions that allow the attainment of health by all its members.

Caring, holism and ecology are essential issues in developing strategies for health promotion. Therefore, those involved should take as a guiding principle that, in each phase of planning, implementation and evaluation of health promotion activities, women and men should become equal partners.

Commitment to Health Promotion

The participants in this Conference pledge:

- to move into the arena of healthy public policy, and to advocate a clear political commitment to health and equity in all sectors;
- to counteract the pressures towards harmful products, resource depletion, unhealthy living conditions and environments, and bad nutrition; and to focus attention on public health issues such as pollution, occupational hazards, housing and settlements;
- to respond to the health gap within and between societies, and to tackle the inequities in health produced by the rules and practices of these societies;
- to acknowledge people as the main health resource; to support and enable them to keep themselves, their families and friends healthy through financial and other means, and to accept the community as the essential voice in matters of its health, living conditions and well-being;
- to reorient health services and their resources towards the promotion of health; and to share power with other sectors, other disciplines and, most importantly, with people themselves;
- to recognize health and its maintenance as a major social investment and challenge; and to address the overall ecological issue of our ways of living.

The Conference urges all concerned to join them in their commitment to a strong public health alliance.

Call for International Action

The Conference calls on the World Health Organization and other international organizations to advocate the promotion of health in all appropriate forums and to support countries in setting up strategies and programmes for health promotion.

The Conference is firmly convinced that if people in all walks of life, nongovernmental and voluntary organizations, governments, the World Health Organization and all other bodies concerned join forces in introducing strategies for health promotion, in line with the moral and social values that form the basis of this CHARTER, Health For All by the year 2000 will become a reality.

CHARTER ADOPTED AT AN INTERNATIONAL CONFERENCE ON HEALTH PROMOTION[1] The move towards a new public health, November 17–21, 1986 Ottawa, Ontario, Canada.

Health Promotion Emblem

A brief explanation of the logo used by WHO since the First International Conference on Health Promotion held in Ottawa, Canada, in 1986. Select an element of the logo for the specific explanation of that part or simply read on for the complete explanation.

This logo was created for the First International Conference on Health Promotion held in Ottawa, Canada, in 1986. At that conference, the Ottawa Charter for Health Promotion was launched. Since then, WHO kept this symbol as the Health Promotion logo (HP logo), as it stands for the approach to health promotion as outlined in the Ottawa Charter.

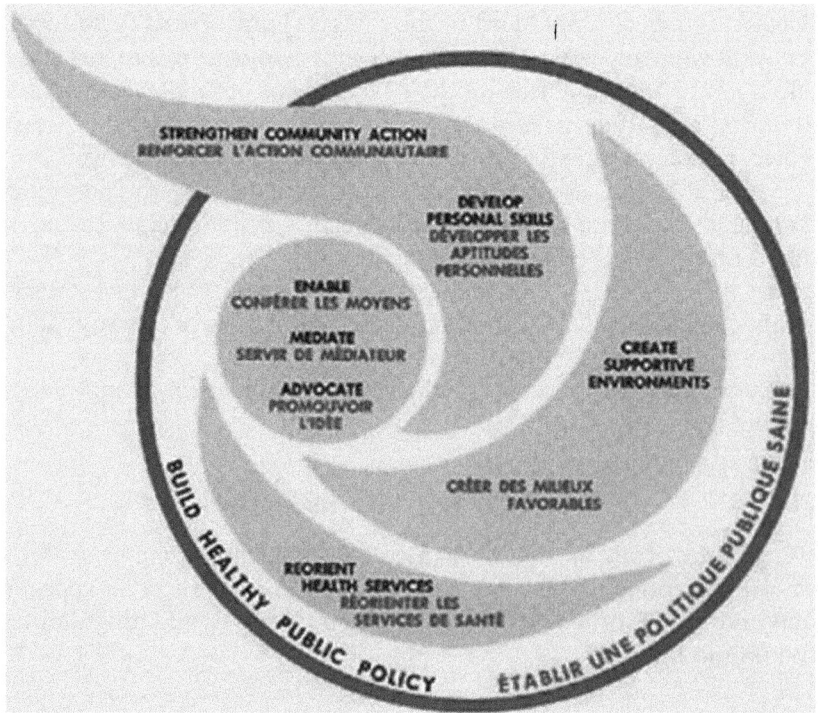

The logo represents a circle with 3 wings. It incorporates five key action areas in Health Promotion (build healthy public policy, create supportive environments for health, strengthen community action for health, develop personal skills, and re-orient health services) and three basic HP strategies (to enable, mediate, and advocate).

The main graphic elements of the HP logo are:

a. one outside circle,
b. one round spot within the circle, and
c. three wings that originate from this inner spot, one of which is breaking the outside circle.

a) The outside circle, originally in red colour, is representing the goal of "Building Healthy Public Policies", therefore symbolising the need for policies to "hold things together". This circle is encompassing the three wings, symbolising the need to address all five key action areas of health promotion identified in the Ottawa Charter in an integrated and complementary manner.

b) The round spot within the circle stands for the three basic strategies for health promotion, "enabling, mediating, and advocacy", which are needed and applied to all health promotion action areas. (Complete definitions of these terms can be found in the Health Promotion Glossary, WHO/HPR/HEP/98.1)

c) The three wings represent (and contain the words of) the five key action areas for health promotion that were identified in the Ottawa Charter for Health Promotion in 1986 and were reconfirmed in the Jakarta Declaration on Leading Health Promotion into the 21st Century in 1997.

More specifically:

• the upper wing that is breaking the circle represents that action is needed to "strengthen community action" and to "develop personal skills". This wing is breaking the circle to symbolise that society and communities as well as individuals are constantly changing and, therefore, the policy sphere has to constantly react and develop to reflect these changes: a "Healthy Public Policy" is needed;
• the middle wing on the right side represents that action is needed to "create supportive environments for health"
• the bottom wing represents that action is needed to "reorient health services" towards preventing diseases and promoting health.

Overall, the logo visualises the idea that Health Promotion is a comprehensive, multi-strategy approach. HP applies diverse strategies and methods in an integrated manner – one of the preconditions "for Health Promotion to be effective" (Jakarta Declaration 1997). Health Promotion addresses the key action areas identified in the Ottawa Charter in an integrated and coherent way.

The term Health Promotion (HP) was, and still today is sometimes, narrowly used as equivalent for Health Education (HE). But HE is one of several key components and action areas of HP as illustrated by the HP logo(see the key action area of "develop personal skills").

The HP logo and approach were reinforced at the second and third conferences on Health Promotion that took place in Sundsvall and in Adelaide.

In the light of the venue of the Fourth International Conference on Health Promotion, that was held in Jakarta, Indonesia, in July 1997, the design of the Ottawa logo was slightly modified to reflect culture and atmosphere of the host country of the conference, making sure that the shape and elements of the original logo were preserved, together with its inner meaning.

The Jakarta Conference logo is a more open and slightly more abstract version of the original HP logo from Ottawa. The three wings, that are now in brick-red colour, still represent the key HP action areas. The outside circle and the inner spot of the Ottawa logo are merged into a unique blue spot from where the three wings originate. This still symbolises that HP addresses its action areas with an integrated multi-strategic approach. Overall, the design of the HP logo adapted for the Conference in Jakarta is more open and lively; all the wings are now reaching out of the circle. This, visualizes the fact that the field of HP has grown and developed, and that today and in the future HP is outreaching to new players and partners, at all levels of society, from local to global level.

Note

1. Co-sponsored by the Canadian Public Health Association, Health and Welfare Canada, and the World Health Organization.

3

Beijing Declaration and Platform for Action at the Fourth World Conference on Women, Beijing 4–15th September

The United Nations (1995)

Resolution 1

Beijing Declaration and Platform for Action[1]

The Fourth World Conference on Women, Having met in Beijing from 4 to 15 September 1995,

1. *Adopts* the Beijing Declaration and Platform for Action, which are annexed to the present resolution;
2. *Recommends* to the General Assembly of the United Nations at its fiftieth session that it endorse the Beijing Declaration and Platform for Action as adopted by the Conference.

Annex I: Beijing Declaration

1. We, the Governments participating in the Fourth World Conference on Women,
2. Gathered here in Beijing in September 1995, the year of the fiftieth anniversary of the founding of the United Nations,
3. Determined to advance the goals of equality, development and peace for all women everywhere in the interest of all humanity,

---

Source: Report of the Fourth World Conference on Women Beijing, 4–15 September 1995.

4. Acknowledging the voices of all women everywhere and taking note of the diversity of women and their roles and circumstances, honouring the women who paved the way and inspired by the hope present in the world's youth,

5. Recognize that the status of women has advanced in some important respects in the past decade but that progress has been uneven, inequalities between women and men have persisted and major obstacles remain, with serious consequences for the well-being of all people,

6. Also recognize that this situation is exacerbated by the increasing poverty that is affecting the lives of the majority of the world's people, in particular women and children, with origins in both the national and international domains,

7. Dedicate ourselves unreservedly to addressing these constraints and obstacles and thus enhancing further the advancement and empowerment of women all over the world, and agree that this requires urgent action in the spirit of determination, hope, cooperation and solidarity, now and to carry us forward into the next century.

We reaffirm our commitment to:

8. The equal rights and inherent human dignity of women and men and other purposes and principles enshrined in the Charter of the United Nations, to the Universal Declaration of Human Rights and other international human rights instruments, in particular the Convention on the Elimination of All Forms of Discrimination against Women and the Convention on the Rights of the Child, as well as the Declaration on the Elimination of Violence against Women and the Declaration on the Right to Development;

9. Ensure the full implementation of the human rights of women and of the girl child as an inalienable, integral and indivisible part of all human rights and fundamental freedoms;

10. Build on consensus and progress made at previous United Nations conferences and summits – on women in Nairobi in 1985, on children in New York in 1990, on environment and development in Rio de Janeiro in 1992, on human rights in Vienna in 1993, on population and development in Cairo in 1994 and on social development in Copenhagen in 1995 with the objective of achieving equality, development and peace;

11. Achieve the full and effective implementation of the Nairobi Forward-looking Strategies for the Advancement of Women;

12. The empowerment and advancement of women, including the right to freedom of thought, conscience, religion and belief, thus contributing to the moral, ethical, spiritual and intellectual needs of women and men, individually or in community with others and thereby guaranteeing them the possibility of realizing their full potential in society and shaping their lives in accordance with their own aspirations.

We are convinced that:

13. Women's empowerment and their full participation on the basis of equality in all spheres of society, including participation in the decision-making process and access to power, are fundamental for the achievement of equality, development and peace;
14. Women's rights are human rights;
15. Equal rights, opportunities and access to resources, equal sharing of responsibilities for the family by men and women, and a harmonious partnership between them are critical to their well-being and that of their families as well as to the consolidation of democracy;
16. Eradication of poverty based on sustained economic growth, social development, environmental protection and social justice requires the involvement of women in economic and social development, equal opportunities and the full and equal participation of women and men as agents and beneficiaries of people-centred sustainable development;
17. The explicit recognition and reaffirmation of the right of all women to control all aspects of their health, in particular their own fertility, is basic to their empowerment;
18. Local, national, regional and global peace is attainable and is inextricably linked with the advancement of women, who are a fundamental force for leadership, conflict resolution and the promotion of lasting peace at all levels;
19. It is essential to design, implement and monitor, with the full participation of women, effective, efficient and mutually reinforcing gender-sensitive policies and programmes, including development policies and programmes, at all levels that will foster the empowerment and advancement of women;
20. The participation and contribution of all actors of civil society, particularly women's groups and networks and other non-governmental organizations and community-based organizations, with full respect for their autonomy, in cooperation with Governments, are important to the effective implementation and follow-up of the Platform for Action;
21. The implementation of the Platform for Action requires commitment from Governments and the international community. By making national and international commitments for action, including those made at the Conference, Governments and the international community recognize the need to take priority action for the empowerment and advancement of women.

We are determined to:

22. Intensify efforts and actions to achieve the goals of the Nairobi Forwardlooking Strategies for the Advancement of Women by the end of this century;
23. Ensure the full enjoyment by women and the girl child of all human rights and fundamental freedoms and take effective action against violations of these rights and freedoms;

24. Take all necessary measures to eliminate all forms of discrimination against women and the girl child and remove all obstacles to gender equality and the advancement and empowerment of women;
25. Encourage men to participate fully in all actions towards equality;
26. Promote women's economic independence, including employment, and eradicate the persistent and increasing burden of poverty on women by addressing the structural causes of poverty through changes in economic structures, ensuring equal access for all women, including those in rural areas, as vital development agents, to productive resources, opportunities and public services;
27. Promote people-centred sustainable development, including sustained economic growth, through the provision of basic education, life-long education, literacy and training, and primary health care for girls and women;
28. Take positive steps to ensure peace for the advancement of women and, recognizing the leading role that women have played in the peace movement, work actively towards general and complete disarmament under strict and effective international control, and support negotiations on the conclusion, without delay, of a universal and multilaterally and effectively verifiable comprehensive nuclear-test-ban treaty which contributes to nuclear disarmament and the prevention of the proliferation of nuclear weapons in all its aspects;
29. Prevent and eliminate all forms of violence against women and girls;
30. Ensure equal access to and equal treatment of women and men in education and health care and enhance women's sexual and reproductive health as well as education;
31. Promote and protect all human rights of women and girls;
32. Intensify efforts to ensure equal enjoyment of all human rights and fundamental freedoms for all women and girls who face multiple barriers to their empowerment and advancement because of such factors as their race, age, language, ethnicity, culture, religion, or disability, or because they are indigenous people;
33. Ensure respect for international law, including humanitarian law, in order to protect women and girls in particular;
34. Develop the fullest potential of girls and women of all ages, ensure their full and equal participation in building a better world for all and enhance their role in the development process.

We are determined to:

35. Ensure women's equal access to economic resources, including land, credit, science and technology, vocational training, information, communication and markets, as a means to further the advancement and empowerment of women and girls, including through the enhancement of their capacities to enjoy the benefits of equal access to these resources, inter alia, by means of international cooperation;

36. Ensure the success of the Platform for Action, which will require a strong commitment on the part of Governments, international organizations and institutions at all levels. We are deeply convinced that economic development, social development and environmental protection are interdependent and mutually reinforcing components of sustainable development, which is the framework for our efforts to achieve a higher quality of life for all people. Equitable social development that recognizes empowering the poor, particularly women living in poverty, to utilize environmental resources sustainably is a necessary foundation for sustainable development. We also recognize that broad-based and sustained economic growth in the context of sustainable development is necessary to sustain social development and social justice. The success of the Platform for Action will also require adequate mobilization of resources at the national and international levels as well as new and additional resources to the developing countries from all available funding mechanisms, including multilateral, bilateral and private sources for the advancement of women; financial resources to strengthen the capacity of national, subregional, regional and international institutions; a commitment to equal rights, equal responsibilities and equal opportunities and to the equal participation of women and men in all national, regional and international bodies and policy-making processes; and the establishment or strengthening of mechanisms at all levels for accountability to the world's women;

37. Ensure also the success of the Platform for Action in countries with economies in transition, which will require continued international cooperation and assistance;

38. We hereby adopt and commit ourselves as Governments to implement the following Platform for Action, ensuring that a gender perspective is reflected in all our policies and programmes. We urge the United Nations system, regional and international financial institutions, other relevant regional and international institutions and all women and men, as well as non-governmental organizations, with full respect for their autonomy, and all sectors of civil society, in cooperation with Governments, to fully commit themselves and contribute to the implementation of this Platform for Action.

Note

1. Adopted at the 16th plenary metting, on 15 September 1995; for the discussion, see chapter V.

4

Transforming Our World: The 2030 Agenda for Sustainable Development

The United Nations (2016)

Preamble

This Agenda is a plan of action for people, planet and prosperity. It also seeks to strengthen universal peace in larger freedom. We recognize that eradicating poverty in all its forms and dimensions, including extreme poverty, is the greatest global challenge and an indispensable requirement for sustainable development.

All countries and all stakeholders, acting in collaborative partnership, will implement this plan. We are resolved to free the human race from the tyranny of poverty and want and to heal and secure our planet. We are determined to take the bold and transformative steps which are urgently needed to shift the world on to a sustainable and resilient path. As we embark on this collective journey, we pledge that no one will be left behind.

The 17 Sustainable Development Goals and 169 targets which we are announcing today demonstrate the scale and ambition of this new universal Agenda. They seek to build on the Millennium Development Goals and complete what they did not achieve. They seek to realize the human rights of all and to achieve gender equality and the empowerment of all women and girls. They are integrated and indivisible and balance the three dimensions of sustainable development: the economic, social and environmental.

The Goals and targets will stimulate action over the next 15 years in areas of critical importance for humanity and the planet.

Source: Transforming Our World: The 2030 Agenda for Sustainable Development, A/RES/70/1 (2016).

People

We are determined to end poverty and hunger, in all their forms and dimensions, and to ensure that all human beings can fulfil their potential in dignity and equality and in a healthy environment.

Planet

We are determined to protect the planet from degradation, including through sustainable consumption and production, sustainably managing its natural resources and taking urgent action on climate change, so that it can support the needs of the present and future generations.

Prosperity

We are determined to ensure that all human beings can enjoy prosperous and fulfilling lives and that economic, social and technological progress occurs in harmony with nature.

Peace

We are determined to foster peaceful, just and inclusive societies which are free from fear and violence. There can be no sustainable development without peace and no peace without sustainable development.

Partnership

We are determined to mobilize the means required to implement this Agenda through a revitalized Global Partnership for Sustainable Development, based on a spirit of strengthened global solidarity, focused in particular on the needs of the poorest and most vulnerable and with the participation of all countries, all stakeholders and all people.

The interlinkages and integrated nature of the Sustainable Development Goals are of crucial importance in ensuring that the purpose of the new Agenda is realized. If we realize our ambitions across the full extent of the Agenda, the lives of all will be profoundly improved and our world will be transformed for the better.

Declaration

Introduction

1. We, the Heads of State and Government and High Representatives, meeting at United Nations Headquarters in New York from 25 to 27 September 2015 as the Organization celebrates its seventieth anniversary, have decided today on new global Sustainable Development Goals.

2. On behalf of the peoples we serve, we have adopted a historic decision on a comprehensive, far-reaching and people-centred set of universal and transformative Goals and targets. We commit ourselves to working tirelessly for the full implementation of this Agenda by 2030. We recognize that eradicating poverty in all its forms and dimensions, including extreme poverty, is the greatest global challenge and an indispensable requirement for sustainable development. We are committed to achieving sustainable development in its three dimensions – economic, social and environmental – in a balanced and integrated manner. We will also build upon the achievements of the Millennium Development Goals and seek to address their unfinished business.

3. We resolve, between now and 2030, to end poverty and hunger everywhere; to combat inequalities within and among countries; to build peaceful, just and inclusive societies; to protect human rights and promote gender equality and the empowerment of women and girls; and to ensure the lasting protection of the planet and its natural resources. We resolve also to create conditions for sustainable, inclusive and sustained economic growth, shared prosperity and decent work for all, taking into account different levels of national development and capacities.

4. As we embark on this great collective journey, we pledge that no one will be left behind. Recognizing that the dignity of the human person is fundamental, we wish to see the Goals and targets met for all nations and peoples and for all segments of society. And we will endeavour to reach the furthest behind first.

5. This is an Agenda of unprecedented scope and significance. It is accepted by all countries and is applicable to all, taking into account different national realities, capacities and levels of development and respecting national policies and priorities. These are universal goals and targets which involve the entire world, developed and developing countries alike. They are integrated and indivisible and balance the three dimensions of sustainable development.

6. The Goals and targets are the result of over two years of intensive public consultation and engagement with civil society and other stakeholders around the world, which paid particular attention to the voices of the poorest and most vulnerable. This consultation included valuable work done by the Open Working Group of the General Assembly on Sustainable Development Goals and by the United Nations, whose Secretary-General provided a synthesis report in December 2014.

Our Vision

7. In these Goals and targets, we are setting out a supremely ambitious and transformational vision. We envisage a world free of poverty, hunger, disease and want, where all life can thrive. We envisage a world free of fear

and violence. A world with universal literacy. A world with equitable and universal access to quality education at all levels, to health care and social protection, where physical, mental and social well-being are assured. A world where we reaffirm our commitments regarding the human right to safe drinking water and sanitation and where there is improved hygiene; and where food is sufficient, safe, affordable and nutritious. A world where human habitats are safe, resilient and sustainable and where there is universal access to affordable, reliable and sustainable energy.

8. We envisage a world of universal respect for human rights and human dignity, the rule of law, justice, equality and non-discrimination; of respect for race, ethnicity and cultural diversity; and of equal opportunity permitting the full realization of human potential and contributing to shared prosperity. A world which invests in its children and in which every child grows up free from violence and exploitation. A world in which every woman and girl enjoys full gender equality and all legal, social and economic barriers to their empowerment have been removed. A just, equitable, tolerant, open and socially inclusive world in which the needs of the most vulnerable are met.

9. We envisage a world in which every country enjoys sustained, inclusive and sustainable economic growth and decent work for all. A world in which consumption and production patterns and use of all natural resources – from air to land, from rivers, lakes and aquifers to oceans and seas – are sustainable. One in which democracy, good governance and the rule of law, as well as an enabling environment at the national and international levels, are essential for sustainable development, including sustained and inclusive economic growth, social development, environmental protection and the eradication of poverty and hunger. One in which development and the application of technology are climate-sensitive, respect biodiversity and are resilient. One in which humanity lives in harmony with nature and in which wildlife and other living species are protected.

Our Shared Principles and Commitments

10. The new Agenda is guided by the purposes and principles of the Charter of the United Nations, including full respect for international law. It is grounded in the Universal Declaration of Human Rights, international human rights treaties, the Millennium Declaration and the 2005 World Summit Outcome. It is informed by other instruments such as the Declaration on the Right to Development.

11. We reaffirm the outcomes of all major United Nations conferences and summits which have laid a solid foundation for sustainable development and have helped to shape the new Agenda. These include the Rio

Declaration on Environment and Development, the World Summit on Sustainable Development, the World Summit for Social Development, the Programme of Action of the International Conference on Population and Development, the Beijing Platform for Action and the United Nations Conference on Sustainable Development. We also reaffirm the follow-up to these conferences, including the outcomes of the Fourth United Nations Conference on the Least Developed Countries, the third International Conference on Small Island Developing States, the second United Nations Conference on Landlocked Developing Countries and the Third United Nations World Conference on Disaster Risk Reduction.

12. We reaffirm all the principles of the Rio Declaration on Environment and Development, including, inter alia, the principle of common but differentiated responsibilities, as set out in principle 7 thereof.

13. The challenges and commitments identified at these major conferences and summits are interrelated and call for integrated solutions. To address them effectively, a new approach is needed. Sustainable development recognizes that eradicating poverty in all its forms and dimensions, combating inequality within and among countries, preserving the planet, creating sustained, inclusive and sustainable economic growth and fostering social inclusion are linked to each other and are interdependent.

Our World Today

14. We are meeting at a time of immense challenges to sustainable development. Billions of our citizens continue to live in poverty and are denied a life of dignity. There are rising inequalities within and among countries. There are enormous disparities of opportunity, wealth and power. Gender inequality remains a key challenge. Unemployment, particularly youth unemployment, is a major concern. Global health threats, more frequent and intense natural disasters, spiralling conflict, violent extremism, terrorism and related humanitarian crises and forced displacement of people threaten to reverse much of the development progress made in recent decades. Natural resource depletion and adverse impacts of environmental degradation, including desertification, drought, land degradation, freshwater scarcity and loss of biodiversity, add to and exacerbate the list of challenges which humanity faces. Climate change is one of the greatest challenges of our time and its adverse impacts undermine the ability of all countries to achieve sustainable development. Increases in global temperature, sea level rise, ocean acidification and other climate change impacts are seriously affecting coastal areas and low-lying coastal countries, including many least developed countries and small island developing States. The survival of many societies, and of the biological support systems of the planet, is at risk.

15. It is also, however, a time of immense opportunity. Significant progress has been made in meeting many development challenges. Within the past generation, hundreds of millions of people have emerged from extreme poverty. Access to education has greatly increased for both boys and girls. The spread of information and communications technology and global interconnectedness has great potential to accelerate human progress, to bridge the digital divide and to develop knowledge societies, as does scientific and technological innovation across areas as diverse as medicine and energy.

16. Almost 15 years ago, the Millennium Development Goals were agreed. These provided an important framework for development and significant progress has been made in a number of areas. But the progress has been uneven, particularly in Africa, least developed countries, landlocked developing countries and small island developing States, and some of the Millennium Development Goals remain off-track, in particular those related to maternal, newborn and child health and to reproductive health. We recommit ourselves to the full realization of all the Millennium Development Goals, including the off-track Millennium Development Goals, in particular by providing focused and scaled-up assistance to least developed countries and other countries in special situations, in line with relevant support programmes. The new Agenda builds on the Millennium Development Goals and seeks to complete what they did not achieve, particularly in reaching the most vulnerable.

17. In its scope, however, the framework we are announcing today goes far beyond the Millennium Development Goals. Alongside continuing development priorities such as poverty eradication, health, education and food security and nutrition, it sets out a wide range of economic, social and environmental objectives. It also promises more peaceful and inclusive societies. It also, crucially, defines means of implementation. Reflecting the integrated approach that we have decided on, there are deep interconnections and many cross-cutting elements across the new Goals and targets.

The New Agenda

18. We are announcing today 17 Sustainable Development Goals with 169 associated targets which are integrated and indivisible. Never before have world leaders pledged common action and endeavour across such a broad and universal policy agenda. We are setting out together on the path towards sustainable development, devoting ourselves collectively to the pursuit of global development and of "win-win" cooperation which can bring huge gains to all countries and all parts of the world. We reaffirm that every State has, and shall freely exercise, full permanent sovereignty over all its wealth, natural resources and economic activity.

We will implement the Agenda for the full benefit of all, for today's generation and for future generations. In doing so, we reaffirm our commitment to international law and emphasize that the Agenda is to be implemented in a manner that is consistent with the rights and obligations of States under international law.

19. We reaffirm the importance of the Universal Declaration of Human Rights, as well as other international instruments relating to human rights and international law. We emphasize the responsibilities of all States, in conformity with the Charter of the United Nations, to respect, protect and promote human rights and fundamental freedoms for all, without distinction of any kind as to race, colour, sex, language, religion, political or other opinion, national or social origin, property, birth, disability or other status.

20. Realizing gender equality and the empowerment of women and girls will make a crucial contribution to progress across all the Goals and targets. The achievement of full human potential and of sustainable development is not possible if one half of humanity continues to be denied its full human rights and opportunities. Women and girls must enjoy equal access to quality education, economic resources and political participation as well as equal opportunities with men and boys for employment, leadership and decision-making at all levels. We will work for a significant increase in investments to close the gender gap and strengthen support for institutions in relation to gender equality and the empowerment of women at the global, regional and national levels. All forms of discrimination and violence against women and girls will be eliminated, including through the engagement of men and boys. The systematic mainstreaming of a gender perspective in the implementation of the Agenda is crucial.

21. The new Goals and targets will come into effect on 1 January 2016 and will guide the decisions we take over the next 15 years. All of us will work to implement the Agenda within our own countries and at the regional and global levels, taking into account different national realities, capacities and levels of development and respecting national policies and priorities. We will respect national policy space for sustained, inclusive and sustainable economic growth, in particular for developing States, while remaining consistent with relevant international rules and commitments. We acknowledge also the importance of the regional and subregional dimensions, regional economic integration and interconnectivity in sustainable development. Regional and subregional frameworks can facilitate the effective translation of sustainable development policies into concrete action at the national level.

22. Each country faces specific challenges in its pursuit of sustainable development. The most vulnerable countries and, in particular, African countries, least developed countries, landlocked developing countries and

small island developing States, deserve special attention, as do countries in situations of conflict and post-conflict countries. There are also serious challenges within many middle-income countries.

23. People who are vulnerable must be empowered. Those whose needs are reflected in the Agenda include all children, youth, persons with disabilities (of whom more than 80 percent live in poverty), people living with HIV/AIDS, older persons, indigenous peoples, refugees and internally displaced persons and migrants. We resolve to take further effective measures and actions, in conformity with international law, to remove obstacles and constraints, strengthen support and meet the special needs of people living in areas affected by complex humanitarian emergencies and in areas affected by terrorism.

24. We are committed to ending poverty in all its forms and dimensions, including by eradicating extreme poverty by 2030. All people must enjoy a basic standard of living, including through social protection systems. We are also determined to end hunger and to achieve food security as a matter of priority and to end all forms of malnutrition. In this regard, we reaffirm the important role and inclusive nature of the Committee on World Food Security and welcome the Rome Declaration on Nutrition and the Framework for Action. We will devote resources to developing rural areas and sustainable agriculture and fisheries, supporting smallholder farmers, especially women farmers, herders and fishers in developing countries, particularly least developed countries.

25. We commit to providing inclusive and equitable quality education at all levels – early childhood, primary, secondary, tertiary, technical and vocational training. All people, irrespective of sex, age, race or ethnicity, and persons with disabilities, migrants, indigenous peoples, children and youth, especially those in vulnerable situations, should have access to life-long learning opportunities that help them to acquire the knowledge and skills needed to exploit opportunities and to participate fully in society. We will strive to provide children and youth with a nurturing environment for the full realization of their rights and capabilities, helping our countries to reap the demographic dividend, including through safe schools and cohesive communities and families.

26. To promote physical and mental health and well-being, and to extend life expectancy for all, we must achieve universal health coverage and access to quality health care. No one must be left behind. We commit to accelerating the progress made to date in reducing newborn, child and maternal mortality by ending all such preventable deaths before 2030. We are committed to ensuring universal access to sexual and reproductive health-care services, including for family planning, information and education. We will equally accelerate the pace of progress made in fighting malaria, HIV/AIDS, tuberculosis, hepatitis, Ebola and other communicable diseases and epidemics, including by addressing growing

anti-microbial resistance and the problem of unattended diseases affecting developing countries. We are committed to the prevention and treatment of non-communicable diseases, including behavioural, developmental and neurological disorders, which constitute a major challenge for sustainable development.

27. We will seek to build strong economic foundations for all our countries. Sustained, inclusive and sustainable economic growth is essential for prosperity. This will only be possible if wealth is shared and income inequality is addressed. We will work to build dynamic, sustainable, innovative and people-centred economies, promoting youth employment and women's economic empowerment, in particular, and decent work for all. We will eradicate forced labour and human trafficking and end child labour in all its forms. All countries stand to benefit from having a healthy and well-educated workforce with the knowledge and skills needed for productive and fulfilling work and full participation in society. We will strengthen the productive capacities of least developed countries in all sectors, including through structural transformation. We will adopt policies which increase productive capacities, productivity and productive employment; financial inclusion; sustainable agriculture, pastoralist and fisheries development; sustainable industrial development; universal access to affordable, reliable, sustainable and modern energy services; sustainable transport systems; and quality and resilient infrastructure.

28. We commit to making fundamental changes in the way that our societies produce and consume goods and services. Governments, international organizations, the business sector and other non-State actors and individuals must contribute to changing unsustainable consumption and production patterns, including through the mobilization, from all sources, of financial and technical assistance to strengthen developing countries' scientific, technological and innovative capacities to move towards more sustainable patterns of consumption and production. We encourage the implementation of the 10-Year Framework of Programmes on Sustainable Consumption and Production Patterns. All countries take action, with developed countries taking the lead, taking into account the development and capabilities of developing countries.

29. We recognize the positive contribution of migrants for inclusive growth and sustainable development. We also recognize that international migration is a multidimensional reality of major relevance for the development of countries of origin, transit and destination, which requires coherent and comprehensive responses. We will cooperate internationally to ensure safe, orderly and regular migration involving full respect for human rights and the humane treatment of migrants regardless of migration status, of refugees and of displaced persons. Such cooperation should also strengthen the resilience of communities hosting refugees, particularly in developing countries. We underline the right of migrants

to return to their country of citizenship, and recall that States must ensure that their returning nationals are duly received.

30. States are strongly urged to refrain from promulgating and applying any unilateral economic, financial or trade measures not in accordance with international law and the Charter of the United Nations that impede the full achievement of economic and social development, particularly in developing countries.

31. We acknowledge that the United Nations Framework Convention on Climate Change is the primary international, intergovernmental forum for negotiating the global response to climate change. We are determined to address decisively the threat posed by climate change and environmental degradation. The global nature of climate change calls for the widest possible international cooperation aimed at accelerating the reduction of global greenhouse gas emissions and addressing adaptation to the adverse impacts of climate change. We note with grave concern the significant gap between the aggregate effect of parties' mitigation pledges in terms of global annual emissions of greenhouse gases by 2020 and aggregate emission pathways consistent with having a likely chance of holding the increase in global average temperature below 2 degrees Celsius or 1.5 degrees Celsius above pre-industrial levels.

32. Looking ahead to the twenty-first session of the Conference of the Parties in Paris, we underscore the commitment of all States to work for an ambitious and universal climate agreement. We reaffirm that the protocol, another legal instrument or agreed outcome with legal force under the Convention applicable to all parties shall address in a balanced manner, inter alia, mitigation, adaptation, finance, technology development and transfer and capacity-building; and transparency of action and support.

33. We recognize that social and economic development depends on the sustainable management of our planet's natural resources. We are therefore determined to conserve and sustainably use oceans and seas, freshwater resources, as well as forests, mountains and drylands and to protect biodiversity, ecosystems and wildlife. We are also determined to promote sustainable tourism, to tackle water scarcity and water pollution, to strengthen cooperation on desertification, dust storms, land degradation and drought and to promote resilience and disaster risk reduction. In this regard, we look forward to the thirteenth meeting of the Conference of the Parties to the Convention on Biological Diversity to be held in Mexico.

34. We recognize that sustainable urban development and management are crucial to the quality of life of our people. We will work with local authorities and communities to renew and plan our cities and human settlements so as to foster community cohesion and personal security and to stimulate innovation and employment. We will reduce the negative impacts of urban activities and of chemicals which are hazardous for

human health and the environment, including through the environmentally sound management and safe use of chemicals, the reduction and recycling of waste and the more efficient use of water and energy. And we will work to minimize the impact of cities on the global climate system. We will also take account of population trends and projections in our national rural and urban development strategies and policies. We look forward to the upcoming United Nations Conference on Housing and Sustainable Urban Development to be held in Quito.

35. Sustainable development cannot be realized without peace and security; and peace and security will be at risk without sustainable development. The new Agenda recognizes the need to build peaceful, just and inclusive societies that provide equal access to justice and that are based on respect for human rights (including the right to development), on effective rule of law and good governance at all levels and on transparent, effective and accountable institutions. Factors which give rise to violence, insecurity and injustice, such as inequality, corruption, poor governance and illicit financial and arms flows, are addressed in the Agenda. We must redouble our efforts to resolve or prevent conflict and to support post-conflict countries, including through ensuring that women have a role in peacebuilding and State-building. We call for further effective measures and actions to be taken, in conformity with international law, to remove the obstacles to the full realization of the right of self-determination of peoples living under colonial and foreign occupation, which continue to adversely affect their economic and social development as well as their environment.

36. We pledge to foster intercultural understanding, tolerance, mutual respect and an ethic of global citizenship and shared responsibility. We acknowledge the natural and cultural diversity of the world and recognize that all cultures and civilizations can contribute to, and are crucial enablers of, sustainable development.

37. Sport is also an important enabler of sustainable development. We recognize the growing contribution of sport to the realization of development and peace in its promotion of tolerance and respect and the contributions it makes to the empowerment of women and of young people, individuals and communities as well as to health, education and social inclusion objectives.

38. We reaffirm, in accordance with the Charter of the United Nations, the need to respect the territorial integrity and political independence of States.

Means of Implementation

39. The scale and ambition of the new Agenda requires a revitalized Global Partnership to ensure its implementation. We fully commit to this. This Partnership will work in a spirit of global solidarity, in particular

solidarity with the poorest and with people in vulnerable situations. It will facilitate an intensive global engagement in support of implementation of all the Goals and targets, bringing together Governments, the private sector, civil society, the United Nations system and other actors and mobilizing all available resources.

40. The means of implementation targets under Goal 17 and under each Sustainable Development Goal are key to realizing our Agenda and are of equal importance with the other Goals and targets. The Agenda, including the Sustainable Development Goals, can be met within the framework of a revitalized Global Partnership for Sustainable Development, supported by the concrete policies and actions as outlined in the outcome document of the third International Conference on Financing for Development, held in Addis Ababa from 13 to 16 July 2015. We welcome the endorsement by the General Assembly of the Addis Ababa Action Agenda, which is an integral part of the 2030 Agenda for Sustainable Development. We recognize that the full implementation of the Addis Ababa Action Agenda is critical for the realization of the Sustainable Development Goals and targets.

41. We recognize that each country has primary responsibility for its own economic and social development. The new Agenda deals with the means required for implementation of the Goals and targets. We recognize that these will include the mobilization of financial resources as well as capacity-building and the transfer of environmentally sound technologies to developing countries on favourable terms, including on concessional and preferential terms, as mutually agreed. Public finance, both domestic and international, will play a vital role in providing essential services and public goods and in catalysing other sources of finance. We acknowledge the role of the diverse private sector, ranging from micro-enterprises to cooperatives to multinationals, and that of civil society organizations and philanthropic organizations in the implementation of the new Agenda.

42. We support the implementation of relevant strategies and programmes of action, including the Istanbul Declaration and Programme of Action, the SIDS Accelerated Modalities of Action (SAMOA) Pathway and the Vienna Programme of Action for Landlocked Developing Countries for the Decade 2014–2024, and reaffirm the importance of supporting the African Union's Agenda 2063 and the programme of the New Partnership for Africa's Development, all of which are integral to the new Agenda. We recognize the major challenge to the achievement of durable peace and sustainable development in countries in conflict and post-conflict situations.

43. We emphasize that international public finance plays an important role in complementing the efforts of countries to mobilize public resources domestically, especially in the poorest and most vulnerable countries with limited domestic resources. An important use of international public finance, including official development assistance (ODA), is to

catalyse additional resource mobilization from other sources, public and private. ODA providers reaffirm their respective commitments, including the commitment by many developed countries to achieve the target of 0.7 per cent of gross national income for official development assistance (ODA/GNI) to developing countries and 0.15 per cent to 0.2 per cent of ODA/GNI to least developed countries.

44. We acknowledge the importance for international financial institutions to support, in line with their mandates, the policy space of each country, in particular developing countries. We recommit to broadening and strengthening the voice and participation of developing countries – including African countries, least developed countries, landlocked developing countries, small island developing States and middle-income countries – in international economic decision-making, norm-setting and global economic governance.

45. We acknowledge also the essential role of national parliaments through their enactment of legislation and adoption of budgets and their role in ensuring accountability for the effective implementation of our commitments. Governments and public institutions will also work closely on implementation with regional and local authorities, subregional institutions, international institutions, academia, philanthropic organizations, volunteer groups and others.

46. We underline the important role and comparative advantage of an adequately resourced, relevant, coherent, efficient and effective United Nations system in supporting the achievement of the Sustainable Development Goals and sustainable development. While stressing the importance of strengthened national ownership and leadership at the country level, we express our support for the ongoing dialogue in the Economic and Social Council on the longer-term positioning of the United Nations development system in the context of this Agenda.

Follow-Up and Review

47. Our Governments have the primary responsibility for follow-up and review, at the national, regional and global levels, in relation to the progress made in implementing the Goals and targets over the coming 15 years. To support accountability to our citizens, we will provide for systematic follow-up and review at the various levels, as set out in this Agenda and the Addis Ababa Action Agenda. The high-level political forum under the auspices of the General Assembly and the Economic and Social Council will have the central role in overseeing follow-up and review at the global level.

48. Indicators are being developed to assist this work. Quality, accessible, timely and reliable disaggregated data will be needed to help with the measurement of progress and to ensure that no one is left behind. Such data is key to decision-making. Data and information from existing

reporting mechanisms should be used where possible. We agree to intensify our efforts to strengthen statistical capacities in developing countries, particularly African countries, least developed countries, land-locked developing countries, small island developing States and middle-income countries. We are committed to developing broader measures of progress to complement gross domestic product.

A Call for Action to Change Our World

49. Seventy years ago, an earlier generation of world leaders came together to create the United Nations. From the ashes of war and division they fashioned this Organization and the values of peace, dialogue and international cooperation which underpin it. The supreme embodiment of those values is the Charter of the United Nations.

50. Today we are also taking a decision of great historic significance. We resolve to build a better future for all people, including the millions who have been denied the chance to lead decent, dignified and rewarding lives and to achieve their full human potential. We can be the first generation to succeed in ending poverty; just as we may be the last to have a chance of saving the planet. The world will be a better place in 2030 if we succeed in our objectives.

51. What we are announcing today – an Agenda for global action for the next 15 years – is a charter for people and planet in the twenty-first century. Children and young women and men are critical agents of change and will find in the new Goals a platform to channel their infinite capacities for activism into the creation of a better world.

52. "We the peoples" are the celebrated opening words of the Charter of the United Nations. It is "we the peoples" who are embarking today on the road to 2030. Our journey will involve Governments as well as parliaments, the United Nations system and other international institutions, local authorities, indigenous peoples, civil society, business and the private sector, the scientific and academic community – and all people. Millions have already engaged with, and will own, this Agenda. It is an Agenda of the people, by the people and for the people – and this, we believe, will ensure its success.

53. The future of humanity and of our planet lies in our hands. It lies also in the hands of today's younger generation who will pass the torch to future generations. We have mapped the road to sustainable development; it will be for all of us to ensure that the journey is successful and its gains irreversible.

Sustainable Development Goals and Targets

54. Following an inclusive process of intergovernmental negotiations, and based on the proposal of the Open Working Group on Sustainable

Development Goals,[1] which includes a chapeau contextualizing the latter, set out below are the Goals and targets which we have agreed.

55. The Sustainable Development Goals and targets are integrated and indivisible, global in nature and universally applicable, taking into account different national realities, capacities and levels of development and respecting national policies and priorities. Targets are defined as aspirational and global, with each Government setting its own national targets guided by the global level of ambition but taking into account national circumstances. Each Government will also decide how these aspirational and global targets should be incorporated into national planning processes, policies and strategies. It is important to recognize the link between sustainable development and other relevant ongoing processes in the economic, social and environmental fields.

56. In deciding upon these Goals and targets, we recognize that each country faces specific challenges to achieve sustainable development, and we underscore the special challenges facing the most vulnerable countries and, in particular, African countries, least developed countries, landlocked developing countries and small island developing States, as well as the specific challenges facing the middle-income countries. Countries in situations of conflict also need special attention.

57. We recognize that baseline data for several of the targets remains unavailable, and we call for increased support for strengthening data collection and capacity-building in Member States, to develop national and global baselines where they do not yet exist. We commit to addressing this gap in data collection so as to better inform the measurement of progress, in particular for those targets below which do not have clear numerical targets.

58. We encourage ongoing efforts by States in other forums to address key issues which pose potential challenges to the implementation of our Agenda, and we respect the independent mandates of those processes. We intend that the Agenda and its implementation would support, and be without prejudice to, those other processes and the decisions taken therein.

59. We recognize that there are different approaches, visions, models and tools available to each country, in accordance with its national circumstances and priorities, to achieve sustainable development; and we reaffirm that planet Earth and its ecosystems are our common home and that "Mother Earth" is a common expression in a number of countries and regions.

Goal 1: End Poverty in All Its Forms Everywhere

1.1 By 2030, eradicate extreme poverty for all people everywhere, currently measured as people living on less than $1.25 a day

1.2 By 2030, reduce at least by half the proportion of men, women and children of all ages living in poverty in all its dimensions according to national definitions

---

Sustainable Development Goals

Goal 1. End poverty in all its forms everywhere

Goal 2. End hunger, achieve food security and improved nutrition and promote sustainable agriculture

Goal 3. Ensure healthy lives and promote well-being for all at all ages

Goal 4. Ensure inclusive and equitable quality education and promote lifelong learning opportunities for all

Goal 5. Achieve gender equality and empower all women and girls

Goal 6. Ensure availability and sustainable management of water and sanitation for all

Goal 7. Ensure access to affordable, reliable, sustainable and modern energy for all

Goal 8. Promote sustained, inclusive and sustainable economic growth, full and productive employment and decent work for all

Goal 9. Build resilient infrastructure, promote inclusive and sustainable industrialization and foster innovation

Goal 10. Reduce inequality within and among countries

Goal 11. Make cities and human settlements inclusive, safe, resilient and sustainable

Goal 12. Ensure sustainable consumption and production patterns

Goal 13. Take urgent action to combat climate change and its impacts[2]

Goal 14. Conserve and sustainably use the oceans, seas and marine resources for sustainable development

Goal 15. Protect, restore and promote sustainable use of terrestrial ecosystems, sustainably manage forests, combat desertification, and halt and reverse land degradation and halt biodiversity loss

Goal 16. Promote peaceful and inclusive societies for sustainable development, provide access to justice for all and build effective, accountable and inclusive institutions at all levels

Goal 17. Strengthen the means of implementation and revitalize the Global Partnership for Sustainable Development

---

1.3 Implement nationally appropriate social protection systems and measures for all, including floors, and by 2030 achieve substantial coverage of the poor and the vulnerable

1.4 By 2030, ensure that all men and women, in particular the poor and the vulnerable, have equal rights to economic resources, as well as access to basic services, ownership and control over land and other forms of property, inheritance, natural resources, appropriate new technology and financial services, including microfinance

1.5 By 2030, build the resilience of the poor and those in vulnerable situations and reduce their exposure and vulnerability to climate-related extreme events and other economic, social and environmental shocks and disasters

1.a Ensure significant mobilization of resources from a variety of sources, including through enhanced development cooperation, in order to provide adequate and predictable means for developing countries, in particular least developed countries, to implement programmes and policies to end poverty in all its dimensions

1.b Create sound policy frameworks at the national, regional and international levels, based on pro-poor and gender-sensitive development strategies, to support accelerated investment in poverty eradication actions

Goal 2: End Hunger, Achieve Food Security and Improved Nutrition and Promote Sustainable Agriculture

2.1 By 2030, end hunger and ensure access by all people, in particular the poor and people in vulnerable situations, including infants, to safe, nutritious and sufficient food all year round

2.2 By 2030, end all forms of malnutrition, including achieving, by 2025, the internationally agreed targets on stunting and wasting in children under 5 years of age, and address the nutritional needs of adolescent girls, pregnant and lactating women and older persons

2.3 By 2030, double the agricultural productivity and incomes of small-scale food producers, in particular women, indigenous peoples, family farmers, pastoralists and fishers, including through secure and equal access to land, other productive resources and inputs, knowledge, financial services, markets and opportunities for value addition and non-farm employment

2.4 By 2030, ensure sustainable food production systems and implement resilient agricultural practices that increase productivity and production, that help maintain ecosystems, that strengthen capacity for adaptation to climate change, extreme weather, drought, flooding and other disasters and that progressively improve land and soil quality

2.5 By 2020, maintain the genetic diversity of seeds, cultivated plants and farmed and domesticated animals and their related wild species, including through soundly managed and diversified seed and plant banks at the national, regional and international levels, and promote access to and fair and equitable sharing of benefits arising from the utilization of genetic resources and associated traditional knowledge, as internationally agreed

2.a Increase investment, including through enhanced international cooperation, in rural infrastructure, agricultural research and extension services, technology development and plant and livestock gene banks in order to enhance agricultural productive capacity in developing countries, in particular least developed countries

2.b Correct and prevent trade restrictions and distortions in world agricultural markets, including through the parallel elimination of all forms of agricultural export subsidies and all export measures with equivalent effect, in accordance with the mandate of the Doha Development Round

2.c Adopt measures to ensure the proper functioning of food commodity markets and their derivatives and facilitate timely access to market information, including on food reserves, in order to help limit extreme food price volatility

Goal 3: Ensure Healthy Lives and Promote Well-Being for All at All Ages

3.1 By 2030, reduce the global maternal mortality ratio to less than 70 per 100,000 live births

3.2 By 2030, end preventable deaths of newborns and children under 5 years of age, with all countries aiming to reduce neonatal mortality to at least as low as 12 per 1,000 live births and under-5 mortality to at least as low as 25 per 1,000 live births

3.3 By 2030, end the epidemics of AIDS, tuberculosis, malaria and neglected tropical diseases and combat hepatitis, water-borne diseases and other communicable diseases

3.4 By 2030, reduce by one third premature mortality from non-communicable diseases through prevention and treatment and promote mental health and well-being

3.5 Strengthen the prevention and treatment of substance abuse, including narcotic drug abuse and harmful use of alcohol

3.6 By 2020, halve the number of global deaths and injuries from road traffic accidents

3.7 By 2030, ensure universal access to sexual and reproductive health-care services, including for family planning, information and education, and the integration of reproductive health into national strategies and programmes

3.8 Achieve universal health coverage, including financial risk protection, access to quality essential health-care services and access to safe, effective, quality and affordable essential medicines and vaccines for all

3.9 By 2030, substantially reduce the number of deaths and illnesses from hazardous chemicals and air, water and soil pollution and contamination

3.a Strengthen the implementation of the World Health Organization Framework Convention on Tobacco Control in all countries, as appropriate

3.b Support the research and development of vaccines and medicines for the communicable and non-communicable diseases that primarily affect developing countries, provide access to affordable essential medicines and vaccines, in accordance with the Doha Declaration on the TRIPS Agreement and Public Health, which affirms the right of developing countries to use to the full the provisions in the Agreement on Trade-Related Aspects of Intellectual Property Rights regarding flexibilities to protect public health, and, in particular, provide access to medicines for all

3.c Substantially increase health financing and the recruitment, development, training and retention of the health workforce in developing countries, especially in least developed countries and small island developing States

3.d Strengthen the capacity of all countries, in particular developing countries, for early warning, risk reduction and management of national and global health risks

Goal 4: Ensure Inclusive and Equitable Quality Education and Promote Lifelong Learning Opportunities for All

4.1 By 2030, ensure that all girls and boys complete free, equitable and quality primary and secondary education leading to relevant and effective learning outcomes

4.2 By 2030, ensure that all girls and boys have access to quality early childhood development, care and pre-primary education so that they are ready for primary education

4.3 By 2030, ensure equal access for all women and men to affordable and quality technical, vocational and tertiary education, including university

4.4 By 2030, substantially increase the number of youth and adults who have relevant skills, including technical and vocational skills, for employment, decent jobs and entrepreneurship

4.5 By 2030, eliminate gender disparities in education and ensure equal access to all levels of education and vocational training for the vulnerable, including persons with disabilities, indigenous peoples and children in vulnerable situations

4.6 By 2030, ensure that all youth and a substantial proportion of adults, both men and women, achieve literacy and numeracy

4.7 By 2030, ensure that all learners acquire the knowledge and skills needed to promote sustainable development, including, among others, through education for sustainable development and sustainable lifestyles, human rights, gender equality, promotion of a culture of peace and non-violence, global citizenship and appreciation of cultural diversity and of culture's contribution to sustainable development

4.a Build and upgrade education facilities that are child, disability and gender sensitive and provide safe, non-violent, inclusive and effective learning environments for all

4.b By 2020, substantially expand globally the number of scholarships available to developing countries, in particular least developed countries, small island developing States and African countries, for enrolment in higher education, including vocational training and information and communications technology, technical, engineering and scientific programmes, in developed countries and other developing countries

4.c By 2030, substantially increase the supply of qualified teachers, including through international cooperation for teacher training in developing countries, especially least developed countries and small island developing States

Goal 5: Achieve Gender Equality and Empower All Women and Girls

5.1 End all forms of discrimination against all women and girls everywhere

5.2 Eliminate all forms of violence against all women and girls in the public and private spheres, including trafficking and sexual and other types of exploitation

5.3 Eliminate all harmful practices, such as child, early and forced marriage and female genital mutilation

5.4 Recognize and value unpaid care and domestic work through the provision of public services, infrastructure and social protection policies and the promotion of shared responsibility within the household and the family as nationally appropriate

5.5 Ensure women's full and effective participation and equal opportunities for leadership at all levels of decision-making in political, economic and public life

5.6 Ensure universal access to sexual and reproductive health and reproductive rights as agreed in accordance with the Programme of Action of the International Conference on Population and Development and the Beijing Platform for Action and the outcome documents of their review conferences

5.a Undertake reforms to give women equal rights to economic resources, as well as access to ownership and control over land and other forms of property, financial services, inheritance and natural resources, in accordance with national laws

5.b Enhance the use of enabling technology, in particular information and communications technology, to promote the empowerment of women

5.c Adopt and strengthen sound policies and enforceable legislation for the promotion of gender equality and the empowerment of all women and girls at all levels

Goal 6: Ensure Availability and Sustainable Management of Water and Sanitation for All

6.1 By 2030, achieve universal and equitable access to safe and affordable drinking water for all

6.2 By 2030, achieve access to adequate and equitable sanitation and hygiene for all and end open defecation, paying special attention to the needs of women and girls and those in vulnerable situations

6.3 By 2030, improve water quality by reducing pollution, eliminating dumping and minimizing release of hazardous chemicals and materials, halving the proportion of untreated wastewater and substantially increasing recycling and safe reuse globally

6.4 By 2030, substantially increase water-use efficiency across all sectors and ensure sustainable withdrawals and supply of freshwater to address water scarcity and substantially reduce the number of people suffering from water scarcity

6.5 By 2030, implement integrated water resources management at all levels, including through transboundary cooperation as appropriate

6.6 By 2020, protect and restore water-related ecosystems, including mountains, forests, wetlands, rivers, aquifers and lakes

6.a By 2030, expand international cooperation and capacity-building support to developing countries in water- and sanitation-related activities and programmes, including water harvesting, desalination, water efficiency, wastewater treatment, recycling and reuse technologies

6.b Support and strengthen the participation of local communities in improving water and sanitation management

Goal 7: Ensure Access to Affordable, Reliable, Sustainable and Modern Energy for All

7.1 By 2030, ensure universal access to affordable, reliable and modern energy services

7.2 By 2030, increase substantially the share of renewable energy in the global energy mix

7.3 By 2030, double the global rate of improvement in energy efficiency

7.a By 2030, enhance international cooperation to facilitate access to clean energy research and technology, including renewable energy, energy efficiency and advanced and cleaner fossil-fuel technology, and promote investment in energy infrastructure and clean energy technology

7.b By 2030, expand infrastructure and upgrade technology for supplying modern and sustainable energy services for all in developing countries, in particular least developed countries, small island developing States and landlocked developing countries, in accordance with their respective programmes of support

Goal 8: Promote Sustained, Inclusive and Sustainable Economic Growth, Full and Productive Employment and Decent Work for All

8.1 Sustain per capita economic growth in accordance with national circumstances and, in particular, at least 7 per cent gross domestic product growth per annum in the least developed countries

8.2 Achieve higher levels of economic productivity through diversification, technological upgrading and innovation, including through a focus on high-value added and labour-intensive sectors

8.3   Promote development-oriented policies that support productive activities, decent job creation, entrepreneurship, creativity and innovation, and encourage the formalization and growth of micro-, small- and medium-sized enterprises, including through access to financial services

8.4   Improve progressively, through 2030, global resource efficiency in consumption and production and endeavour to decouple economic growth from environmental degradation, in accordance with the 10-Year Framework of Programmes on Sustainable Consumption and Production, with developed countries taking the lead

8.5   By 2030, achieve full and productive employment and decent work for all women and men, including for young people and persons with disabilities, and equal pay for work of equal value

8.6   By 2020, substantially reduce the proportion of youth not in employment, education or training

8.7   Take immediate and effective measures to eradicate forced labour, end modern slavery and human trafficking and secure the prohibition and elimination of the worst forms of child labour, including recruitment and use of child soldiers, and by 2025 end child labour in all its forms

8.8   Protect labour rights and promote safe and secure working environments for all workers, including migrant workers, in particular women migrants, and those in precarious employment

8.9   By 2030, devise and implement policies to promote sustainable tourism that creates jobs and promotes local culture and products

8.10 Strengthen the capacity of domestic financial institutions to encourage and expand access to banking, insurance and financial services for all

8.a   Increase Aid for Trade support for developing countries, in particular least developed countries, including through the Enhanced Integrated Framework for Trade-related Technical Assistance to Least Developed Countries

8.b   By 2020, develop and operationalize a global strategy for youth employment and implement the Global Jobs Pact of the International Labour Organization

Goal 9: Build Resilient Infrastructure, Promote Inclusive and Sustainable Industrialization and Foster Innovation

9.1   Develop quality, reliable, sustainable and resilient infrastructure, including regional and transborder infrastructure, to support economic development and human well-being, with a focus on affordable and equitable access for all

9.2   Promote inclusive and sustainable industrialization and, by 2030, significantly raise industry's share of employment and gross domestic product, in line with national circumstances, and double its share in least developed countries

9.3 Increase the access of small-scale industrial and other enterprises, in particular in developing countries, to financial services, including affordable credit, and their integration into value chains and markets

9.4 By 2030, upgrade infrastructure and retrofit industries to make them sustainable, with increased resource-use efficiency and greater adoption of clean and environmentally sound technologies and industrial processes, with all countries taking action in accordance with their respective capabilities

9.5 Enhance scientific research, upgrade the technological capabilities of industrial sectors in all countries, in particular developing countries, including, by 2030, encouraging innovation and substantially increasing the number of research and development workers per 1 million people and public and private research and development spending

9.a Facilitate sustainable and resilient infrastructure development in developing countries through enhanced financial, technological and technical support to African countries, least developed countries, landlocked developing countries and small island developing States

9.b Support domestic technology development, research and innovation in developing countries, including by ensuring a conducive policy environment for, inter alia, industrial diversification and value addition to commodities

9.c Significantly increase access to information and communications technology and strive to provide universal and affordable access to the Internet in least developed countries by 2020

Goal 10: Reduce Inequality Within and Among Countries

10.1 By 2030, progressively achieve and sustain income growth of the bottom 40 per cent of the population at a rate higher than the national average

10.2 By 2030, empower and promote the social, economic and political inclusion of all, irrespective of age, sex, disability, race, ethnicity, origin, religion or economic or other status

10.3 Ensure equal opportunity and reduce inequalities of outcome, including by eliminating discriminatory laws, policies and practices and promoting appropriate legislation, policies and action in this regard

10.4 Adopt policies, especially fiscal, wage and social protection policies, and progressively achieve greater equality

10.5 Improve the regulation and monitoring of global financial markets and institutions and strengthen the implementation of such regulations

10.6 Ensure enhanced representation and voice for developing countries in decision-making in global international economic and financial institutions in order to deliver more effective, credible, accountable and legitimate institutions

10.7 Facilitate orderly, safe, regular and responsible migration and mobility of people, including through the implementation of planned and well-managed migration policies

10.a Implement the principle of special and differential treatment for developing countries, in particular least developed countries, in accordance with World Trade Organization agreements

10.b Encourage official development assistance and financial flows, including foreign direct investment, to States where the need is greatest, in particular least developed countries, African countries, small island developing States and landlocked developing countries, in accordance with their national plans and programmes

10.c By 2030, reduce to less than 3 per cent the transaction costs of migrant remittances and eliminate remittance corridors with costs higher than 5 per cent

Goal 11: Make Cities and Human Settlements Inclusive, Safe, Resilient and Sustainable

11.1 By 2030, ensure access for all to adequate, safe and affordable housing and basic services and upgrade slums

11.2 By 2030, provide access to safe, affordable, accessible and sustainable transport systems for all, improving road safety, notably by expanding public transport, with special attention to the needs of those in vulnerable situations, women, children, persons with disabilities and older persons

11.3 By 2030, enhance inclusive and sustainable urbanization and capacity for participatory, integrated and sustainable human settlement planning and management in all countries

11.4 Strengthen efforts to protect and safeguard the world's cultural and natural heritage

11.5 By 2030, significantly reduce the number of deaths and the number of people affected and substantially decrease the direct economic losses relative to global gross domestic product caused by disasters, including water-related disasters, with a focus on protecting the poor and people in vulnerable situations

11.6 By 2030, reduce the adverse per capita environmental impact of cities, including by paying special attention to air quality and municipal and other waste management

11.7 By 2030, provide universal access to safe, inclusive and accessible, green and public spaces, in particular for women and children, older persons and persons with disabilities

11.a Support positive economic, social and environmental links between urban, peri-urban and rural areas by strengthening national and regional development planning

11.b By 2020, substantially increase the number of cities and human settlements adopting and implementing integrated policies and plans towards inclusion, resource efficiency, mitigation and adaptation to climate change, resilience to disasters, and develop and implement, in line with the Sendai Framework for Disaster Risk Reduction 2015–2030, holistic disaster risk management at all levels

11.c Support least developed countries, including through financial and technical assistance, in building sustainable and resilient buildings utilizing local materials

Goal 12: Ensure Sustainable Consumption and Production Patterns

12.1 Implement the 10-Year Framework of Programmes on Sustainable Consumption and Production Patterns, all countries taking action, with developed countries taking the lead, taking into account the development and capabilities of developing countries

12.2 By 2030, achieve the sustainable management and efficient use of natural resources

12.3 By 2030, halve per capita global food waste at the retail and consumer levels and reduce food losses along production and supply chains, including post-harvest losses

12.4 By 2020, achieve the environmentally sound management of chemicals and all wastes throughout their life cycle, in accordance with agreed international frameworks, and significantly reduce their release to air, water and soil in order to minimize their adverse impacts on human health and the environment

12.5 By 2030, substantially reduce waste generation through prevention, reduction, recycling and reuse

12.6 Encourage companies, especially large and transnational companies, to adopt sustainable practices and to integrate sustainability information into their reporting cycle

12.7 Promote public procurement practices that are sustainable, in accordance with national policies and priorities

12.8 By 2030, ensure that people everywhere have the relevant information and awareness for sustainable development and lifestyles in harmony with nature

12.a Support developing countries to strengthen their scientific and technological capacity to move towards more sustainable patterns of consumption and production

12.b Develop and implement tools to monitor sustainable development impacts for sustainable tourism that creates jobs and promotes local culture and products

12.c Rationalize inefficient fossil-fuel subsidies that encourage wasteful consumption by removing market distortions, in accordance with national circumstances, including by restructuring taxation and phasing out those harmful subsidies, where they exist, to reflect their environmental impacts, taking fully into account the specific needs and conditions of developing countries and minimizing the possible adverse impacts on their development in a manner that protects the poor and the affected communities

Goal 13: Take Urgent Action to Combat Climate Change and Its Impacts[3]

13.1 Strengthen resilience and adaptive capacity to climate-related hazards and natural disasters in all countries

13.2 Integrate climate change measures into national policies, strategies and planning

13.3 Improve education, awareness-raising and human and institutional capacity on climate change mitigation, adaptation, impact reduction and early warning

13.a Implement the commitment undertaken by developed-country parties to the United Nations Framework Convention on Climate Change to a goal of mobilizing jointly $100 billion annually by 2020 from all sources to address the needs of developing countries in the context of meaningful mitigation actions and transparency on implementation and fully operationalize the Green Climate Fund through its capitalization as soon as possible

13.b Promote mechanisms for raising capacity for effective climate change-related planning and management in least developed countries and small island developing States, including focusing on women, youth and local and marginalized communities

Goal 14: Conserve and Sustainably Use the Oceans, Seas and Marine Resources for Sustainable Development

14.1 By 2025, prevent and significantly reduce marine pollution of all kinds, in particular from land-based activities, including marine debris and nutrient pollution

14.2 By 2020, sustainably manage and protect marine and coastal ecosystems to avoid significant adverse impacts, including by strengthening their resilience, and take action for their restoration in order to achieve healthy and productive oceans

14.3 Minimize and address the impacts of ocean acidification, including through enhanced scientific cooperation at all levels

14.4 By 2020, effectively regulate harvesting and end overfishing, illegal, unreported and unregulated fishing and destructive fishing practices and implement science-based management plans, in order to restore fish stocks in the shortest time feasible, at least to levels that can produce maximum sustainable yield as determined by their biological characteristics

14.5 By 2020, conserve at least 10 per cent of coastal and marine areas, consistent with national and international law and based on the best available scientific information

14.6 By 2020, prohibit certain forms of fisheries subsidies which contribute to overcapacity and overfishing, eliminate subsidies that contribute to illegal, unreported and unregulated fishing and refrain from introducing new such subsidies, recognizing that appropriate and effective special and differential treatment for developing and least developed countries should be an integral part of the World Trade Organization fisheries subsidies negotiation[4]

14.7 By 2030, increase the economic benefits to small island developing States and least developed countries from the sustainable use of marine resources, including through sustainable management of fisheries, aquaculture and tourism

14.a Increase scientific knowledge, develop research capacity and transfer marine technology, taking into account the Intergovernmental Oceanographic Commission Criteria and Guidelines on the Transfer of Marine Technology, in order to improve ocean health and to enhance the contribution of marine biodiversity to the development of developing countries, in particular small island developing States and least developed countries

14.b Provide access for small-scale artisanal fishers to marine resources and markets

14.c Enhance the conservation and sustainable use of oceans and their resources by implementing international law as reflected in the United Nations Convention on the Law of the Sea, which provides the legal framework for the conservation and sustainable use of oceans and their resources, as recalled in paragraph 158 of "The future we want"

Goal 15: Protect, Restore and Promote Sustainable Use of Terrestrial Ecosystems, Sustainably Manage Forests, Combat Desertification, and Halt and Reverse Land Degradation and Halt Biodiversity Loss

15.1 By 2020, ensure the conservation, restoration and sustainable use of terrestrial and inland freshwater ecosystems and their services, in particular forests, wetlands, mountains and drylands, in line with obligations under international agreements

15.2 By 2020, promote the implementation of sustainable management of all types of forests, halt deforestation, restore degraded forests and substantially increase afforestation and reforestation globally

15.3 By 2030, combat desertification, restore degraded land and soil, including land affected by desertification, drought and floods, and strive to achieve a land degradation-neutral world

15.4 By 2030, ensure the conservation of mountain ecosystems, including their biodiversity, in order to enhance their capacity to provide benefits that are essential for sustainable development

15.5 Take urgent and significant action to reduce the degradation of natural habitats, halt the loss of biodiversity and, by 2020, protect and prevent the extinction of threatened species

15.6 Promote fair and equitable sharing of the benefits arising from the utilization of genetic resources and promote appropriate access to such resources, as internationally agreed

15.7 Take urgent action to end poaching and trafficking of protected species of flora and fauna and address both demand and supply of illegal wildlife products

15.8 By 2020, introduce measures to prevent the introduction and significantly reduce the impact of invasive alien species on land and water ecosystems and control or eradicate the priority species

15.9 By 2020, integrate ecosystem and biodiversity values into national and local planning, development processes, poverty reduction strategies and accounts

15.a Mobilize and significantly increase financial resources from all sources to conserve and sustainably use biodiversity and ecosystems

15.b Mobilize significant resources from all sources and at all levels to finance sustainable forest management and provide adequate incentives to developing countries to advance such management, including for conservation and reforestation

15.c Enhance global support for efforts to combat poaching and trafficking of protected species, including by increasing the capacity of local communities to pursue sustainable livelihood opportunities

Goal 16: Promote Peaceful and Inclusive Societies for Sustainable Development, Provide Access to Justice for All and Build Effective, Accountable and Inclusive Institutions at All Levels

16.1 Significantly reduce all forms of violence and related death rates everywhere

16.2 End abuse, exploitation, trafficking and all forms of violence against and torture of children

16.3 Promote the rule of law at the national and international levels and ensure equal access to justice for all

16.4 By 2030, significantly reduce illicit financial and arms flows, strengthen the recovery and return of stolen assets and combat all forms of organized crime

16.5 Substantially reduce corruption and bribery in all their forms

16.6 Develop effective, accountable and transparent institutions at all levels

16.7 Ensure responsive, inclusive, participatory and representative decision-making at all levels

16.8 Broaden and strengthen the participation of developing countries in the institutions of global governance

16.9 By 2030, provide legal identity for all, including birth registration

16.10 Ensure public access to information and protect fundamental freedoms, in accordance with national legislation and international agreements

16.a Strengthen relevant national institutions, including through international cooperation, for building capacity at all levels, in particular in developing countries, to prevent violence and combat terrorism and crime

16.b Promote and enforce non-discriminatory laws and policies for sustainable development

Goal 17: Strengthen the Means of Implementation and Revitalize the Global Partnership for Sustainable Development

Finance

17.1 Strengthen domestic resource mobilization, including through international support to developing countries, to improve domestic capacity for tax and other revenue collection

17.2 Developed countries to implement fully their official development assistance commitments, including the commitment by many developed countries to achieve the target of 0.7 per cent of gross national income for official development assistance (ODA/GNI) to developing countries and 0.15 to 0.20 per cent of ODA/GNI to least developed countries; ODA providers are encouraged to consider setting a target to provide at least 0.20 per cent of ODA/GNI to least developed countries

17.3 Mobilize additional financial resources for developing countries from multiple sources

17.4 Assist developing countries in attaining long-term debt sustainability through coordinated policies aimed at fostering debt financing, debt relief and debt restructuring, as appropriate, and address the external debt of highly indebted poor countries to reduce debt distress

17.5 Adopt and implement investment promotion regimes for least developed countries

Technology

17.6 Enhance North-South, South-South and triangular regional and international cooperation on and access to science, technology and innovation and enhance knowledge sharing on mutually agreed terms, including through improved coordination among existing mechanisms, in particular at the United Nations level, and through a global technology facilitation mechanism

17.7 Promote the development, transfer, dissemination and diffusion of environmentally sound technologies to developing countries on favourable terms, including on concessional and preferential terms, as mutually agreed

17.8 Fully operationalize the technology bank and science, technology and innovation capacity-building mechanism for least developed countries by 2017 and enhance the use of enabling technology, in particular information and communications technology

Capacity-Building

17.9 Enhance international support for implementing effective and targeted capacity-building in developing countries to support national plans to implement all the Sustainable Development Goals, including through North-South, South-South and triangular cooperation

Trade

17.10 Promote a universal, rules-based, open, non-discriminatory and equitable multilateral trading system under the World Trade Organization, including through the conclusion of negotiations under its Doha Development Agenda

17.11 Significantly increase the exports of developing countries, in particular with a view to doubling the least developed countries' share of global exports by 2020

17.12 Realize timely implementation of duty-free and quota-free market access on a lasting basis for all least developed countries, consistent with World Trade Organization decisions, including by ensuring that preferential rules of origin applicable to imports from least developed countries are transparent and simple, and contribute to facilitating market access

Systemic Issues

Policy and Institutional Coherence

17.13 Enhance global macroeconomic stability, including through policy coordination and policy coherence

17.14 Enhance policy coherence for sustainable development

17.15 Respect each country's policy space and leadership to establish and implement policies for poverty eradication and sustainable development

Multi-stakeholder Partnerships

17.16 Enhance the Global Partnership for Sustainable Development, complemented by multi-stakeholder partnerships that mobilize and share knowledge, expertise, technology and financial resources, to support the achievement of the Sustainable Development Goals in all countries, in particular developing countries

17.17 Encourage and promote effective public, public-private and civil society partnerships, building on the experience and resourcing strategies of partnerships

Data, Monitoring and Accountability

17.18 By 2020, enhance capacity-building support to developing countries, including for least developed countries and small island developing States, to increase significantly the availability of high-quality, timely and reliable data disaggregated by income, gender, age, race, ethnicity, migratory status, disability, geographic location and other characteristics relevant in national contexts

17.19 By 2030, build on existing initiatives to develop measurements of progress on sustainable development that complement gross domestic product, and support statistical capacity-building in developing countries

Means of Implementation and the Global Partnership

60. We reaffirm our strong commitment to the full implementation of this new Agenda. We recognize that we will not be able to achieve our ambitious Goals and targets without a revitalized and enhanced Global Partnership and comparably ambitious means of implementation. The revitalized Global Partnership will facilitate an intensive global engagement in support of implementation of all the Goals and targets, bringing together Governments, civil society, the private sector, the United Nations system and other actors and mobilizing all available resources.

61. The Agenda's Goals and targets deal with the means required to realize our collective ambitions. The means of implementation targets under each Sustainable Development Goal and Goal 17, which are referred to above, are key to realizing our Agenda and are of equal importance with the other Goals and argets. We shall accord them equal priority in our implementation efforts and in the global indicator framework for monitoring our progress.

62. This Agenda, including the Sustainable Development Goals, can be met within the framework of a revitalized Global Partnership for Sustainable Development, supported by the concrete policies and actions outlined in the Addis Ababa Action Agenda,[5] which is an integral part of the 2030 Agenda for Sustainable Development. The Addis Ababa Action Agenda supports, complements and helps to contextualize the 2030 Agenda's means of implementation targets. It relates to domestic public resources, domestic and international private business and finance, international development cooperation, international trade as an engine for development, debt and debt sustainability, addressing systemic issues and science, technology, innovation and capacity-building, and data, monitoring and follow-up.

63. Cohesive nationally owned sustainable development strategies, supported by integrated national financing frameworks, will be at the heart of our efforts. We reiterate that each country has primary responsibility for its own economic and social development and that the role of national policies and development strategies cannot be overemphasized. We will respect each country's policy space and leadership to implement policies for poverty eradication and sustainable development, while remaining consistent with relevant international rules and commitments. At the same time, national development efforts need to be supported by an enabling international economic environment, including coherent and mutually supporting world trade, monetary and financial systems, and strengthened and enhanced global economic governance. Processes to develop and facilitate the availability of appropriate knowledge and technologies globally, as well as capacity-building, are also critical. We commit to pursuing policy coherence and an enabling environment for sustainable development at all levels and by all actors, and to reinvigorating the Global Partnership for Sustainable Development.

64. We support the implementation of relevant strategies and programmes of action, including the Istanbul Declaration and Programme of Action, the SIDS Accelerated Modalities of Action (SAMOA) Pathway and the Vienna Programme of Action for Landlocked Developing Countries for the Decade 2014–2024, and reaffirm the importance of supporting the African Union's Agenda 2063 and the programme of the New Partnership for Africa's Development, all of which are integral to the new Agenda. We recognize the major challenge to the achievement of durable peace and sustainable development in countries in conflict and post-conflict situations.

65. We recognize that middle-income countries still face significant challenges to achieve sustainable development. In order to ensure that achievements made to date are sustained, efforts to address ongoing challenges should be strengthened through the exchange of experiences, improved coordination, and better and focused support of the United Nations development system, the international financial institutions, regional organizations and other stakeholders.

66. We underscore that, for all countries, public policies and the mobilization and effective use of domestic resources, underscored by the principle of national ownership, are central to our common pursuit of sustainable development, including achieving the Sustainable Development Goals. We recognize that domestic resources are first and foremost generated by economic growth, supported by an enabling environment at all levels.

67. Private business activity, investment and innovation are major drivers of productivity, inclusive economic growth and job creation. We acknowledge the diversity of the private sector, ranging from micro-enterprises to cooperatives to multinationals. We call upon all businesses to apply their creativity and innovation to solving sustainable development challenges. We will foster a dynamic and well-functioning business sector, while protecting labour rights and environmental and health standards in accordance with relevant international standards and agreements and other ongoing initiatives in this regard, such as the Guiding Principles on Business and Human Rights and the labour standards of the International Labour Organization, the Convention on the Rights of the Child and key multilateral environmental agreements, for parties to those agreements.

68. International trade is an engine for inclusive economic growth and poverty reduction, and contributes to the promotion of sustainable development. We will continue to promote a universal, rules-based, open, transparent, predictable, inclusive, non-discriminatory and equitable multilateral trading system under the World Trade Organization, as well as meaningful trade liberalization. We call upon all members of the World Trade Organization to redouble their efforts to promptly conclude the negotiations on the Doha Development Agenda. We attach great importance to providing trade-related capacity-building for developing countries, including African countries, least developed countries, landlocked developing countries, small island developing States and middle-income countries, including for the promotion of regional economic integration and interconnectivity.

69. We recognize the need to assist developing countries in attaining long-term debt sustainability through coordinated policies aimed at fostering debt financing, debt relief, debt restructuring and sound debt management, as appropriate. Many countries remain vulnerable to debt crises and some are in the midst of crises, including a number of least developed countries, small island developing States and some developed countries. We reiterate that debtors and creditors must work together to prevent and

resolve unsustainable debt situations. Maintaining sustainable debt levels is the responsibility of the borrowing countries; however we acknowledge that lenders also have a responsibility to lend in a way that does not undermine a country's debt sustainability. We will support the maintenance of debt sustainability of those countries that have received debt relief and achieved sustainable debt levels.

70. We hereby launch a Technology Facilitation Mechanism which was established by the Addis Ababa Action Agenda in order to support the Sustainable Development Goals. The Technology Facilitation Mechanism will be based on a multi-stakeholder collaboration between Member States, civil society, the private sector, the scientific community, United Nations entities and other stakeholders and will be composed of a United Nations inter-agency task team on science, technology and innovation for the Sustainable Development Goals, a collaborative multi-stakeholder forum on science, technology and innovation for the Sustainable Development Goals and an online platform.

- The United Nations inter-agency task team on science, technology and innovation for the Sustainable Development Goals will promote coordination, coherence and cooperation within the United Nations system on science, technology and innovation-related matters, enhancing synergy and efficiency, in particular to enhance capacity-building initiatives. The task team will draw on existing resources and will work with 10 representatives from civil society, the private sector and the scientific community to prepare the meetings of the multi-stakeholder forum on science, technology and innovation for the Sustainable Development Goals, as well as in the development and operationalization of the online platform, including preparing proposals for the modalities for the forum and the online platform. The 10 representatives will be appointed by the Secretary-General, for periods of two years. The task team will be open to the participation of all United Nations agencies, funds and programmes and the functional commissions of the Economic and Social Council and it will initially be composed of the entities that currently integrate the informal working group on technology facilitation, namely, the Department of Economic and Social Affairs of the Secretariat, the United Nations Environment Programme, the United Nations Industrial Development Organization, the United Nations Educational, Scientific and Cultural Organization, the United Nations Conference on Trade and Development, the International Telecommunication Union, the World Intellectual Property Organization and the World Bank.
- The online platform will be used to establish a comprehensive mapping of, and serve as a gateway for, information on existing science, technology and innovation initiatives, mechanisms and programmes, within and beyond the United Nations. The online platform will facilitate access to information, knowledge and experience, as well as best practices and lessons learned, on science, technology and innovation facilitation initiatives

and policies. The online platform will also facilitate the dissemination of relevant open access scientific publications generated worldwide. The online platform will be developed on the basis of an independent technical assessment which will take into account best practices and lessons learned from other initiatives, within and beyond the United Nations, in order to ensure that it will complement, facilitate access to and provide adequate information on existing science, technology and innovation platforms, avoiding duplications and enhancing synergies.

- The multi-stakeholder forum on science, technology and innovation for the Sustainable Development Goals will be convened once a year, for a period of two days, to discuss science, technology and innovation cooperation around thematic areas for the implementation of the Sustainable Development Goals, congregating all relevant stakeholders to actively contribute in their area of expertise. The forum will provide a venue for facilitating interaction, matchmaking and the establishment of networks between relevant stakeholders and multi-stakeholder partnerships in order to identify and examine technology needs and gaps, including on scientific cooperation, innovation and capacity-building, and also in order to help to facilitate development, transfer and dissemination of relevant technologies for the Sustainable Development Goals. The meetings of the forum will be convened by the President of the Economic and Social Council before the meeting of the high-level political forum under the auspices of the Council or, alternatively, in conjunction with other forums or conferences, as appropriate, taking into account the theme to be considered and on the basis of a collaboration with the organizers of the other forums or conferences. The meetings of the forum will be co-chaired by two Member States and will result in a summary of discussions elaborated by the two co-Chairs, as an input to the meetings of the high-level political forum, in the context of the follow-up and review of the implementation of the post-2015 development agenda.
- The meetings of the high-level political forum will be informed by the summary of the multi-stakeholder forum. The themes for the subsequent multi-stakeholder forum on science, technology and innovation for the Sustainable Development Goals will be considered by the high-level political forum on sustainable development, taking into account expert inputs from the task team.

71. We reiterate that this Agenda and the Sustainable Development Goals and targets, including the means of implementation, are universal, indivisible and interlinked.

Follow-Up and Review

72. We commit to engaging in systematic follow-up and review of the implementation of this Agenda over the next 15 years. A robust,

voluntary, effective, participatory, transparent and integrated follow-up and review framework will make a vital contribution to implementation and will help countries to maximize and track progress in implementing this Agenda in order to ensure that no one is left behind.

73. Operating at the national, regional and global levels, it will promote accountability to our citizens, support effective international cooperation in achieving this Agenda and foster exchanges of best practices and mutual learning. It will mobilize support to overcome shared challenges and identify new and emerging issues. As this is a universal Agenda, mutual trust and understanding among all nations will be important.

74. Follow-up and review processes at all levels will be guided by the following principles:

    (a) They will be voluntary and country-led, will take into account different national realities, capacities and levels of development and will respect policy space and priorities. As national ownership is key to achieving sustainable development, the outcome from national-level processes will be the foundation for reviews at the regional and global levels, given that the global review will be primarily based on national official data sources.

    (b) They will track progress in implementing the universal Goals and targets, including the means of implementation, in all countries in a manner which respects their universal, integrated and interrelated nature and the three dimensions of sustainable development.

    (c) They will maintain a longer-term orientation, identify achievements, challenges, gaps and critical success factors and support countries in making informed policy choices. They will help to mobilize the necessary means of implementation and partnerships, support the identification of solutions and best practices and promote the coordination and effectiveness of the international development system.

    (d) They will be open, inclusive, participatory and transparent for all people and will support reporting by all relevant stakeholders.

    (e) They will be people-centred, gender-sensitive, respect human rights and have a particular focus on the poorest, most vulnerable and those furthest behind.

    (f) They will build on existing platforms and processes, where these exist, avoid duplication and respond to national circumstances, capacities, needs and priorities. They will evolve over time, taking into account emerging issues and the development of new methodologies, and will minimize the reporting burden on national administrations.

    (g) They will be rigorous and based on evidence, informed by country-led evaluations and data which is high-quality, accessible, timely, reliable and disaggregated by income, sex, age, race, ethnicity, migration status, disability and geographic location and other characteristics relevant in national contexts.

(h)   They will require enhanced capacity-building support for developing countries, including the strengthening of national data systems and evaluation programmes, particularly in African countries, least developed countries, small island developing States, landlocked developing countries and middle-income countries.

(i)   They will benefit from the active support of the United Nations system and other multilateral institutions.

75.   The Goals and targets will be followed up and reviewed using a set of global indicators. These will be complemented by indicators at the regional and national levels which will be developed by Member States, in addition to the outcomes of work undertaken for the development of the baselines for those targets where national and global baseline data does not yet exist. The global indicator framework, to be developed by the Inter-Agency and Expert Group on Sustainable Development Goal Indicators, will be agreed by the Statistical Commission by March 2016 and adopted thereafter by the Economic and Social Council and the General Assembly, in line with existing mandates. This framework will be simple yet robust, address all Sustainable Development Goals and targets, including for means of implementation, and preserve the political balance, integration and ambition contained therein.

76.   We will support developing countries, particularly African countries, least developed countries, small island developing States and landlocked developing countries, in strengthening the capacity of national statistical offices and data systems to ensure access to high-quality, timely, reliable and disaggregated data. We will promote transparent and accountable scaling-up of appropriate public-private cooperation to exploit the contribution to be made by a wide range of data, including earth observation and geospatial information, while ensuring national ownership in supporting and tracking progress.

77.   We commit to fully engage in conducting regular and inclusive reviews of progress at the subnational, national, regional and global levels. We will draw as far as possible on the existing network of follow-up and review institutions and mechanisms. National reports will allow assessments of progress and identify challenges at the regional and global level. Along with regional dialogues and global reviews, they will inform recommendations for follow-up at various levels.

National Level

78.   We encourage all Member States to develop as soon as practicable ambitious national responses to the overall implementation of this Agenda. These can support the transition to the Sustainable Development Goals and build on existing planning instruments, such as national development and sustainable development strategies, as appropriate.

79. We also encourage Member States to conduct regular and inclusive reviews of progress at the national and subnational levels which are country-led and country-driven. Such reviews should draw on contributions from indigenous peoples, civil society, the private sector and other stakeholders, in line with national circumstances, policies and priorities. National parliaments as well as other institutions can also support these processes.

Regional level

80. Follow-up and review at the regional and subregional levels can, as appropriate, provide useful opportunities for peer learning, including through voluntary reviews, sharing of best practices and discussion on shared targets. We welcome in this respect the cooperation of regional and subregional commissions and organizations. Inclusive regional processes will draw on national-level reviews and contribute to follow-up and review at the global level, including at the high-level political forum on sustainable development.

81. Recognizing the importance of building on existing follow-up and review mechanisms at the regional level and allowing adequate policy space, we encourage all Member States to identify the most suitable regional forum in which to engage. United Nations regional commissions are encouraged to continue supporting Member States in this regard.

Global Level

82. The high-level political forum will have a central role in overseeing a network of follow-up and review processes at the global level, working coherently with the General Assembly, the Economic and Social Council and other relevant organs and forums, in accordance with existing mandates. It will facilitate sharing of experiences, including successes, challenges and lessons learned, and provide political leadership, guidance and recommendations for follow-up. It will promote system-wide coherence and coordination of sustainable development policies. It should ensure that the Agenda remains relevant and ambitious and should focus on the assessment of progress, achievements and challenges faced by developed and developing countries as well as new and emerging issues. Effective linkages will be made with the follow-up and review arrangements of all relevant United Nations conferences and processes, including on least developed countries, small island developing States and landlocked developing countries.

83. Follow-up and review at the high-level political forum will be informed by an annual progress report on the Sustainable Development Goals to be prepared by the Secretary-General in cooperation with the United Nations system, based on the global indicator framework and data produced by national statistical systems and information collected at the

regional level. The high-level political forum will also be informed by the *Global Sustainable Development Report*, which shall strengthen the science-policy interface and could provide a strong evidence-based instrument to support policymakers in promoting poverty eradication and sustainable development. We invite the President of the Economic and Social Council to conduct a process of consultations on the scope, methodology and frequency of the global report as well as its relation to the progress report, the outcome of which should be reflected in the ministerial declaration of the session of the high-level political forum in 2016.

84. The high-level political forum, under the auspices of the Economic and Social Council, shall carry out regular reviews, in line with General Assembly resolution 67/290 of 9 July 2013. Reviews will be voluntary, while encouraging reporting, and include developed and developing countries as well as relevant United Nations entities and other stakeholders, including civil society and the private sector. They shall be State-led, involving ministerial and other relevant high-level participants. They shall provide a platform for partnerships, including through the participation of major groups and other relevant stakeholders.

85. Thematic reviews of progress on the Sustainable Development Goals, including cross-cutting issues, will also take place at the high-level political forum. These will be supported by reviews by the functional commissions of the Economic and Social Council and other intergovernmental bodies and forums which should reflect the integrated nature of the Goals as well as the interlinkages between them. They will engage all relevant stakeholders and, where possible, feed into, and be aligned with, the cycle of the high-level political forum.

86. We welcome, as outlined in the Addis Ababa Action Agenda, the dedicated follow-up and review for the financing for development outcomes as well as all the means of implementation of the Sustainable Development Goals which is integrated with the follow-up and review framework of this Agenda. The intergovernmentally agreed conclusions and recommendations of the annual Economic and Social Council forum on financing for development will be fed into the overall follow-up and review of the implementation of this Agenda in the high-level political forum.

87. Meeting every four years under the auspices of the General Assembly, the high-level political forum will provide high-level political guidance on the Agenda and its implementation, identify progress and emerging challenges and mobilize further actions to accelerate implementation. The next high-level political forum under the auspices of the General Assembly will be held in 2019, with the cycle of meetings thus reset, in order to maximize coherence with the quadrennial comprehensive policy review process.

88. We also stress the importance of system-wide strategic planning, implementation and reporting in order to ensure coherent and integrated

support to the implementation of the new Agenda by the United Nations development system. The relevant governing bodies should take action to review such support to implementation and to report on progress and obstacles. We welcome the ongoing dialogue in the Economic and Social Council on the longer-term positioning of the United Nations development system and look forward to taking action on these issues, as appropriate.

89. The high-level political forum will support participation in follow-up and review processes by the major groups and other relevant stakeholders in line with resolution 67/290. We call upon those actors to report on their contribution to the implementation of the Agenda.

90. We request the Secretary-General, in consultation with Member States, to prepare a report, for consideration at the seventieth session of the General Assembly in preparation for the 2016 meeting of the high-level political forum, which outlines critical milestones towards coherent, efficient and inclusive follow-up and review at the global level. The report should include a proposal on the organizational arrangements for State-led reviews at the high-level political forum under the auspices of the Economic and Social Council, including recommendations on voluntary common reporting guidelines. It should clarify institutional responsibilities and provide guidance on annual themes, on a sequence of thematic reviews, and on options for periodic reviews for the high-level political forum.

91. We reaffirm our unwavering commitment to achieving this Agenda and utilizing it to the full to transform our world for the better by 2030.

Notes

1. Contained in the report of the Open Working Group of the General Assembly on Sustainable Development Goals (A/68/970 and Corr.1; see also A/68/970/Add.1 and 2).
2. Acknowledging that the United Nations Framework Convention on Climate Change is the primary international, intergovernmental forum for negotiating the global response to climate change.
3. Acknowledging that the United Nations Framework Convention on Climate Change is the primary international, intergovernmental forum for negotiating the global response to climate change.
4. Taking into account ongoing World Trade Organization negotiations, the Doha Development Agenda and the Hong Kong ministerial mandate.
5. The Addis Ababa Action Agenda of the Third International Conference on Financing for Development (Addis Ababa Action Agenda), adopted by the General Assembly on 27 July 2015 (resolution 69/313).

Global Health and Global Governance

5

Action on Global Health: Addressing Global Health Governance Challenges

Ilona Kickbusch

A Change in Perspective

Despite progress in many areas of global health action there is an increasing consensus that it is insufficient and too slow. It was hoped throughout the last 10 years that an approach targeting specific diseases would move global health a significant step forward, but we are again at a stage where it is becoming clear that our response must be based on another intellectual premise. This insight is best expressed by Inge Kaul, in her seminal work on global public goods: *the pervasiveness of today's crises suggests that they might all suffer from a common cause, such as a common flaw in policy making, rather than from issue specific problems. If so, issue specific responses, typical to date, would be insufficient – allowing global crisis to persist and even multiply.*[1]

A change in perspective implies that we approach the global health crisis[2] not primarily as a set of disease-based challenges but as a challenge on how to ensure health as a global public good through a reliable mechanism of global governance. This means we need to gain an understanding of the political determinants of global health and seek to analyze the 'common flaws' most relevant to policy making in the global health arena. Two such flaws gain particular prominence:

- the lack of a long term commitment – at all levels of governance – to institution building for health, including sustainable financing and strengthening

Source: *Public Health: Journal of The Royal Institute of Public Health*, 119(11) (2005): 969–973.

of human resources as part of the health development agenda; this reflects the need to reform the practices of the donor community,

- the weak political commitment and fragile institutional and organisational base – at all levels of governance – to address the global production of health risks as part of a new global health policy agenda; this reflects the need to develop new mechanisms which support the production, protection and financing of global public goods.

At present, the dominant perspective applied to global health action is organized around issue specific problems, usually specific diseases such as AIDS, malaria and tuberculosis. The unresolved debate within the global health world whether to put the priority on issue specific (vertical) or systemic (horizontal) measures, has come back into focus as major disease based initiatives are faced with the lack of human resources and delivery systems in the developing world.

On the one hand, some disease based approaches have shown extraordinary success, the most well known of course being the eradication of smallpox. A recent publication of the What Works Working Group[3] analyzes 17 selected global health success stories throughout the developing world ranging from eliminating polio in Latin America to improving the health of the poor in Mexico and preventing iodine deficiency disease in China. Piecemeal projects, argues the economist William Easterly, have led to significant improvements in reducing infant mortality and increasing access to clean water at the community level.[4]

On the other hand, even the What Works Working Group expresses frustration with the fact that many of the successes have not been replicated and scaled up to the extent that would seem reasonable and expected given their high impact on the health of populations. Just recently, the World Health Organization and UNAIDS were forced to announce that their goals of '3 by 5' – to treat three million people with HIV by the end of 2005 – would not be reached, mainly due to shortages of health workers and problems with drug supplies.[5]

Frustration has turned to pessimism and anger in many other areas of global health. For example, the broad public awareness of the significance of health action in the developing world has been created largely through the information on HIV/AIDS and its devastating consequences in the poorest countries of Africa. But after 20 years of global advocacy, high political profile – even making HIV/AIDS the first ever health issue to be discussed at the UN Security Council – and millions of dollars in aid, the lack of progress in fighting the epidemic is obvious. In view of the fact that there are an estimated 25 million HIV positive people in Africa and that the disease is wiping out many of the development gains of recent decades, Stephen Lewis, the UN Secretary General's special envoy on HIV/AIDS in Africa, has expressed his sentiment in the following way: There are no excuses left, no rationalizations to hide behind, no murky slanders to justify indifference – there will only be the mass graves of the betrayed.[6]

A New Political Ecosystem

In order to understand some of the governance issues that need to be addressed we need to consider the new political ecosystem of health with its new set of actors and redirection of functions, the new political space it embodies, and the new quality of its politics. This system of global health governance deserves closer analysis than it has received so far, not only in the public health world but also in the field of international relations, as it constitutes an excellent example of the development of transnational society at the end of the 20th century. Public health research will increasingly need to incorporate not only epidemiological evidence but move into compiling new types of evidence related to policy and implementation – this inevitably relates to the analysis of the distribution of power and resources within and between countries and different actors.

A Redirection of Global Health Functions

The most striking development in the international arena has been the increase in the number of transnational actors. Global health is different from international health in that it defines itself through a complex interplay between state and non-state actors and through new organisational mechanisms that allow for their involvement. The new global health system is clearly pluralistic and characterized by increasing privatisation: functions which in the international health system would have been dealt with through interstate mechanisms or/and international/multilateral organisations such as the WHO. They are now more commonly conducted by other types of organisations such as an increasing array of public–private partnerships. A prime example is the important role played by the Bill and Melinda Gates Foundation, which pump as much money into the global health system as the regular budget of the WHO. The Initiative on Public–Private Partnerships for Health (IPPPH) lists 80 public–private partnerships on its website[7] but informal discussions with staff of the initiative suggest around 150 public–private partnerships in the global health arena.

This redirection of functions and programmes formerly in the public/governmental/interstate arena towards private organisations has significant consequences in terms of priority-setting, ownership, control and accountability. It has clearly reduced the importance of the UN system – some would say it is a reflection of its failures to respond adequately to the challenges at hand. William Easterly[4] sees the present development as a positive move away from what he deems to be the weak accountability of international agencies for their 'big plans' and programmes towards a focused intervention-based accountability of 'visible piece–meal steps'. He also suggests that a more pluralist system will provide developing countries with intervention choices that they do not have when confronted with the 'Cartel of Good Intentions' of development agencies and their comprehensive plans.[8]

What he does not address is the downside of this development, which also leads to what can be termed the Balkanisation of global health. The term is used to describe the process of fragmentation or division of a region into many smaller regions that are often hostile or non-cooperative with each other.[9] In the global health arena a multitude of actors now competes for funds, media attention and legitimacy – and while many of these initiatives will have helped make a difference on the ground, they have not helped to create reliable infrastructures for health and have frequently put additional burdens on developing countries through complex application and reporting procedures. Nancy Birdsall[10] in a recent analysis of the 'seven deadly sins' of donor behaviour provides the example of Tanzania, which in the period from 2001 to 2002 had 1300 foreign aid based projects, 1000 donor meetings a year, and had to produce 2400 donor reports every quarter. She underlines that such 'collusion and coordination failures' undermine the governance capacity of countries and indicate the need for 'ambitious and structural changes in the overall aid architecture'.

A New Political Space

Not only have the actors in health increased exponentially, the situation is rendered more complex by the fact that within the last decade health has entered a new political space. It is present in a broad range of governance arenas and global health is no longer in the remit of only the ministers of health as part of their responsibilities within the deliberations of the WHO. The recognition and acceptance that health is not only an outcome of development but contributes significantly to economic growth, social stability and individual life chances, as put forward for example by the Commission on Macroeconomics and Health,[11] has made health part of many larger, more comprehensive policy initiatives. The recognition of new infectious disease threats through outbreaks or bioterrorism has raised the profile of health in security policy. The relevance of epidemics in relation to failed states and loss of social cohesion has made it a concern of foreign policy. Intellectual property agreements have put health on the agenda of trade policies. Global health advocates are now active in writing briefs for meetings such as the World Economic Forum, the National Economic Partnership for African Development (NEPAD) or the G8 meetings.[12]

The Millennium Development Declaration and its goals which were agreed in 2000 by the heads of state in the context of the United Nations give health high prominence.[13] Three of the goals refer explicitly to health outcomes: reducing under five mortality, reducing maternal mortality, and reversing the spread of HIV/AIDS, malaria and TB. Health features in some of the sub-targets, in particular the target related to water and sanitation and the target on access to essential drugs.

As global health moves out of the technical domain into other policy arenas the ideological differences in the basic premise of global health governance approaches become more prominent and we can witness a shift from the technical analysis of health risks into more politically driven assessments from viewpoints such as national security, market opportunities or global social justice.

A Different Quality of International Health Politics

But beyond the sheer quantitative increase in the number of actors and the new policy space we also experience a changed quality in the global policy arena.

First, it becomes more and more difficult to draw borders between different fields of action, which until now have been assigned to different ministries at the national level and to different organisations at the international level. Indeed, one consequence has been that it is more difficult to define what is domestic and what foreign policy. But, even more importantly as some would argue, health has not only moved into other 'harder' policy arenas, but it has also gained a new type of strength and relevance. As an increasing number of health issues now have relevance to foreign and security policy, health moves from being defined as low politics to high politics.[14]

Second, the very nature of politics has changed from elites to one that involves ordinary people. Global Health issues have become an ethical driving force promoting the idea of a common humanity. The women's health movement and the HIV/AIDS movement were at the forefront of insisting that health is indivisible and that every life, no matter where on earth, has the same value. This has taken global health into the realm of popular global culture where pop stars have played a prominent role in setting global agendas such as 'Make poverty history'. While many are skeptical, if not downright cynical, about such efforts they do reflect a new level of awareness of global health issues that was not present in earlier generations. Barry Buzan,[15] in his analysis of the social structure of globalization, has termed this the interhuman domain, in which ordinary citizens accept a global frame of reference.

Globalisation has also provided 'opportunities for groups such as women, lesbians and gay men, disabled persons, indigenous people to mobilize to a degree that was generally unavailable to them in . . . territorial politics'.[16] Increasingly, therefore, the interstate system is cross cut by an array of networks, alliances and partnerships focused on 'identity politics . . . based on ethnicity, class, religion, sex, sexuality or other criteria'.[17] The movements for reproductive rights or HIV/AIDS are obvious examples – and they have clearly changed the priorities set in the global health arena.

Third, these movements are usually loosely coupled networks, driven by moral politics and supported by modern information technology. Manual Castels has defined networks as the organisational form of the 21st century and sees them as the appropriate instruments for a capitalist economy based on innovation, globalisation, and decentralised concentration.[18] Networks also respond to what has been called the globalisation paradox: expanding global governance capacity without centralising policy-making power.[19]

The Role of Public Health Associations

Nation states are usually seen to have four major functions: security, rule of law, social welfare, and identity and participation. These functions have clear equivalents at the level of global governance: human security and human rights, international rule of law/global ethics, fairness in global distribution, common identity as global citizens and a global voice and channels of participation. Health risks in the 21st century are obviously transnational – both as regards infectious and non-infectious diseases as well as the determinants of health. Indeed transboundary issues and collective human security issues gain increasing importance. Global risk production is localised through the globalisation of everyday life – no aspect of how we 'live, love, work and play'[20] is free from a global dimension – be it the food we eat, the advertising we see, the information we access or the fears we have. As a consequence, the development of international norms and standards as well as compliance, transparency and accountability (CTA) gain increasing importance.

Examples in the health arena are the International Health Regulations and the first ever treaty signed in international health, the Framework Convention on Tobacco Control. What emerges is that accountability in health is widened to include the national and local constituency as well as the global constituency, a fact well illustrated on the occasion of the recent SARS epidemic. But the notion of CTA needs to be widened beyond nation states and international organisations to include the wide range of actors in the global health arena. Very first steps are underway in the corporate social responsibility arena, supported by the UN Global Compact.

Networks – both advocacy networks and technical networks – contribute significantly to the compliance with global norms through network effects that support the transfer of 'rules, practices, institutional structures'.[21] But this is not sufficient. At present there is no mechanism whereby the various actors within the pluralist global health system are held to account. A first analytical attempt is being made by the people's Health Movement through the Global Health Watch alternative health report to be published in July 2005.

National public health associations (NPHA), as important technical networks, will need to move their activity into the CTA arena and be at the forefront of explaining and exploring the interface of national and global public health agendas. A global domestic policy agenda could include:

- Reform and strengthen global institutions and international law for health
- Control unsafe goods and products, ensure corporate social responsibility
- Ensure access to essential medicines, vaccines and health knowledge and research
- Increase human capacity and health literacy
- Create primary health care and public health infrastructures, surveillance and information systems
- Create professional capacity and ensure human resources
- Fight major diseases and defined global health emergencies including rapid response.

One mechanism to do this could be to organise national Global Health Summits based on a national global health strategy and a national global health report based on CTA indicators. In this manner NPHA could set out the dimensions and parameters of such a global governance debate. NPHA will need to engage in the policy arenas into which health has moved and seek to understand their goals and ensure that public health principles and values remain present. As global health becomes both high politics and a key concern of global citizenship,[22,23] NPHA have a unique historical opportunity to step out of the shadows and develop intellectual leadership in the five action areas of the Brighton declaration:[24]

- health as a global public good
- health as a key component of collective human security
- health a key factor of global governance of interdependence
- health as responsible business practice and social responsibility
- health as global citizenship.

References

1. Kaul I, Grunberg I, Stern M, editors. *Global public goods: international cooperation in the 21st century*. New York: UNDP; 1999.
2. Kickbusch I. The Leavell lecture – the end of public health as we know it: constructing global public health in the 21st century. *Public Health* 2004;463–9.
3. Levine R. *Millions saved. Proven successes in global health*. Washington D.C.: Center for Global Development; 2004.
4. Easterly W. *Tone deaf on Africa The New York times*, 3 July 2005.
5. Altman LK, McNeil DG. *Health groups expect to miss AIDS target The New York times, 30 June* 2005.
6. McNeil DG. *Plan to battle AIDS worldwide is falling short New York times, 28 March* 2004.
7. http://www.ipph.org.
8. Easterly W. *The cartel of good intentions Foreign policy, July–August* 2002.

9. http://www.en/wikipedia.org/wiki/Balkanization.
10. Birdsall N. *Seven deadly sins: reflections on donor failings Working paper number 50, center for global development December* 2004.
11. *Macroeconomics and health: investing in health for economic development, WHO commission on macroeconomics and health, December* 2001.
12. Labonte R, Schrecker T, Sanders D, Meeus W. *Fatal indifference The G8, Africa and global health*. Ottawa: UCT Press; 2004.
13. http://www.un.org/millenniumgoals/
14. Fidler D. *Health as foreign policy: harnessing globalization for health, background paper for WHO conference on health promotion in Bangkok Thailand, draft, May* 2005.
15. Buzan B. *From international to world society*. Cambridge: Cambridge University Press; 2004.
16. Scholte JA. *Globalization: a critical introduction*. London: Macmillan; 2000.
17. http://wwwen/wikipedia.org/wiki/identity politics
18. Castells M. *The rise of the network society*. 2 ed. Oxford: Blackwell Publishers; 2000.
19. Slaughter AM. *A new world order*. Princeton: Princeton University Press; 2004.
20. *The Ottawa charter*, http://www.ldb.org/iuhpe/ottawa.htm.
21. Raustiala K. The architecture of international cooperation: transgovernmental networks and the future of international law. *V J Int Law* 2002;**43**:1.
22. Dower N. *An introduction to global citizenship*. Edinburgh: Edinburgh University Press; 2003.
23. Singer P. *One world: the ethics of globalization*. New Haven: Yale University Press; 2002.
24. *The Brighton declaration*, http://www.publichealthnews.com/pdf/UKPHA20040422.pdf.

6

Global Health Inequalities: An International Comparison

J.P. Ruger and H.-J. Kim

Interest in health inequality among countries is growing.[1] The World Health Organization (WHO),[2-4] World Bank,[5 6] Unicef,[7 8] Pan American Health Organization, United Nations Development Programme,[9-12] the UK Department of International Development[13] and the broader global health community[14-16] have made this issue a priority. One book, *Challenging inequities in health: from ethics to action,* published results from the Global Health Equity Initiative examining social inequalities in health in 15 countries.[1 16-18] On 18 March 2005, the WHO Commission on Social Determinants of Health was formed and charged with identifying interventions and policies to reduce global health inequalities. However, the formulation of a global policy for closing the gap between industrialised and developing countries is hampered by the lack of a sound knowledge base. Although much work has focused on the determinants of average health,[1] the conceptualisation and measurement of poverty worldwide,[19 20] social determinants of, and social inequalities in, health within and between industrialised countries,[21-28] and comparing inequalities between socioeconomic groups within countries,[1 29-37] there have been few studies on health disparities between countries. This paper aims to study cross-national inequalities in adult and child mortality using a novel approach, with clustering techniques to stratify countries into mortality groups (better-off, worse-off, mid-level) and to examine associated risk factors for inequality. Stratifying countries into mortality groups of different levels of health, which, to our knowledge, has yet to be carried out,

Source: *Journal of Epidemiology & Community Health,* 60(11) (2006): 928–936.

better enables multilateral institutions like the WHO and the World Bank to devise policies and interventions to reduce the gap in mortality between countries. This type of study is particularly relevant for organisations such as the World Bank, which work in several policy domains outside the health sector.[38] This study uses data from the World Development Indicators (WDI), enabling the study of numerous economic and social variables unavailable in other international datasets which also include health.

Methods

Data

WDI 2003[39] is a cross-national database of >500 time-series indicators (variables) for 207 countries and 18 country groups, covering 1960–2001, for which data are available. The 94 indicators include size, growth and structure of a country's population; health; labour force; education and illiteracy; natural resources and land usage; income and poverty; expenditure on food, housing, fuel and power; transport and communication; urbanisation and pollution; national accounts, debt and trade; exchange rates, prices, taxation, levels of aid; and healthcare and education. The World Bank also classifies countries on the basis of annual gross national product per capita. As far as possible, data included in the WDI conform to the United Nations System of National Accounts and the methods of specialised agencies of the United Nations.

Measures

The two primary outcome measures in our study were underfive mortality (the probability that a child born in the indicated year would die before age 5 years, using current age-specific mortality data) and adult mortality (the probability of dying between ages 15 and 60 years, using current age-specific mortality data for men between these ages). We did not choose life expectancy as our main dependent variable, because it applies to a hypothetical cohort born each year and does not reflect real-life cohorts at different ages. It also combines adult, infant and child mortality, precluding assessment across the life span.

The main covariates in our bivariate and multivariate analyses stem from several key topics. Gross national income (GNI) per capita, purchasing power parity (PPP; $) is (the sum of gross national product and the terms of trade adjustment) converted to international dollars (using PPP rates). The population below $1 a day is the percentage of the population living on <$1.08 a day at 1993 international prices. The percentage living in the lowest income quintile is the income share that accrues to subgroups of the population indicated by quintiles. The Gini Index measures the degree of income inequality within a country.[39]

Adult illiteracy rate for men or women is the percentage of people aged ≥15 years who cannot, with understanding, read and write a short, simple statement about their everyday life. Total health expenditure is the sum of public and private health expenditure (preventive and curative health services, family planning, nutrition and emergency medical aid). Private health expenditure includes out-of-pocket spending by households, private insurance, spending by non-profit institutions serving households (other than social insurance) and direct service payments by private corporations. Public health expenditure includes government spending (central and local), external loans and grants, and social health insurance funds. Doctors are graduates of any faculty or school of medicine who are working in the country in any medical field. Hospital beds include those for acute or chronic care in public, private, general and specialised hospitals and rehabilitation centres. Outpatient visits per capita is the number of visits to healthcare facilities per capita, including repeat visits.[39]

As per disease-prevention strategies in environmental health, an improved water source is considered to be a household connection, public standpipe, borehole, protected well or spring, or rainwater collection. Access to water is the availability of ≥20 l/person/day from a source within 1 km of the dwelling. Improved sanitation facilities (private or shared, not public) are those that permit excreta to be disposed away from contact with humans, animals and insects, ranging from protected pit latrines to flush toilets.[39]

For health conditions and interventions, the incidence of tuberculosis is the estimated number of new cases (pulmonary, smear positive, extrapulmonary). The prevalence of HIV is the percentage of the population infected. The child immunisation rate is the percentage of children <1 year adequately vaccinated for measles (one dose) and diphtheria, pertussis and tetanus (three doses). The prevalence of smoking is the percentage of men and women >18 years (15 in some countries) who smoke cigarettes.[39]

Rural population refers to the percentage of the total population living in rural areas in a given country. Electric power consumption measures the production of power plants and heat/power plants excluding transmission, distribution and transformation losses.[39]

Household final consumption expenditure per capita is the market value of all goods and services purchased by households. Gross domestic product (GDP) at purchasing power prices is the sum of the gross value added by all resident producers in the economy and product taxes (excluding subsidies) not included in the valuation of output.[39] Trade in goods as a share of GDP is the sum of merchandise exports and imports divided by the value of GDP ($US). Food production and consumption price indices measure changes in prices.

Total debt service is the sum of principal repayments and interest paid in foreign currency, goods or services on long-term debt, and interest paid on short-term debt and repayments to the International Monetary Fund.

Finally, military expenditures include all current and capital expenditures on the armed forces, including peacekeeping forces, defence ministries and other relevant government agencies, paramilitary forces and military space activities.[39] For information and technology, internet users are people with access to the world wide web. Personal computers are those designed to be used by one person.

As missing data was an issue for all covariates, we followed the World Bank recommendations regarding the selection of variables and years, with a 5-year range, using the most recent data available (1996–2000).

Statistical Analysis

As we were interested in measuring inequalities in health between population groups, as opposed to measuring interindividual health inequalities, we wanted to identify groups, assess intergroup disparities and examine risk factors associated with such disparities. To group countries by mortality, we used cluster analysis, which tests whether individual members (countries in our analysis) within clusters are similar to each other (homogeneous or compact with respect to certain characteristics) and different from members of other clusters (with respect to the same characteristics). Cluster analysis is a statistical method well suited for classifying data into cluster groups,[40 41] and has several basic science and medical applications, such as the classification of elements of the periodic table and of diseases for research on aetiology and treatment.[40 41] We used the K-means clustering technique to minimise variability within clusters and to maximise variability between clusters. For adult and child mortality in 2000, the 3-means cluster analysis produced more discriminatory results than 2-means or 4–5-means clusters.

We especially wanted to identify a least healthy (worseoff) group, a most healthy (better-off) group and other groups in between. In choosing groups for comparison, we tried to balance the need to demonstrate the magnitude of health inequality across groups with the need to have groups large enough to produce significant results.[17 24] Should it be desirable to disproportionately weigh the improvements in health among the least healthy (and, to a certain extent, those with mid-level health), more than improvements among the healthiest,[9 12 17] our analyses could be extended to assess shortfalls in health achievement (mortality), using the most healthy as the reference group.[9 17 42 43]

We assessed overall sample characteristics using univariate analysis. Bivariate analysis examined unadjusted relationships between mortality groups and major country-level factors: poverty and income; education; health expenditure; disease prevention; health risk factors; environment; energy production and use; economy; trade; monetary; external debt; military spending; communication; and information technology. x^2 tests allowed comparisons among groups (categorical variables). Bivariate correlations between continuous variables were analysed with the t test. Analysis of variance among multiple groups evaluated differences in unadjusted mean values between each pair of means for each group, and confidence intervals (CIs) were calculated for each group mean to permit groupwise comparisons.

Multivariate analyses were carried out using multinomial logistic regression. We stratified adult mortality into three levels per 1000: 80–250; 258–449; and 460–725. The strata for under-five mortality for 1000 were 3.9–60; 66–156; and 160–316. Multinomial logistic regression analyses were used to estimate associations between the mortality groups and independent variables; to avoid potential collinearity, we used the stepwise procedure of entering several variables at a time. Separate sets of models were estimated for under-five and adult mortality. In each case, the healthiest group, group 1, was the reference group. For each predictor in the model, we estimated one parameter that represented the effect of a 1-unit increase on the logit (log odds) scale (a 10-unit increase produced little change in the log odds). Our first model included only income, education, health expenditure, disease prevention and health risk factor variables. In the second model, we added indicators on the environment, economy, monetary, communication, and information technology. Multinomial regression was validated using the Wald test and log likelihood ratios. Two-tailed p values or 95% CIs are reported for all analyses. We used SAS, Stata and EXCEL statistical software for analyses.

Results

Appendix A (available at http://jech.bmjjournals.com/supplemental) shows data on under-five and adult mortality by country. Cluster analysis identified three distinct groups for under-five mortality. Group 1 (lowest mortality) included 117 countries with a mean mortality per 1000 live births of 20.6 and a range of 3.9–60 (fig 1A). Group 2 (mid-level) included 45 countries with a mean mortality per 1000 live births of 105.6 and a range of 66–156. Group 3 (highest mortality) included 23 countries with a mean mortality per 1000 live births of 207.3 and a range of 160–316.

Cluster analysis also identified three distinct categories for adult mortality (men; fig 1B). Group 1 included 111 countries with a mean mortality per 1000 adult men of 173.5 and a range of 80.4–250. Group 2 included 49 countries with a mortality per 1000 adult men of 331.8 and a range of 258–449. Group 3 included 29 countries with a mean rate per 1000 adult men of 583.7 and a range of 460–725.

Figure 1A shows that the 23 countries with the highest under-five mortality are in western and sub-Saharan Africa and Afghanistan. Figure 1B shows even more strikingly that the 29 countries with the highest adult mortalities are also in western and sub-Saharan Africa.

Figure 2A shows time-series trend data on adult mortality. Between 1960 and 1990, all three groups experienced a decline in mortality. Since 1990, however, adult mortality has actually increased for the worse-off group.

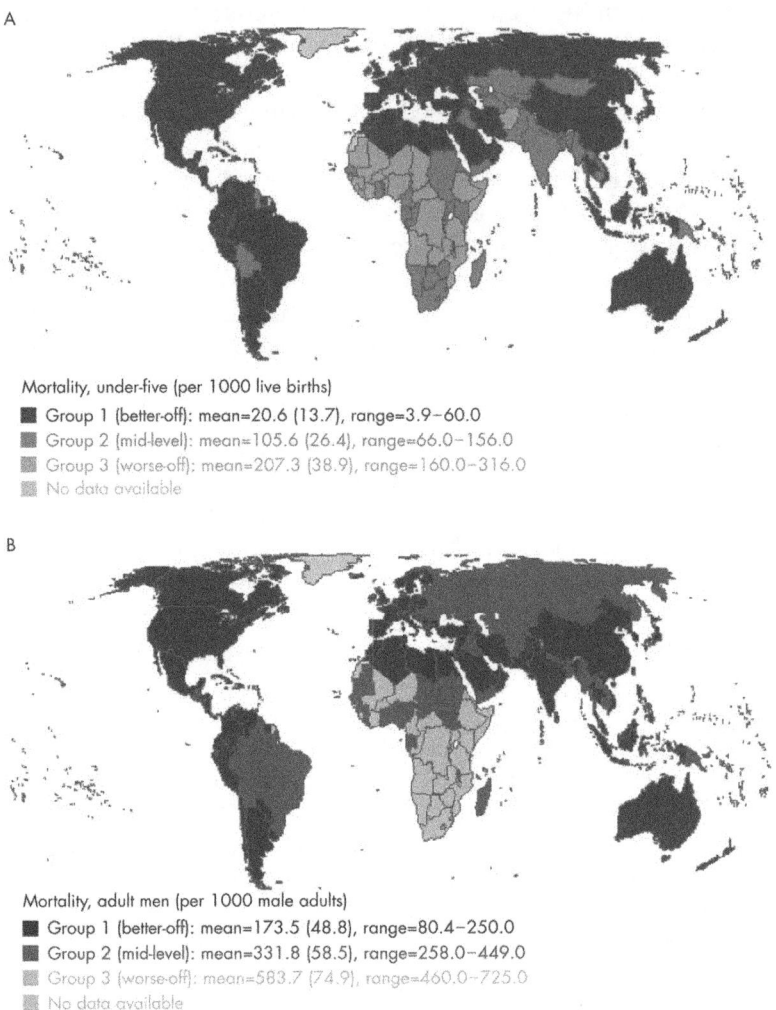

Figure 1: (A) Global distribution of disparities in under-five mortality. (B) Global distribution of disparities in adult mortality (men).

Figure 2B shows the rate of change in adult mortality between 1960 and 2000. It shows that gaps in mortality are becoming wider and that the rate of change in mortality has become markedly different, especially among the countries with the highest mortality, which have switched from a reduction to an increase. Between 1990 and 2000, adult male mortality increased by roughly 30% among the worse-off countries, whereas it stayed roughly the same or even fell slightly for the better-off and mid-level groups.

Figure 3A shows time-series trend data on under-five mortality between 1960 and 2000. Under-five mortality declined for all three groups. Figure 3B shows rates of change in under-five mortality between 1960 and 2000.

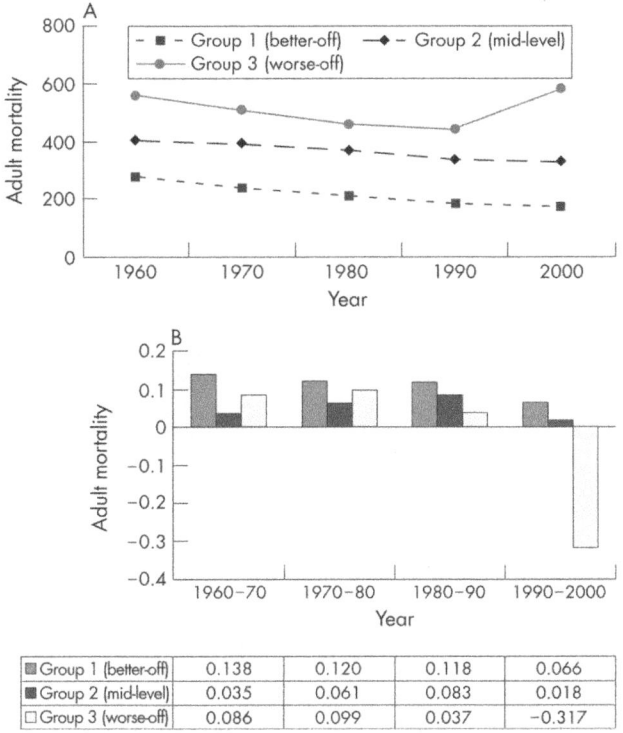

	1960–70	1970–80	1980–90	1990–2000
▣ Group 1 (better-off)	0.138	0.120	0.118	0.066
▪ Group 2 (mid-level)	0.035	0.061	0.083	0.018
□ Group 3 (worse-off)	0.086	0.099	0.037	−0.317

Figure 2: (A) Time-series trend of adult mortality. (B) Rate of change in adult mortality.

The rate of change for the worse-off group slowed down more (from 0.124 between 1960 and 1970 to 0.056 between 1990 and 2000) than, for example, the rate for the better-off group (from 0.292 from 1960 to 1970 to 0.326 between 1990 and 2000). The worse-off group therefore made slower progress in reducing under-five mortality than the better-off group.

Table 1 compares the country indicators of the adult and under-five mortality groups. For under-five mortality, the worse-off group had a mean GNI per capita in international dollars of $1011 as compared with $12 086 for the better-off group – a 10-fold difference. This relationship held true for GNI per capita calculated by the Atlas method as well. Additionally, the percentage of the population living on <$1 per day was considerably greater in the worse-off group (53%) than in the better-off group (8%). Educational outcomes were also unequal, with a fourfold difference in the female illiteracy rate (63% for the worse-off group v 14% for the better-off group). Differences in health expenditures and risk factors showed considerably more investment in overall spending ($650 v 10$ per capita), human resources such as doctors (7 v 1 per capita) and health-related capital (eg, 5 v 0.7 hospital beds per capita) for the better-off as compared with the worse-off group. An interesting finding was the degree of difference in the consumer

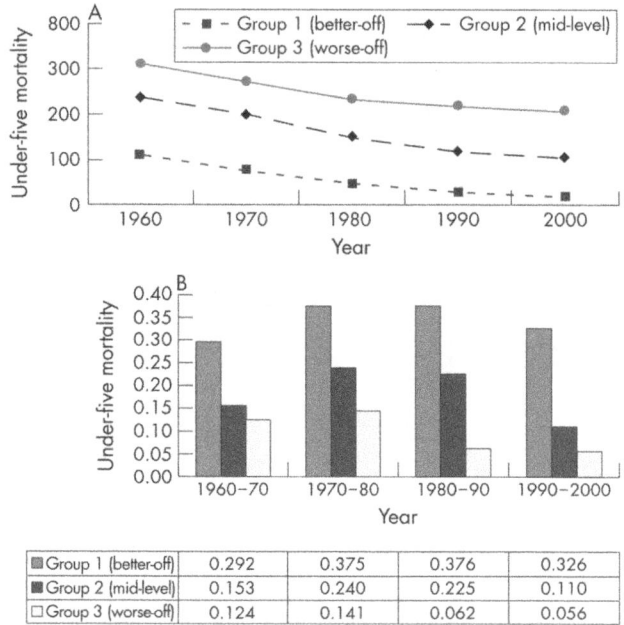

Figure 3: (A) Time-series trend of child mortality. (B) Rate of change in child mortality.

price index (CPI) between the worse-off (CPI of 25 035) and better-off (CPI of 251) groups, indicating the influence of considerable instability and inflation in general prices. The better-off group fared much better on all measures of communication (fixed line and mobile phone subscribers) and information technology (computer and internet).

In general, these relationships held true for the adult mortality groups (table 1). The better-off group had higher per capita income; less poverty (percentage of the population living on <1$ per day); better educational outcomes, more expenditure on health and human and physical resources; better access to sanitation and safe water; lower rates of diseases such as HIV/AIDS and tuberculosis; higher percentage of GDP in trade; and greater access to communication and information technology. For both indicators of mortality, therefore, the better-off groups of countries had better resources, investments and outcomes than the worse-off groups, suggesting that multiple factors influence a country's level of health.

Table 2 presents unadjusted and adjusted odds ratios (ORs) for associations between country indicators and mortality groups. For adult mortality (table 2), associations among several independent variables (GNI per capita, PPP ($); adult illiteracy rate for men and women; health expenditure per capita ($) total; improved water source and improved sanitation facilities; and incidence of tuberculosis) and the comparison between the worse-off and better-off groups lost significance after adjustments for all other variables in

Table 1: Country indicators by mortality groups

Country indicators	Adult mortality (men)			Under-five mortality		
	Group 1 (better-off)	Group 2 (mid-level)	Group 3 (worse-off)	Group 1 (better-off)	Group 2 (mid-level)	Group 3 (worse-off)
Health						
Mortality, per 1000	173.5 (77.2 to 269.8)	331.7‡ (216.4 to 447.0)	583.7‡‡ (435.9 to 731.5)	20.6 (−6.4 to 47.6)	105.6‡ (53.5 to 157.7)	207.2‡ (130.5 to 283.9)
Poverty and income						
GNI per capita, PPP ($)	12 772.1 (10 774.0 to 14.770.2)	3700.7‡ (2636.8 to 4764.6)	2058.5‡ (1113.6 to 3003.4)	12 085.8 (10 229.7 to 13 941.9)	2659.7‡ (1954.9 to 3364.5)	1011.0‡ (810.5 to 1211.5)
GNI per capita, Atlas method	9540.7 (7416.5 to 11 664.9)	1460.8‡ (732.7 to 2188.9)	617.8‡ (303.3 to 932.3)	8848.0 (6861.2 to 10 834.8)	812.3‡ (565.3 to 1059.3)	255.7‡ (190.4 to 321.0)
% Living on <$1 per day	9.7 (5.5 to 13.9)	18.0 (11.1 to 24.9)	46.4‡ (35.3 to 57.5)	8.4 (5.0 to 11.8)	25.5‡ (17.6 to 33.4)	53.3‡ (42.0 to 64.6)
% Living in the lowest income quintile	7.4 (6.3 to 8.5)	6.0 (5.5 to 6.5)	2.4* (2.4 to 2.4)	7.3 (6.1 to 8.5)	4.8* (−0.3 to 9.9)	No data
Gini coefficient	40.7 (37.4 to 44.0)	37.9 (34.2 to 41.6)	46.2 (16.3 to 76.1)	40.2 (37.2 to 43.2)	37.5 (33.8 to 41.2)	46.2 (16.3 to 76.1)
Education outcome						
Adult illiteracy rate, men (%)	10.2 (7.5 to 12.9)	19.9† (13.3 to 26.5)	32.6‡ (25.1 to 40.1)	8.3 (6.1 to 10.5)	24.4‡ (18.9 to 29.9)	41.8‡ (33.4 to 50.2)
Adult illiteracy rate, women (%)	17.0 (12.6 to 21.4)	32.5‡ (22.2 to 42.8)	47.0‡ (36.8 to 57.2)	13.8 (10.3 to 17.3)	38.2‡ (29.4 to 47.0)	63.3‡ (54.8 to 71.8)
Health expenditure, services and use						
Health expenditure per capita ($)	688.7 (505.6 to 871.8)	91.8‡ (45.1 to 138.5)	34.0‡ (10.9 to 57.1)	649.8 (481.6 to 818.0)	56.1‡ (31.9 to 80.3)	9.9‡ (7.8 to 12.0)
Health expenditure, private (% of GDP)	2.4 (2.1 to 2.7)	1.6* (1.2 to 2.0)	2.2 (1.7 to 2.7)	2.4 (2.1 to 2.7)	2.0 (1.5 to 2.5)	1.7 (1.3 to 2.1)
Health expenditure, public (% of GDP)	3.9 (3.6 to 4.2)	2.9† (2.5 to 3.3)	2.3‡ (1.9 to 2.7)	3.9 (3.6 to 4.2)	2.7† (2.2 to 3.2)	2.0‡ (1.7 to 2.3)
Births attended by skilled health staff (% of total)	90.3 (83.5 to 97.1)	57.5 (42.0 to 73.0)	42.5 (30.4 to 54.6)	88.7 (81.0 to 96.4)	62.0 (48.0 to 76.0)	33.8 (25.5 to 42.1)
Inpatient admission rate	13.0 (10.9 to 15.1)	18.0 (13.6 to 22.4)	10.1 (−27.9 to 48.1)	14.2 (12.2 to 16.2)	13.9 (−4.4 to 32.2)	10.1 (−27.9 to 48.1)
Outpatient visits	4.9 (3.7 to 6.1)	5.5 (3.0 to 8.0)	0.6 (−0.6 to 1.8)	5.5 (4.4 to 6.6)	0.6 (−0.1 to 1.3)	0.6 (−0.6 to 1.8)
Doctors	6.0 (4.8 to 7.2)	8.1 (6.0 to 10.2)	1.3 (0.9 to 1.7)	7.1 (5.9 to 8.3)	5.4 (2.8 to 8.0)	1.2 (0.9 to 1.5)
Hospital beds	4.3 (3.5 to 5.1)	6.7 (4.7 to 8.7)	0.7 (0.0 to 1.4)	4.9 (4.1 to 5.7)	4.1 (1.1 to 7.1)	0.7 (−0.4 to 1.8)

(Continued)

Table 1: (Continued)

Country indicators	Adult mortality (men)			Under-five mortality		
	Group 1 (better-off)	Group 2 (mid-level)	Group 3 (worse-off)	Group 1 (better-off)	Group 2 (mid-level)	Group 3 (worse-off)
Environmental health factors						
Improved water source (% with access)	89.8 (87.1 to 92.5)	70.5‡ (63.2 to 77.8)	60.6‡ (53.6 to 67.6)	89.9 (87.3 to 92.5)	70.5‡ (64.4 to 76.6)	51.8‡ (43.8 to 59.8)
Improved sanitation facilities (% with access)	87.3 (82.7 to 91.9)	60.5‡ (51.0 to 70.0)	50.4‡ (38.9 to 61.9)	88.5 (84.6 to 92.4)	58.6‡ (49.7 to 67.5)	42.8‡ (30.9 to 54.7)
Health conditions and interventions						
Incidence of TB	56.9 (42.9 to 70.9)	189.0‡ (156.6 to 221.4)	396.0‡ (348.6 to 443.4)	59.5 (46.9 to 72.1)	288.4‡ (232.4 to 344.4)	325.9‡ (295.4 to 356.4)
Prevalence of HIV	0.3 (0.2 to 0.4)	1.0 (0.6 to 1.4)	11.2‡ (7.3 to 15.1)	0.3 (0.2 to 0.4)	6.1‡ (2.7 to 9.5)	5.5‡ (3.7 to 7.3)
Immunisation, measles (% of children)	88.5 (SD 2.2) (86.3 to 90.7)	77.1‡ (SD 6.1) (71.0 to 83.2)	61.5‡ (SD 6.9) (54.6 to 68.4)	89.9 (SD 1.9) (88.0 to 91.8)	72.4‡ (SD 5.7) (66.7 to 78.1)	55.3‡ (SD 7.7) (47.6 to 63.0)
Prevalence of smoking, men (%)	39.1 (36.2 to 42.0)	45.6 (40.8 to 50.4)	37.6 (22.4 to 52.8)	40.6 (37.8 to 43.4)	43.9 (37.9 to 49.9)	29.2 (7.6 to 50.8)
Prevalence of smoking, women (%)	16.5 (14.0 to 19.0)	15.7 (11.7 to 19.7)	12.5 (4.9 to 20.1)	17.0 (14.7 to 19.3)	13.4 (9.1 to 17.7)	12.7 (−3.7 to 29.1)
Environment						
Rural population (% of total population)	35.9 (31.9 to 39.9)	54.6‡ (48.9 to 60.3)	67.2‡ (60.6 to 73.8)	35.3 (31.7 to 38.9)	61.5‡ (56.2 to 66.8)	70.2‡ (64.5 to 75.9)
Energy production and use						
Electric power consumption (kWh per capita)	4675 (3570.0 to 5780.0)	1316.2† (870.2 to 1762.2)	441.5† (−389.5 to 1272.5)	4457.8 (3442.2 to 5473.4)	830.4† (392.8 to 1268.0)	57.6* (36.6 to 78.6)
Economy						
Household final consumption expenditure per capita growth (%)	2.5 (1.2 to 3.8)	2.8 (−1.1 to 6.7)	1.0 (−3.4 to 5.4)	3.1 (1.8 to 4.4)	2.6 (−1.3 to 6.5)	−2.2 (−7.7 to 3.3)
GDP per capita, PPP ($)	13 066.8 (11 016.5 to 15 117.1)	3831.7‡ (2730.3 to 4933.1)	2096.0‡ (1138.6 to 3053.4)	12 305.1 (10 410.2 to 14 200.0)	2745.6 (2015.9 to 3475.3)	1058.0 (837.4 to 1278.6)
Trade						
Trade in goods (% of GDP)	68.3 (59.7 to 76.9)	76.1 (65.1 to 87.1)	55.3 (42.8 to 67.8)	68.9 (61.4 to 76.4)	70.8 (59.8 to 81.8)	50.0 (35.1 to 64.9)

Monetary						
Food production index 1989–91 = 100	119.6 (112.9 to 126.3)	116.7 (104.8 to 128.6)	118.1 (109.8 to 126.4)	115.2 (108.2 to 22.2)	123.9 (115.2 to 132.6)	127.6 (115.5 to 139.7)
CPI	193.5 (111.6 to 275.4)	357.8 (78.3 to 637.3)	18 333.4 (−14 710.2 to 51 377.0)	251.4 (132.7 to 370.1)	178.4 (132.5 to 224.3)	25 035.2* (−20 992.4 to 71 062.8)
External debt						
Total debt service (% of GNI)	5.7 (4.9 to 6.5)	5.9 (4.5 to 7.3)	4.1 (2.2 to 6.0)	6.3 (5.5 to 7.1)	4.5 (3.3 to 5.7)	4.8 (2.4 to 7.2)
Military expenditure						
Military expenditure (% of GDP)	2.7 (2.2 to 3.2)	2.2 (1.4 to 3.0)	3.1 (2.1 to 4.1)	2.7 (2.2 to 3.2)	2.1 (1.7 to 2.5)	3.2 (1.8 to 4.6)
Communication						
Fixed line and mobile phone subscribers (per 1000 people)	532.7 (447.7 to 617.7)	147.2‡ (88.6 to 205.8)	36.9‡ (8.5 to 65.3)	519.0 (440.7 to 597.3)	60.1‡ (39.4 to 80.8)	8.4‡ (3.6 to 13.2)
Information Technology						
Internet users	3 563 084.4 (989 602.9 to 6 136 565.9)	256 830.4 (7677.7 to 505 983.1)	108 724.2 (−59 695.6 to 277 144.0)	3 368 419.7 (952 600.4 to 5 784 239.0)	217 637.5 (−53 247.8 to 488 522.8)	17 650.1 (5180.8 to 30 119.4)
Personal computers (per 1000 people)	149.9 (118.3 to 181.5)	27.5‡ (14.1 to 40.9)	9.7‡ (2.2 to 17.2)	143.1 (113.4 to 172.8)	13.2‡ (7.7 to 18.7)	2.8‡ (1.4 to 4.2)

Values are given in mean or percentage (95% CI).
CPI, consumer price index; GDP, gross domestic product; GNI, gross national income; PPP, purchasing power parity; TB, tuberculosis.
*Significant at p < 0.05 when compared with group 1; †significant at p < 0.01 when compared with group 1; ‡significant at p < 0.001 when compared with group 1.

Table 2: Multivariate logistic regression relating country indicators to mortality groups

	Adult mortality, men				Child mortality, under-five			
	Mid-level (group 2) v better-off (group 1)		Worse-off (group 3) v better-off (group 1)		Mid-level (group 2) v better-off (group 1)		Worse-off (group 3) v better-off (group 1)	
Group Covariates	Unadjusted OR (95% CI)	Adjusted OR (95% CI)	Unadjusted OR (95% CI)	Adjusted OR (95% CI)	Unadjusted OR (95% CI)	Adjusted OR (95% CI)	Unadjusted OR (95% CI)	Adjusted OR (95% CI)
Income category								
GNI per capita, PPP ($)	0.99§ (0.99 to 0.99)	1.00 (0.99 to 1.00)	0.99§ (0.99 to 0.99)	0.99 (0.98 to 1.00)	0.99§ (0.99 to 0.99)	1.00 (0.99 to 1.00)	0.99§ (0.99 to 0.99)	0.76 (0)
Education outcome								
Adult illiteracy rate, men (%)	1.05‡ (1.01 to 1.08)	1.06 (0.87 to 1.30)	1.09§ (1.05 to 1.13)	1.15 (0.73 to 1.80)	1.11§ (1.06 to 1.16)	0.89 (0.64 to 1.24)	1.19§ (1.12 to 1.27)	$5.49 \times 10^{0.07}$ (0)
Adult illiteracy rate, women (%)	1.03‡ (1.01 to 1.05)	0.97 (0.84 to 1.10)	1.05§ (1.02 to 1.07)	1.05 (0.74 to 1.47)	1.06§ (1.03 to 1.09)	1.15 (0.92 to 1.43)	1.12§ (1.07 to 1.17)	0.16 (0)
Health expenditure								
Health expenditure per capita ($)	0.99§ (0.99 to 0.99)	0.98 (0.96 to 1.00)	0.98§ (0.97 to 0.99)	0.97 (0.81 to 1.17)	0.98§ (0.98 to 0.99)	0.89* (0.77 to 1.02)	0.81§ (0.72 to 0.90)	2.90×10^{-8} (0)
Health expenditure, public (%)	0.71‡ (0.57 to 0.88)	2.78‡ (1.11 to 6.92)	0.55§ (0.40 to 0.76)	2.33* (0.06 to 8.81)	0.65§ (0.51 to 0.83)	10.75 (0.85 to 135.69)	0.43§ (0.28 to 0.65)	0.00 (0)
Environmental health factor								
Improved water source (%)	0.93§ (0.90 to 0.95)	0.95 (0.89 to 1.02)	0.91§ (0.88 to 0.94)	1.07 (0.80 to 1.43)	0.92§ (0.89 to 0.95)	0.98 (0.84 to 1.13)	0.87§ (0.83 to 0.91)	1.69×10^{-6} (0)
Improved sanitation facilities (%)	0.94§ (0.92 to 0.97)	1.00 (0.94 to 1.05)	0.93§ (0.91 to 0.96)	1.01 (0.92 to 1.12)	0.93§ (0.91 to 0.96)	0.97 (0.89 to 1.07)	0.91§ (0.88 to 0.94)	89.78 (0)
Health condition								
TB incidence (%)	1.01§ (1.01 to 1.02)	1.00* (0.99 to 1.02)	1.04§ (1.02 to 1.05)	1.04 (0.96 to 1.12)	1.02§ (1.01 to 1.03)	1.01* (0.99 to 1.03)	1.02§ (1.01 to 1.03)	0.61 (0.99 to 1.04)
HIV prevalence (%)	2.71§ (1.51 to 4.88)	1.41 (0.60 to 3.31)	6.43§ (3.16 to 13.08)	18.62* (0.30 to 1135)	3.12§ (1.86 to 5.25)	2.24* (0.75 to 6.71)	3.10§ (1.84 to 5.21)	8.83×10^{19} (0)
n	82		82		82		82	
Log likelihood	-30.039442		-30.039442		-9.0381474		-9.0381474	
Pseudo-R²	0.6572		0.6572		0.8927		0.8927	

GNI, gross national income; PPP, purchasing power parity; TB, tuberculosis.
*Approaching borderline †significance at p < 0.1; ‡significant at p < 0.05 level; §significant at p < 0.01 level; §significant at p < 0.001 level.

What This Paper Adds

- This paper deals with gaps in the literature on health inequalities in work on intercountry inequalities, on the use of a threshold or norm (established by clustering techniques) and on the identification of "health gaps" for development policy purposes.
- Most of the prior work has focused on international comparisons of intra-country inequalities, with the use of the application of econometric techniques in inequality assessment to health as a variable (Gini coefficient, etc).

---

the table. The exception was public health expenditure (%; adjusted OR 2.3; 95% CI 0.06 to 8.8) and prevalence of HIV infection (adjusted OR 18.6; 95% CI 0.30 to 1135), which approached borderline significance.

A similar trend was found for under-five mortality (table 2): associations between all the independent variables and the comparison between the worse-off and better-off groups were no longer significant after adjustments for all other variables shown in the table. The exception was seen in the comparison between the mid-level and better-off groups for total health expenditure per capita ($; adjusted OR 0.89; 95% CI 0.77 to 1.02), incidence of tuberculosis (adjusted OR 1.01; 95% CI 0.99 to 1.03) and prevalence of HIV (adjusted OR 2.24; 95% CI 0.75 to 6.71) approaching borderline significance. This logistic model had fairly good predictive power, with a pseudo-R^2 of 0.65 (the pseudo-R^2 is the log likelihood on a scale where 0 corresponds to the constant-only model and 1 corresponds to perfect prediction for discrete models).[44]

Discussion

To our knowledge, this is the first systematic study of cross-national inequalities in adult and child mortality to identify mortality groups (most healthy, least healthy, mid-level health) by cluster analysis, and examine risk factors associated with inequality in mortality. We found that inequalities in child and adult mortality are wide, are growing and are influenced by several economic, social and health-sector variables. On average, groups of countries with high adult or under-five mortality also had lower average incomes, more extreme poverty, lower levels of investment in human and physical resources, higher inflation and less trade, less effective disease prevention, and worse educational outcomes and health risk factors. Global health inequalities should be studied in conjunction with levels of development and social and economic inequalities. Global efforts to deal with inequalities in health require attention to the worse-off countries, geographic concentrations and multidimensional approach to development.[45]

A particularly disturbing finding from this study was that countries with high under-five mortality are making slower progress than countries with

---

Policy Implications

- Stratifying countries into mortality groups of different levels of health, which to our knowledge has yet to be carried out, better enables multilateral institutions like the World Health Organization and the World Bank to devise policies and interventions to reduce the gap in mortality between countries.
- This type of study is particularly relevant for organisations such as the World Bank, which work in several policy domains outside the health sector.
- The study uses data from the World Development Indicators (WDI), enabling the study of numerous economic and social variables unavailable in other international datasets which also include health.

---

lower rates. Moreover, gaps in adult mortality are becoming wider and countries with the highest mortality have actually shifted from reduced mortality to increased mortality. Studies comparing inequalities in child health within countries have found similar trends.[38] This finding suggests that the widening inequality in both child and adult mortality may reflect of growing gaps in living conditions and standards between geographical areas and between rich and poor countries.

Our study showed that the countries with the highest adult and under-five mortality have multiple deprivations. For example, they have a fourfold higher percentage of people living on <$1 per day; more than a twofold higher female illiteracy rate, and less than one sixth the GNI per capita in international dollars; and one fifth the outpatient visits, hospital beds and doctors as their low-mortality counterparts. The gap in total expenditure on health per capita is even greater: a 20-fold difference between adult mortality groups and a 50-fold difference between child mortality groups. Additionally, this study confirms what is known about the HIV/AIDS epidemic, as countries with high adult mortality had a higher prevalence of HIV infection (OR per 1% increase 18.6; 95% CI 0.3 to 1135). It is therefore important to deal with health needs on multiple fronts with an integrated set of strategies, especially in western and sub-Saharan Africa and Afghanistan. These results are particularly relevant for multinational organisations, such as the World Bank, which work in multiple policy domains affecting health inequalities.

Our study has some limitations.

1. Although the World Bank reviews social and economic indicators for reliability and validity, the definitions and methods underlying indicators can vary from country to country and over time. Data may also be collected differently in different countries. Therefore, we have focused on broad trends rather than small differences among indicators.

2. Social indicators may refer to different years. Because social changes tend to occur slowly and many factors contribute to a single indicator, the World Bank does not generally recommend that values be imputed for missing data.
3. Although WDI data are collected from supposedly authoritative sources, many statistical systems in developing economies are weak, which affects the availability, reliability and margin of error associated with individual data. However, the WDI dataset is considered one of the most reliable and valid of all international databases that include health, social and economic indicators.
4. We have tried to control for multicollinearity in our study, but as commonly recognised in multivariate modelling, we note inevitably high levels of correlations between variables, in some cases resulting in wide CIs.
5. Numerous historical and political factors, including coloniser status and years of colonisation, composition of military power, political orientation and political total years of dictatorship, and human rights are not included in the dataset but may influence both health and development.

In conclusion, this study identified three distinct mortality groups worldwide (worse-off, better-off, mid-level) and showed that key associated factors to health disparities among countries include both factors within the health sector and factors related to a country's overall level of development. Thus, this analysis could be extended by multinational development actors to assess shortfalls in mortality using better-off as the reference group. It is important to note, however, that global health policy focused on narrowing the mortality gap between countries is not simply a matter of poverty reduction or development. It requires a commitment to social justice.[46–48]

References

1. Evans T, Whitehead M, Diderichsen F, *et al*. Introduction. In: Evans T, Whitehead M, Diderichsen F, Bhuiya A, Wirth M, eds. *Challenging inequalities in health: from ethics to action*. Oxford: Oxford University Press, 2001:3–11.
2. World Health Organization. *The World Health Report 1999: making the difference*. Geneva, Switzerland: World Health Organization, 1999; http://www.who.int/whr2001/2001/archives/1999/en/pdf/whr99.pdf (accessed 14 Jan 2004).
3. World Health Organization. *The World Health Report 2000: health systems: improving performance*. Geneva, Switzerland: World Health Organization, 2000; http://www.who.int/whr2001/2001/archives/2000/en/index.htm (accessed 14 Jan 2004).
4. Jong-wook L. Global health improvement and WHO: shaping the future. *Lancet* 2003;362:2083–8.
5. World Bank. *Health, nutrition and population sector strategy*. Washington, DC: World Bank, 1997.

6. Gwatkin DR. Health inequalities and the health of the poor: what do we know? What can we do? *Bull World Health Organ* 2000;78:3–18.
7. United Nations Children's Fund. *Unicef annual report*. New York, NY: United Nations Children's Fund, 1997.
8. United Nations Children's Fund. *The progress of nations 1999*. New York, NY: United Nations Children's Fund, 1999.
9. Anand S, Sen A. *Gender inequality in human development: theories and measurement*, In: United Nations Development Programme, background papers: human development report 1995. New York, NY: United Nations Development Programme, 1995:1–20.
10. United Nations Development Programme. Human development reports. http://hdr.undp.org (accessed 20 Aug 2003).
11. United Nations Development Programme. *Human development report 1996: economic growth and human development*. New York, NY: Oxford University Press, 1996; http://hdr.undp.org/reports/global/1996/en/ (accessed 14 Jan2004).
12. United Nations Development Programme. *Human development report 2003: millennium development goals: a compact among nations to end poverty*. New York, NY: Oxford University Press, 2003; http://www.undp.org/hdr2003/ (accessed 30 Jul 2003).
13. UK Department for International Development. *Better health for poor people: strategies for achieving the international development targets*. London, UK: Department for International Development, 2001.
14. Foege WH. Global public health: targeting inequities. *JAMA* 1998;279:1931–2.
15. Howson CP, Fineberg HV, Bloom BR. The pursuit of global health: the relevance of engagement for developed countries. *Lancet* 1998;351:586–90.
16. Evans T, Whitehead M, Diderichsen F, *et al*. *Challenging inequities in health: from ethics to action: summary*. New York, USA: Rockefeller Foundation, http://www.rockfound.org/Library/Challenging_Inequalities_In_Health_From_Ethics_to_Action.pdf (accessed 30 Jul 2003).
17. An and S, Diderichsen F, Evans T, *et al*. Measuring disparities in health: methods and indicators (chapter 5). In: Evans T, Whitehead M, Diderichsen F, Bhuiya A, Wirth M, eds. *Challenging inequalities in health: from ethics to action*. Oxford: Oxford University Press, 2001:48–67.
18. Diderichsen F, Whitehead M, Burstrom B, *et al*. Sweden and Britain: the impact of policy context on inequities in health (chapter 17). In: Evans T, Whitehead M, Diderichsen F, Bhuiya A, Wirth M, eds. *Challenging inequalities in health: from ethics to action*. Oxford: Oxford University Press, 2001:240–55.
19. Sen A. *Inequality reexamined*. Cambridge, MA: Harvard University Press, 1992.
20. World Bank. *World Development Report 2000/2001: attacking poverty*. New York, NY: Oxford University Press, 2000.
21. Townsend P, Davidson N. *Inequalities in health: the black report*. Hammonds worth: Penguin, 1982.
22. North F, Syme SL, Feeney A, *et al*. Explaining socioeconomic differences in sickness absence: the Whitehall II study. *BMJ* 1993;306:361–6.
23. Wilkinson RG. *Unhealthy societies: the afflictions of inequality*. London, UK: Routledge, 1996.
24. Mackenbach JP, Kunst AE, Cavelaars AE, *et al*. Socioeconomic inequalities in morbidity and mortality in western Europe. The EU Working Group on Socioeconomic Inequalities in Health. *Lancet* 1997;349:1655–9.
25. Acheson D, Barker D, Chambers J, *et al*. *The report of the independent inquiry into the inequalities in health*. London, UK: The Stationery Office, 1998.
26. Lynch JW, Kaplan GA, Cohen RD, *et al*. Do cardiovascular risk factors explain the relationship between socioeconomic status, risk of all-cause mortality, cardiovascular mortality and acute myocardial infarction? *Am J Epidemiol* 1996;144:934–42.

27. Hallqvist J, Diderichsen F, Theorell T, *et al.* Is the effect of job strain due to interaction between high psychological deman and low decision latitude? *Soc Sci Med* 1998;46:1405–15.
28. Department of Health. *Tackling health inequalities: consultation on a plan for delivery.* London, UK: Department of Health, 2001.
29. Kunst AE, Geurts JJ, van den Berg J. International variation in socioeconomic inequalities in self reported health. *J Epidemiol Community Health* 1995;49:117–23.
30. Kakwani N, Wagstaff A, van Doorslaer E. Socioeconomic inequalities in health: measurement, computation and statistical inference. *J Econom* 1997;77:87–103.
31. Kunst A. *Cross-national comparisons of socioeconomic differences in mortality.* The Hague: CIP-Gegevens Koninklijke Bibltiotheek, 1997.
32. Whitehead M, Diderichsen F. International evidence on social inequalities in health. In: Dever F, Whitehead M, eds. *Health inequalities – decennial supplement.*Series DS No 15.Office for National Statistics. London, UK: The Stationery Office, 1997:44–69.
33. Wagstaff A. Socioeconomic inequalities in child mortality: comparisons across nine developing countries. *Bull World Health Organ* 2000;78:19–29.
34. Wagstaff A, Watanabe N. *Socioeconomic inequalities in child malnutrition in the developing world*, Policy research working paper 2434. Washington, DC: World Bank, September, 2000.
35. Van Doorslaer E, Wagstaff A, Bleichrodt H, *et al.* Income-related inequities in health: some international comparisons. *J Health Econ* 1997;16:93–112.
36. Wagstaff A. *Inequalities in health in developing countries: swimming against the tide?* Policy research working paper 2795. Washington, DC: World Bank, 2002.
37. Victora CG, Wagstaff A, Schellenberg JA, *et al.* Applying an equity lens to child health and mortality: more of the same is not enough. *Lancet* 2003;362:233–41.
38. Ruger JP. Changing Role of the World Bank in Global Health. *American Journal of Public health* 2005;95(1):60–70.
39. World Bank. *World development indicators 2003*. Washington, DC: World Bank, 2003.
40. Everitt BS. *Cluster analysis*, 3rd edn. London, UK: Edward Arnold, 1993.
41. Everitt BS. Commentary: classification and cluster analysis. *BMJ* 1995;311:535–6.
42. Ruger JP. Measuring disparities in healthcare. *BMJ* 2006;333:274.
43. Ruger JP. Toward a theory of a right to health: capability and incompletely theorized agreements. *Yale J Law Human* 2006;18:273–326.
44. Judge GG, Griffiths RC, Lutkepohl H, *et al. The theory and practice of econometrics*, 2nd edn. New York, NY: John Wiley & Sons, 1985.
45. Ruger JP. Global tobacco control: an integrated approach to global health policy. *Development* 2005;48:65–9.
46. Ruger JP. Health and social justice. *Lancet* 2004;364:1075–80.
47. Ruger JP. Ethics of the social determinants of health. *Lancet* 2004;364:1092–7.
48. Ruger JP. Millennium development goals for health: building human capabilities. *Bull World Health Organ* 2004;82:951–2.

7

Stigmatized Ethnicity, Public Health, and Globalization

S. Harris Ali

D uring extreme events and disaster situations, many potentially disruptive aspects of individual and group life that normally remain latent
in day-to-day life may come to the surface. Extreme events, therefore,
provide a unique opportunity – a type of "natural experiment" – to study the
"exception" to better understand the "rule" in terms of gaining insights into
social structure and behaviour more generally (Stallings 2002). The value of
adopting this vantage point is quite evident in the work of sociologists in various traditions, for example, in Durkheim's (1951) classic work on how acts of
deviance are functional to society by providing instances and making explicit
what exactly the moral boundaries are for a given behaviour. This point is
similarly illustrated by Garfinkel's (1967) ethnomethological technique of
intentionally violating social norms through "breaching experiments" and
observing the reactions of the interactants, in order to gain insights into
the nature of our taken-for-granted sense of shared social reality. Extreme
events may also serve as "signal events" (Edelstein 2004, 16) that problema-
tize and subsequently politicize and bring to light previously unrecognized,
normalized, and naturalized circumstances and conditions that are, in fact,
dangerous and require political or social attention. For example, the highly
publicized chemical contamination incident in Love Canal, New York brought
to light the extent to which toxic contamination was a major environmental
health problem throughout the whole of North America (Szasz 1994). The
present analysis considers another kind of extreme event, namely a disease

Source: *Canadian Ethnic Studies/Études ethniques au Canada*, 40(3) (2008): 43–64.

outbreak – Severe Acute Respiratory Syndrome (SARS) – in Toronto. The analysis of this event is used as a springboard for discussion in three ways. The first is to explore the general implications that mechanisms of social control and surveillance (within the context of public health) have for ethnic minorities. The second is to help understand the nature of the interactions and reactions to members of certain ethnic minorities during the outbreak. The role of stigma (Goffman 1963) is used to direct and guide this part of the discussion. Finally, at a more conceptual level, the notion of stigma is situated within the contexts of public health practice and globalization to stimulate theoretical developments in ethnicity studies in new directions.

The study presented in this paper was part of a larger research project on "SARS and the Global City" in which other dimensions of the outbreak were investigated, including aspects related to public health governance and disease vulnerability among marginalized communities. As part of this larger research project, a series of 16 unstructured interviews was conducted in Toronto from September 2005 to March 2006 with local officials from Chinese cultural centres, legal aid clinics with an Asian Canadian client base, community health centres, hospitality workers, and domestics. The analysis developed here draws in part from these interview data, but also relies on the various public inquiry commission reports produced shortly after the outbreak by different levels of government, such as the federal report by the National Advisory Committee on SARS and Public Health (Naylor 2003) and the Ontario Expert Panel on SARS and Infectious Disease Control (Walker 2003). In addition, newspaper articles, as well as other reports, were consulted, and most helpful for our analysis here was one prepared by Leung and Guan (2004) for the Chinese Canadian National Council, entitled "Yellow Peril Revisited: Impact of SARS on the Chinese and Southeast Asian Canadian Communities."

Background: The SARS Outbreak in Toronto

In late February 2003, a physician from Guandong Province in China, who had treated patients with an unclassified form of atypical pneumonia, stayed at the Metropole Hotel in Hong Kong while attending a nephew's wedding. The physician himself became ill and, during his visit, twelve guests of the hotel became infected with the disease later to be identified as SARS. These guests continued their travels to various locations, including Singapore, Hanoi, Toronto, and other areas of Hong Kong. A seventy-eight-year-old Canadian woman was one such guest, and, soon after her return to her Toronto home, her forty-four-year-old son also became ill. On March 5, 2003, the woman died, while her son died two days later. During the son's hospital stay, many patients and staff were unknowingly exposed to the SARS virus through various pathways.

Within the first month of the outbreak, there were, in total, 13 SARS-related deaths, 97 probable cases, and 1,137 suspected cases in the Toronto area. By the following month, the total number of cases appeared to reach a plateau; and on May 16, 2003, the official pronouncement was made that the SARS outbreak was over (Naylor 2003). One week later, however, a second wave of SARS patients appeared in a Toronto-area hospital, where five patients were quarantined. The exact chain of events that led to this second outbreak (referred to as SARS II) remains unknown (ibid.). By the third week of April 2003, the World Health Organization (WHO) issued an unprecedented travel advisory, recommending that visitors postpone all but essential trips to Toronto. One week later, a delegation that included the provincial health minister and the public health commissioner visited WHO headquarters in Geneva. In response to the delegation's plea, the travel advisory to Toronto was lifted after the WHO was given assurances that the city would intensify the screening of travelers to and from Canada. In the meantime, the city's three Chinatowns remained deserted, while the occupancy rate in Toronto hotels fell dramatically with the cancellation of several international conferences and a plummeting decline in tourists (CBC 2003).

Surveillance and SARS

A defining characteristic of public health strategies in late modernity is an emphasis on enumeration and surveillance as a means of maintaining social order in the face of potential disruption from health threats (Petersen and Lupton 1996; Fischer and Poland 1998; Petersen 1996; Lupton 1999; Castel 1991; Elden 2003). Specifically, the use of statistics, population profiles, and monitoring in the service of surveillance have become important public health measures through which "populations" can be regulated and controlled through the attribution of risk (Lupton 1995, 174). That is, such measures are used to define the notions of "normality" and "pathology" as applied to groups within the population. As documented by Michel Foucault (1991), this use of medical surveillance as a method of social control has deep historical roots. Traditionally, the control of populations comes under the purview of policing and governmental regulations, but these became supplanted by the European medical establishment during the period of 1720–1800. The influence of medical surveillance intensified during this period because of the need to identify and monitor the health status of the population as part of the battle against leprosy and the plague. In reviewing the work of Foucault, Elden (2003, 242) notes that two different surveillance and control approaches were used with respect to each disease. In the case of leprosy, social control was exercised through the expulsion of the leper from the shared space of the community with the aim of maintaining the purity of the urban environment. On the other hand, in the case of the plague, the "emergency plan" was employed. Here quarantines were imposed (especially in large port areas) as the town became divided into distinct sectors that were patrolled by street inspectors. Detailed reports on each sector were prepared by these officials and kept in a centralized information system for monitoring purposes. This emergency plan strategy resulted in a system of generalized surveillance based on the compartmentalization and control of the movements and activities of sub-populations defined as "pathological." Elden notes that each type of strategy represented a different way of utilizing and controlling space, with each reflecting a different form of political power. With leprosy, the practice of outright exclusion represented a negative form of political power. In contrast, the disciplinary orientation based on strict spatial partitioning, careful surveillance, detailed inspection, and maintaining social order represented a positive form of political power because this type of orientation retains an inclusion-orientation. (See also Sarasin [2008] who discerns a third form of political power in the work of Foucault involving a mixture of security and tolerance related to Foucaults' accounts of smallpox). The response to the SARS outbreak in Toronto clearly paralleled that found during the historic plague period. An illustration of this is found in considering the mode of risk communication employed during the outbreak. Specifically, Torontonians received reports of newly identified SARS cases and a running tally every evening via televised press conferences held by the provincial health officials. Such a practice appears to be a modern variant of the "bills of mortality" – the weekly reports of the number of deaths in each parish of London caused by the plague in 1665 (Edelstein 2004, 2). The practice also illustrates the modern form of ensuring an inclusion orientation, namely, one predicated upon computerized information systems and communications technologies (King 2002).

In contemporary times, health surveillance involves the tracking of health events or health determinants through the continuous collection of data (Naylor 2003, 92). Once the data are collected, they are analyzed and interpreted, with the results being used to make decisions about issuing advisories, alerts, or warnings. Various strategies were pursued to control the spread of SARS in Toronto, including surveillance and contact tracing, isolation and quarantine, and travel restrictions (Gostin et al. 2003, 3229). From a legal standpoint, the term "isolation" refers to limiting the movement of those who exhibit the symptoms of the disease (i.e., they are patients), while "quarantine" applies to restricting the movement of healthy people (Walker-Renshaw 2003). Once cases were identified by the screening procedure, they were *isolated*, while those found to be in contact with the cases were *quarantined*. In addition, those who had entered the epicenter hospital after March 16 were asked to adhere to a ten-day home quarantine. By the end of the SARS outbreak, Toronto public health had investigated 1,907 separate reports in addition to 220 cases of probable or suspected SARS cases (Naylor 2003, 35).[1]

The use of medical surveillance to identify cases may have important social ramifications for those diseases that differentially affect segments of the population. Of note in this connection is the fact that the surveillance efforts of public health officials may have inadvertently amplified the stigmatization of affected groups (Gostin et al. 2003, 3233).

Stigmatization and SARS

In his seminal work, Goffman (1963) notes that stigma is an attribute that socially discredits the individual or group. It is, therefore, a relational concept, that is, a notion that necessarily involves reference to at least two parties, namely, the stigmatized individual or group – towards which feelings of disgrace and disgust are directed – and the non-stigmatized "normals." Since the time of this pioneering work, the concept has been refined and, in an excellent review, Link and Phelan (2001) summarize five interrelated components or processes that together constitute the process of stigmatization – labeling, stereotyping, separation, status loss, and discrimination. Each of these components takes on a unique significance within the context of stigmatization associated with infectious disease – some of these will be simply introduced at this point and further elaborated upon later.

The first component refers to the process whereby people distinguish and label human differences. This involves the issue of how a particular human difference is selected to matter socially and involves what Goffman (1963) refers to as a "mark," that is, a discrediting feature that is observable to others and interpreted as deviant, flawed, limited, spoiled or generally undesirable.[2] Three stigmatizing marks may be identified in the case of SARS in Toronto, namely: (i) the wearing of respiratory masks to cover the face in public places; (ii) employment in the health care sector; and (iii) Chinese-Canadian ethnicity.

Before proceeding further, it should be noted that in a comparative study of the stigma of HIV/AIDS, tuberculosis, and SARS in Hong Kong, Mak et al. (2006) found that the attribution of stigma was a function of controllability, personal responsibility, and blame. That is, first, a determination is made if the person is held personally responsible for his/her illness or whether he/she is deemed helpless in the circumstance; then an inference will be made as to whether the person should be blamed and the extent to which he/she should be stigmatized. Such criteria clearly have greater relevance for a homogenous population, but, as will be discussed, these attribution features will interact with racially-based attributes in a multicultural society such as Toronto. In other words, SARS was what Yang et al. (2007) refer to as a "situational threat" – one in which stigma results from being placed in a social situation that influences how one is treated.

The second component involves the process of stereotyping that occurs through the linking of the labeled person with the undesirable characteristics that form a culturally held stereotype. For example, Briggs (2005, 272) notes that the title of a *New York Times* article on SARS reads, "From China's Provinces, a Crafty Germ Breaks Out" (Rosenthal 2003, 1). The caption has the effect of associating both the virus and the human being with the stereotype of the Chinese as sneaky and/or cunning. Furthermore, in those cases of stigmatization involving infectious disease, such stereotyping builds on the alleged unhygienic sanitary or dietary practices of a particular group – a prejudice that has deep roots that can be traced to "developmentalism" and colonialism.[3]

Third, an "us-them" separation is socially constructed by placing those labeled into distinct categories; that is, an "othering" process occurs (see, for example, Eichelberger [2007] on this process *within* the Chinese-American community itself during SARS). For the purposes of this paper, as alluded to above, medical surveillance, quarantine, and isolation may result in the setting apart of others in a most formal sense, which will obviously facilitate the "us-them" separation.

Fourth, the labeled person experiences status loss and discrimination (both individual and structural discrimination) that lead to unequal outcomes in terms of disadvantages in life chances such as income, education, housing status, and, most relevant to the case of SARS, in terms of medical treatment, health, and psychological well-being.

Finally, all of the above component processes must occur within a context that is contingent upon access to social, economic, and political power; that is, there must exist a power differential between those imposing the stigmatized status and those that are stigmatized because, as Link and Phelan note, "it takes power to stigmatize" (2001, 375). As will be discussed in the final sections of this paper, the power situation that allows the stigmatization of SARS is today influenced by certain ideologies associated with the hegemonic influences of the "new public health" and globalization.

The "SARS Mask" and the Social Construction of Risk and Stigma

The social construction of risk (Hannigan 1995, 92–108) during the SARS outbreak was undoubtedly influenced by the visual image of individuals donning N95 respiratory masks to protect themselves from the contagious spread of the SARS coronavirus. The surgical-type mask is tied around the back of the head and covers the nose and mouth entirely, leaving only the eyes and forehead exposed. Images of individuals wearing these masks were prominently displayed in much of the television and press media coverage of the SARS epidemic in Toronto (and other cities around the world), and the masked visage soon became a poignant symbol of SARS throughout the world. Notably, this occurred despite the fact that, in actuality, the appearance of individuals wearing masks on the streets of Toronto was very rare. Nevertheless, largely through media representations, the mask, in essence, became a type of universal stigma symbol – a "mark" – that conveyed a "spoiled identity" by suggesting the association of that individual (in whatever capacity) with SARS. At the same time, the masked face served as a foreboding symbol that generally raised public awareness and alarm around SARS, embodying the disease and its spread with ominous characteristics.

The image of an individual wearing the N95 mask in the public space is particularly graphic, evocative, and disturbing because it shatters the everyday taken-forgranted sense of reality that most people have concerning the safety of the spaces they occupy in daily life. The use of a respiratory mask outside the hospital setting is obviously not a common occurrence. Thus, the presence of a person wearing a medical mask in public tends to be perceived as "alien" – and, therefore, disruptive of the conventional and comfortable normality of events in everyday public life. In essence, the N95 mask, within the context of the SARS outbreak coverage, functioned as a "rhetorical idiom" (Ibarra and Kitsuse 1993), that is, as an image cluster that endows claims with moral significance (see also Hannigan 1995, 36). Specifically, it was an example of a symbol imbued with a "sense of endangerment" – a threat to one's health and safety.[4]

Content analysis studies of the visual images of SARS in the national newsprint media (i.e., the *National Post* and *Globe and Mail* newspapers and *Maclean's* and *Time* [Canadian Editions] newsmagazines) found that the vast majority of images contained people wearing masks (82.5% [of 120], 57.1% [of 119], 63% [of 23] and 88.2% [of 17] respectively). Furthermore, 50% of the *Post*, 34.5% of the *Globe*, 22.2% of *Maclean's* and 35.3% of *Time* depicted Asian people with masks (Leung and Guan, 2004). The association of masks with Asian people contributed to the racialization of SARS as a "Chinese" disease, and the study concluded, "By labeling SARS as an Asian virus, media sensationalism and its resultant public panic toward Asian Canadian populations have been rationalized and, therefore, justified" (ibid. 13). I will return to this important theme later in this paper, but another aspect of stigmatization that first needs to be considered involves the experience of those in the health care sector during SARS – an important point to consider in reference to the racialization of the disease because the health care sector is one of the few occupational sectors in Canada in which Asian (particularly Asian women) are proportionately represented (interviews: Ontario Nursing Association official, 16 September 2005; Provincial Chief Nursing officer, 27 March 2006).

The SARS Pariah Syndrome and the Health Care Sector

Although the names of individual hospital employees infected with SARS were not disclosed in Toronto, by implication, every hospital employee was publicly perceived as a health risk. As a consequence, all health care workers found themselves stigmatized to some extent (Gostin et al. 2003, 3230). Interviews with health care workers involved in the SARS crisis consistently revealed the high level of stress and stigmatization experienced. Many noted that people were afraid to associate with them (or even with their spouses), and, in many cases, this led to the experience of acute social isolation and ostracism (Walker 2003). Some even noticed that members of the public would cross the street to avoid walking near their homes. The isolation resulting from this SARS pariah syndrome was especially difficult for staff members who contracted the disease because family members were not allowed to visit them in the hospitals, while upon discharge, they were sent directly into home quarantine. Moreover, frequent visits by public health and medical officials dressed in protective gear frightened neighbours and friends and further added to the feelings of alienation experienced by the stigmatized. One health care worker returned home to find that she was no longer welcome there, as her housemates had placed all her belongings outside. The effects of stigmatization also extended to the children of health care workers who were being called by school officials and told to keep their children at home (Walker 2003).

It should be noted that the experiences resulting from the stigmatization described above represented only one aspect of the kinds of psychosocial impacts endured by health care workers. Other psychosocial impacts that arose in being at the frontline of the SARS battle included: the fear of going to work in a hospital; the fear of taking care of SARS patients, which might lead to feelings of confusion, anger, and guilt; and having to deal with the lingering resentment of colleagues who might not have contributed what was expected (ibid.). Furthermore, these psychosocial impacts had to be dealt with in a crisis situation in which health care workers were working excessive hours and double shifts. Added stress was also caused by concerns that workers had about the improper fitting of the N95 masks that might increase their risk of exposure during their work shifts. The severity of the psychological problems experienced is revealed by the fact that a year after the SARS outbreak, dozens of health-care workers in Toronto are still unable to return to work because of psychological trauma (Palmer 2003).

SARS and the Chinese Canadian Community in Toronto

The popular association of SARS with China led to the experience of stigmatization for members of the Chinese Canadian community. Canadian citizens of Chinese origin comprised about 7.5% (348,010) of the 4.7 million people living in the Toronto metropolitan area in 2001 (Statistics Canada 2005), and the city has become the preferred destination of most immigrants from Asian countries to Canada (Li 1998). Historically, the early Chinese settlers who came to Canada to help complete the national railway system were subjected to overt forms of racism such as the Chinese Head Tax and

Exclusion Act in the early 1900s. During this period, both law makers and the media described the migration of Chinese immigrants to Canada as the "yellow peril" (Leung and Guan 2004). Such attitudes still prevail today, as sociologists have noted that the Chinese have always represented a source of uncertainty and fear in Canada and have never gained full acceptance as "true Canadians" (Li 1994, 1998).

A recent example of the underlying hostility towards this group is given by Hier and Greenberg (2002) who discuss how the arrival of four boats carrying refugees from Fujian Province, China, in 1999 led to a cultural construction of enemy stereotypes. That is, the event was quickly used by the media, government, and police to criticize Canadian immigration policy, fashioned around fabricated charges concerning the health and security risks that these migrants allegedly posed to Canadian citizens. Under such article headings as "These Refugees and Immigrants Can Kill You" (*National Post*, August 21, 1999) and "Canada's Open Door" (*Maclean's*, August 23, 1999), the media portrayal focused attention on particular allegations informed by themes of risk and a rhetoric of endangerment: the risk of disease the migrants would bring; the use of Canada as a stop-over for migrants ultimately destined for the United States (hence the risk of increased U.S.-Canada tensions); and an increased risk of organized crime (Greenberg and Hier 2001). This perception of immigrants as threats to public health undoubtedly served as a backdrop to, and subtly influenced, the response to Chinese Canadians during the SARS crisis. Hier and Greenberg (2002) note that, although expressions of explicitly racist stereotypes appear to have dissipated in Canada, they have, in fact, re-appeared in a variety of new racialized discursive articulations which do not make reference to explicit biological or genetic terms. These new discursive articulations are covert in nature and include such stereotypes as "unneighbourly houses," "unusual aesthetic values," "substandard social integration," and "criminality" (examples given in Li 1994, 1998; Tator et al. 1997; Henry and Tator 2000). Extreme situations such as disease outbreaks may, however, serve as the triggering stimulus for the return of more overt forms of racism (Keil and Ali 2006b; Zheng 2005).

In a submission presented to the provincially commissioned SARS Inquiry Commission, a representative from the Asian Canadian Labour Alliance (ACAL) remarked: "We have witnessed the ugly reality of racism that is always hidden just beneath the surface of our so-called tolerant society, and it is always ready to strike at the first opportunity available" (Walker 2003). Leaders from the Chinese Canadian community noted a growing tide of "anti-Chinese" sentiment in Toronto and in the country as a whole, and called upon federal and provincial officials to take action to stem this tide. Just as the Fujiian-Chinese migrant incident triggered opportunities for racially-based sentiments to come to the surface, the SARS outbreak did so as well. Several incidents raised by the ACAL spokesperson and others in the Chinese Canadian community reveal the extent of these reactions.

The ACAL spokesperson related numerous instances of racially-based actions that had taken place during the SARS crisis: seats beside passengers with an Asian appearance remained vacant in crowded subway trains and buses; several Asian Canadian tenants were told to move out by their landlords; a government official visiting a nursing home caring primarily for Chinese Canadian seniors insisted on wearing a mask during her entire visit to do routine testing and commented to the nursing home staff that she only wore the mask at that particular institution; domestic workers from the Filipino Canadian community were laid off; staff from the Canadian Immigration and Refugee board insisted on wearing masks only during those hearings involving people of Chinese descent, despite the fact that all such applicants had to have been in Canada for at least two years prior to their hearing (this practice was dropped after a formal complaint was lodged by ACAL). Furthermore, the Chinese Council of Canadians noted that it had received several racially-based messages that singled out the Chinese community for the outbreak (Sorensen 2003).

Remarks made by public officials were also tainted by racial bias. For example, one member of the provincial legislature for a Toronto-area constituency openly suggested that the SARS crisis in the city was the result of federal immigration policy, while the provincial Energy Minister mockingly coughed upon entering a media scrum, at which point a journalist jokingly asked if the coughing was part of the government's new communication strategy to "give the media SARS." The politician deadpanned, "I enjoyed my trip to Asia." Although he later apologized for his remark, ACAL noted that the social repercussions of the exchange were already evident (Rider 2003).

Racially-based expressions were also directly experienced by Chinese Canadians at the interpersonal level. For example, a family member of a SARS patient remarked: "People treat us like monsters. They say we eat like rats and that we live like pigs." Another member of the Chinese Canadian community noted: "If you cough, people look at you funny. They're wrapping this thing on the Chinese because it started in China. It is really unfair" (Sorensen 2003). One woman whose mother and brother died of SARS even spoke of experiencing discrimination during the cremation of her brother: the funeral home director refused entry to her husband, but later relented when he realized that he would not otherwise receive payment (ibid.).

Those of Asian background with jobs in the health care sector faced stigmatization on two fronts. As alluded to previously, nursing is one of the few professions in Canada in which people of Chinese and Filipino descent were well-represented. However, some of these nurses and health care workers had to contend with suspicion from their colleagues, to the point that they even felt pressure to cut ties with their own ethnic communities to prove that they were "clean" (interview: Toronto nurse, September 2005).

Stigmatization also had an impact on the economy of Chinatown as restaurants in the area reported a drop of 60–80% during the SARS crisis and for several weeks thereafter (ACAL 2003). It was noted, however, that although a total of $150 million (CDN) had been set aside by the three levels of government for marketing and advertising campaigns to attract tourists back to the stigmatized city, "not one cent has gone to Chinatown to help these workers or businesses" (ibid.).

To counter the SARS-related racial paranoia that developed, ACAL (2003) urged the provincial health ministry to set aside funding for public education campaigns that would provide accurate and consistent information. The emergence of racial paranoia frequently occurs in what are referred to as situations of "moral panic" (Cohen 1972). In moral panic situations, a scapegoat or "folk devil" emerges and serves as an identifiable object onto which social fears and anxieties are projected. Consequently, an individual or a group of individuals is perceived in a pejorative way by the general public because the "folk devil" is held responsible for introducing the threat (Goode and Ben-Yehuda 1994, 22, cited by Ungar 2001, 281). As Hier observes, however, the object of the panic is not the folk devil *per se*. Rather, the folk devil merely symbolizes an "ideological embodiment of deeper anxieties, perceived as 'a problem' only in and through social definition and construction" (2003, 6). The stigmatized as a folk devil in the case of SARS was, therefore, a social construction – one which directs us to the cultural dimension of racism – and is intimately connected to other racialized discursive articulations, such as those associated with an anti-immigration ideology. For example, Martin Collacutt (2002) alleges that family reunification policies give larger ethnic groups (such as Chinese Canadians) an advantage over smaller groups, thus threatening Canada's commitment to diversity – a fallacious argument that ostensibly hides a more general anti-immigration sentiment.

Hier and Walby (2006) note that two analytical paradigms may be discerned in the study of racism in Canada. The first tends to focus on the social politics of redistribution of resources, including the economic. When members of a minority are blocked from receiving, or gaining access to resources, discriminatory practice is likely to play some role. The second paradigm focuses on the cultural politics of recognition and tends to analyze racism in terms of the lived subjectivities and phenomenal reality of everyday life. Although the former does come into play in the SARS outbreak incident in terms of the lack of government funds directed towards Chinese Canadian businesses most affected, most of the impact of the conflation of ethnicity and health stigmatization was felt in the many instances of racism experienced by members of the Asian Canadian community (as briefly reviewed above). In particular, these instances reveal how those racist sentiments that are found acceptable by the dominant group and verbalized during private conversations, but not in the public sphere of everyday interaction (Zamudio and Rios 2006), found explicit expression during the outbreak. It is not unreasonable to expect that such expression was rationalized or legitimated on the basis of a perspective that rationalizes "othering" as a means of self-protection. That is, stemming from the conflated status of ethnicity and disease, a form of "segregationist racism" results, in which marked attempts are made by the dominant group to distance itself physically and socially from visible minorities (ibid.). Segregationist racism acts at the level of what Homi Bhabha (1994, xvi) refers to as "symbolic citizenship" in which the dominant group must have some criteria to discriminate between the "good" migrant and the "bad" migrant, or to decide which cultures are "safe" and which are not. Bhabha notes that this prejudiced discernment is made through a surveillant culture of "security." In this connection, the stigmatization of the Chinese Canadian community and of health care workers during the SARS outbreak can be understood in terms of a social construction that is implicitly and tacitly informed by larger ideological currents of what has been referred to as the "new public health" hegemony (Sanford and Ali 2005; Keil and Ali 2006b), as well as a reaction to perceived threats of globalization in a post-9/11 era.

Stigmatized Ethnicity in the Context of Public Health

Goffman (1963, 139) notes that the stigmatization of individuals can serve as a means of social control of racial, ethnic, and religious groups. Seen in this light, stigmatization, as expressed through racialization, represents one mechanism through which social control is exerted. The connection of racialization to social control raises an important issue related to power and politics that is worth considering from a sociological point of view. Social theorists such as Foucault (1991) and Bourdieu (1984) note that social control may be embedded in established knowledge systems that legitimize structures of social inequality. As such, the members of a dominant group may implicitly make reference to institutionalized and widely held knowledge systems to lend an "official" or authoritative voice to their position and to justify their imposition of a stigmatized status on others. Stigma, therefore, can only be understood in terms of the convergence of culture, power, and difference (Parker and Aggleton 2003). This understanding has particular relevance when considering the interplay among public health, as a "scientifically"-based endeavour having authoritative appeal, infectious disease, and stigmatized ethnicity. Briggs (2005, 276) notes that racialization is hardly an isolated process because, in reality, racial categories "intersect" with other forms of subordination. This intersecting quality of racialization is illustrated, for example, in Packard's (1989) argument that a tuberculosis epidemic – that was rooted in exploitative labour conditions in the South African mining industry – provided an initial rationalization for apartheid. A second example may be found in the history of Vancouver's Chinatown. Anderson (1992) documents how Chinatown was set apart and vilified in the nineteenth and early twentieth centuries as a site of vice, corruption, disease, and danger (see Craddock 2000, and Shah 2001, for similar analyses in the context of San Francisco). As the abandonment of Chinatown during the SARS incident in Toronto attests, this stigmatization of people and place may now occur on the basis of a different rationale – that is, one informed and justified by the "new public health" discourse on risk and surveillance.

Emanating from the sanitarian movement at the turn of the century, the strategies of the "old" public health discourse tended to focus on issues relating to public hygiene – for example, sanitation, overcrowded living conditions, hazardous work environments, maintaining clean streets, and so on (Garrett 2000). In contrast, today these "health promotion" strategies target lifestyle and risk factors in disease onset. This latter emphasis forms the basis of the "new" public health. Formalized in the influential Lalonde Report (*New Perspectives on the Health of Canadians*) in 1974, the "new" public health approach directed attention away from an exclusive and traditional biomedical orientation concerning health care towards one based on the "health fields" of: lifestyle, health care organization, human biology, and the environment. With the institutionalization of the "new" public health approach, however, attention has gradually shifted to an emphasis on changing behaviours on the individual level in order to prevent illness (Hancock 1986). Although recent efforts have been made to address the tacit "blaming the victim" assumption embedded within the new public health philosophy (see, for example, Labonte and Penfold 1981; Baum 1999, 1990), the individualist tendency still persists. In this connection, Armstrong (1995; see also Petersen 1996) notes that because the emphasis is on monitoring individual behaviours and lifestyle, the new public health strategies are simply modern extensions of the types of surveillance tactics noted by Foucault (i.e., the monitoring and social control of populations through individual level data and statistics). Furthermore, Petersen and Lupton (1996) contend that attempts at broadening the focus of the new public health to consider the psychological, social, and biophysical elements of health has led to an increased level of scrutiny wherein very few areas of social life remain immune to regulation of some kind. For example, "life-style surveys" are frequently used to gather data about many aspects of personal life for the calculation of "risk"; this includes data concerning many aspects of life, such as those related to leisure activities; level of exercise; the extent of social and sexual contact; the intake of fats, fibre, sugar, alcohol, and tobacco; body weight; blood pressure; and cholesterol (Bunton 1992; Petersen 1996). New public health strategies now tend to intervene in such areas as community development, personal skills development, the control of advertising "unhealthy" dangerous consumer products, and the regulation of urban space (for example, the "Health City projects"). What should be noted about these efforts are that they offer new opportunities for the monitoring and periodic screening of sub-populations (Petersen 1996, 49). Surveillance forms of this type are essentially risk-profiling techniques which enable the identification and formal recognition of many more categories of "at risk" populations and "risky" situations. That is, the scope of what is to be defined as "pathological" (as opposed to "normal") has in effect been expanded by such techniques.

The separation of those infected or exposed to SARS from the healthy is certainly warranted to avert the risk of transmission of this infectious disease, but, as Gostin et al. observe, such a practice also has the potential for using public health as a "subterfuge for discrimination" (2003, 3233). Thus, for example, one U.S. court invalidated an early twentieth-century quarantine in San Francisco that operated exclusively against the Chinese community, noting that the public health officials had targeted this community with an "evil eye and an unequal hand" (ibid.). Expanding on this notion, Lupton (1995, 117) notes that, historically, the "risky" persona has been cast in public health discourse in terms of attributes such as gender, social class, ethnicity, and sexual identity. Thus, certain social groupings – such as the poor, the working class, women, non-Europeans – as well as certain geographical locations – for example, slums, working-class neighbourhoods, the African continent – have been designated by Western societies as the contaminating "other" towards which public health measures are directed (ibid. 174). According to Petersen and Lupton (1996), the perspective of the "contaminating other" has become consolidated in contemporary times with the advent of the "new public health" and, as will now be discussed, this has implications for the manner in which the experiences of the SARS outbreak were felt by those in the Asian Canadian community.

Stigmatized Ethnicity within the Context of Globalization

The stigmatizing effects of the new public health techniques have a particular currency within the context of globalization – a period in which the migration of different people from around the world is said to have intensified (Shamir 2005). As Briggs notes, "narratives about epidemics make racial inequalities seem natural – as if bacteria and viruses gravitate toward populations and respect social boundaries" (2005, 272). For this reason, it is perhaps not surprising that infectious disease control and immigration regulations have always had a close affinity. This affinity is based on the fact that both public health and immigration are concerned with the effects and regulation of space and movement and the administration, security, and permeability of borders (Bashford 2002, 352) – where borders refers to both the boundaries of the body (with reference to infectious disease spread) and the territorial unit (with reference to the movement of people). Today, the biopolitics of movement within the age of globalization takes on new significance. Dean affirms that today biopolitics is not only about managing health, sanitation, propagation, longevity, and lifestyle – as emphasized in the work of Foucault and his adherents – but now also involves an international dimension. That is, biopolitics should now include that which governs the movement, transitions, settlement, and repatriation of various populations – including refugees, migrants, guest workers, tourists and students (1999, 100, cited by Bashford 2002, 348–349).

Recently, in the post-9/11 era, new opportunities have arisen for the convergence of different types of threats, thus giving rise to new forms of stigmatization that are based on the type of overlap between public health, infectious disease, and ethnicity reviewed in the previous sections. First, the threat of infectious disease has recently garnered much greater public attention. In part, this was an outcome of the interactions among the American scientific community, public health officials, and defence experts who collectively put forth the argument that "new and emerging diseases" represented a serious threat to national security, international development, and global public health (King 2002). This argument in turn reflected "American anxieties about living in a globalizing world, in which the assumptions and institutions of the Cold War era no longer seemed adequate to the task of ensuring the safety and interest of U.S. citizens" (ibid. 764). The resultant emphasis on security that was adopted to address the infectious disease threat was also directed towards the threats that were associated with terrorism and the movement of "outsiders" in a highly globalized and mobile world. In this light, Shamir writes of the development of a pervasive "paradigm of suspicion" that conflates the perceived threats of crime, immigration, and terrorism. For Shamir, the key issue is how individuals and groups are classified according to perceived threats and risks, and especially on the role of technology and statistical means involved in operationalizing these classifications and creating elaborate forms of such distinctions. The use of technical means to track possible security threats may dovetail with similar measures used during disease outbreaks to track a virus (see, for example, van Wagner 2008, for further discussion of this in the context of SARS in Toronto in relation to the associated social and legal ramifications). This conflation of national security and public health has particular implications for the nature of stigmatization under the contemporary conditions of neoliberalization and globalization.

First, it should be noted that the techniques of contact tracing, quarantine/isolation, and surveillance may be thought of as traditional techniques of infectious disease control, dating back centuries to the days when ships where quarantined to port. What is different today is that the employment of these "old" public health techniques have different effects within the context of the new public health and neoliberal globalization. The neoliberal tendency to withdraw various state services originally intended to protect the collective welfare has led to many changes. Most notably, within the realm of public health, it has induced the emergence of a prevailing cultural norm that holds that it is solely the individuals' responsibility to protect themselves from risk, thus absolving the state of any responsibilities in regard to collective health, while at the same time downloading the costs of public health to the individual (Petersen 1996, 4). Such a norm tends to individualize the risk and leads to a more defensive orientation that has significant implications for the stigmatization of others who are conceived of not only as health threats, but as racialized health threats. Thus, for example, it was reported that a Chinese Canadian women was told to "go back to your own country and stop transporting diseases here" (Leung and Guan 2004, 19). Furthermore, as expressed by some members of the Asian community, there was disappointment that the government's SARS recovery plans did not include any anti-racism programs but only emphasized economic recovery – and then only for businesses outside Chinatown (ibid. 22), again illustrating the withdrawal of the state from protective collective welfare functions.

Ho-fung Hung (2004) notes that the organized responses to SARS could be classified along two lines. The first, referred to as the "globalist response," is predicated on the idea that fighting a global disease requires a global response in which nations and officials at all levels must cooperate and coordinate their response. With this orientation, attention becomes focused on the task at hand (i.e., preventing the global transmission of disease) instead of on the political question of where to place the blame. Consequently, the entire chain of transmission warrants public health attention. In contrast, the "anti-globalist" response focuses only on one point in the viral transmission chain, namely, the point of origin. As such, the originating culture and society are blamed for the pandemic, and actions are taken specifically against that culture/society, with the reasoning being that the pandemic can only be stopped by isolating that particular culture/society and avoiding all contact with members belonging to it. Hung (27) goes on to note:

> This suggested solution is antithetical to the globalization process, and akin to global apartheid. For whereas actions that isolate people showing symptoms of a contagious disease can be called a quarantine, actions that indiscriminately isolate a whole culture and society as well as the people from there should be called apartheid.

Therein lies the obfuscated distinction between quarantine as a health measure and avoidance as a racially motivated action.

Issues involving stigmatized ethnicity, especially as they relate to public health threats, will undoubtedly increase in significance in the future with the intensification of globalization (and, most notably, with the pending threat of avian flu – see, for example, Keil and Ali 2006a; Davis 2005). This will be particularly true for large cosmopolitan centres that are often home to people from around the world. It has been found, for example, that the diffusion of the SARS coronavirus followed a circuit of human flows that networked the worlds' "global cities" – all of which are characterized by the presence of numerous diaspora communities (Ali and Keil 2006). In this light, the study of the relationship between ethnicity and infectious disease is one lens through which the politics of race and the workings of multiculturalism in a city may be studied (Keil and Ali 2006b). The transmigration of people and the flow of human beings among societies of the world will likely intensify with increased cultural, economic, and political globalization, thus increasing the potential for disease spread (Barret et al. 1998; Davis 2005; Galea et al. 2005; IOM 1997; McMichael 2001). However, it has been noted by Sarasin (2006, 220) that under such circumstances –

> There is clearly an intensified exchange and global spread of pathogenic microorganisms along the path of migration. Yet, isn't this exchange just as pronounced, or even more so, in worldwide tourist traffic? Can we really be certain immigrants have so many more infectious diseases as to justify our fear of them, whereas business travelers, sex tourists, students, and all the other millions who cross international borders everyday are somehow less affected by them? SARS has taught us differently.

Sarasin further warns us that infectious disease may very well represent the "metaphorical core of globalization" in a post-9/11 world in which, increasingly, the "other" is perceived in terms of a viral invader conceived of, not only as a health threat, but as a threat in many other senses as well, be it defined in terms of terrorism, economics, culture, or politics. In this sense, ethnic stigmatization may not only represent the manifestation of "othering" at the level of social interaction, but as the first step towards a more troubling process of demonization that may ultimately contribute to the potential creation of a moral panic. Thus, the conflation of stigmatization ethnicity and health threats will likely take on a new and unique significance under contemporary globalized circumstances.

Conclusion

The tacit and latent symbolic boundaries separating the dominant group and visible minorities are silently reproduced in the everyday interactions of civil society. Extreme events, however, afford the opportunity for negative sentiments to become visible and to violate these boundaries, as evidenced, for example, by the onset of racially-based actions and words. Such developments are particularly evident during disease outbreaks. One of the features of the epidemic disease is its transgression of boundaries, and, during disease outbreaks, lines are drawn to make these boundaries material in the form of quarantine (Zheng 2005), as has historically been done in Chinatowns throughout North America. Today, the individualizing tendencies of the "new public health," coupled with socially negative reactions to the inclinations towards a "borderless" world brought on by globalization, has led to new justifications for the imposition of stigmatized status, one which further strengthens the association of ethnicity with disease. For these reasons, the relationship among stigmatization, ethnicity, and health must be given much more careful consideration by both analysts and policymakers, lest we risk a return to a more unjust and intolerant society.

Notes

1. It is also worth noting that one of the key recommendations from an influential commission inquiry report into the SARS outbreak in Toronto pertained to the establishment of a central public health agency designed to coordinate efforts across different sectors (Naylor 2003). In response, the newly established Public Health Agency of Canada emphasized an increase in surveillance capacity in many of its' constituent branches, including Infectious Disease and Emergency Preparedness, Health Promotion, and Chronic Disease Prevention, and Public Health Practice and Regional Operations (see www.phac-aspc.gc.ca), with a concomitant loss in collective empowerment and community capacity building.
2. Link and Phelan (2001) prefer the term "label," rather than "mark" or "attribute," because they wish to emphasize that the identification and election of the particular feature for social attention is not exclusively a characteristic of the stigmatized individual, but the product of a relational social process.
3. According to Escobar (1995), the bias of developmentalism is based on political involvement in development and the institutional promotion of development as a means of improving life in poor countries, which in turn is predicated on the notion of modernity in which the West self-characterizes itself in opposition to "others" and "elsewheres" that are imagined to not be modern (Robinson 2006, 4).
4. Interestingly, Baehr (2008) notes that this same mask connotes a different meaning in the context of Hong Kong, where the donning of the mask was conceived of in positive terms as a cultural norm was established that if one were ill, but not wearing a mask in public, than one was not fulfilling his or her responsibility to the community of protecting others. That is, the stigmatized individual would be the ill person who was not wearing a mask to protect others in public.

References

ACAL (Asian Canadian Labour Alliance). 2003. Healing the Scars in post-SARS Toronto. http://www.buzzardpress.com/acla/sars/healing_scars_post-sars.html.

Ali, S. Harris, and Roger Keil. 2006. Global Cities and the Spread of Infectious Disease: The Case of Severe Acute Respiratory Syndrome (SARS) in Toronto, Canada. *Urban Studies* 43.3: 1–19.

Anderson, K. 1992. *Vancouver's Chinatown: Racial Discourse in Canada, 1874–1980.* Montreal-Kingston: McGill-Queen's University Press.

Armstrong, D. 1995. The Rise of Surveillance Medicine. *Sociology of Health and Illness* 17: 393–404.

Baehr, Peter. 2008. City under Siege: Authoritarian Toleration, Mass Culture, and the SARS Crisis in Hong Kong. In *Networked Disease: Emerging Infections in the Global City*, ed. S. Harris Ali and Roger Keil, 138–151. Oxford: Wiley-Blackwell.

Barrett R., C. Kuzawa, T. McDale, and G. Armelagos. 1998. Emerging and Re-emerging Infectious Diseases: The Third Epidemiologic Transition. *Annual Review of Anthropology* 27: 247–71.

Bashford, Alison. 2002. At the Border: Contagion, Immigration, Nation. *Australian Historical Studies* 120: 344–358.

Baum, Frances. 1999. *The New Public Health: An Australian Perspective.* Oxford: Oxford University Press.

———. 1990. The New Public Health: Force for Change or Reaction? *Health Promotion International* 5.2: 145–150.

Bhabha, Homi K. 1994. *The Location of Culture.* New York: Routledge.

Bourdieu, Pierre. 1984. *Distinction: A Social Critique of the Judgement of Taste.* Trans. R. Nice. London: Routledge.

Briggs, Charles L. 2005. Communicability, Racial Discourse, and Disease. *Annual Review of Anthropology* 34: 269–91.

Bunton, R. 1992. More than a Wooly Jumper: Health Promotion as Social Regulation. *Critical Public Health* 3.2: 4–11.

Castel, R. 1991. From Dangerousness to Risk. In *The Foucault Effect: Studies in Governmentality*, ed. G. Burchell, C. Gordon, and P. Miller. Hemel Hempstead: Harvester Wheatsheaf.

CBC (Canadian Broadcasting Corporation). 2003. Toronto hotels lose millions over SARS. http://www.cbc.ca/canada/story/2003/04/03/sars_hotels030403.html.

Cohen, S. 1972. *Folk Devils and Moral Panics: The Creation of the Mods and Rockers*. London: McGibbon and Kee.

Collacutt, Martin. 2002. *Canadian Immigration Policy: The Need for Major Reform*. Public Policy Sources 64. Vancouver: Fraser Institute.

Craddock, Susan. 2000. *City of Plagues: Disease, Poverty, and Deviance in San Francisco*. Minneapolis: University of Minnesota Press.

Davis, Mike. 2005. *The Monster at Our Door: The Global Threat of Avian Flu*. New York: New Press.

Dean, Mitchell. 1999. *Governmentality: Power and Rule in Modern Society*. London: Sage.

Durkheim, Emile. 1951. *Suicide*. New York: Free Press.

Edelstein, Michael. 2004. *Contaminated Communities: Coping with Residential Toxic Exposure*. 2nd ed. Boulder: Westview.

Eichelberger, Laura. 2007. SARS and New York's Chinatown: The Politics of Risk and Blame during an Epidemic of Fear. *Social Science and Medicine* 65: 1284–1295.

Elden, Stuart. 2003. Plague, Panopticon, Police. *Surveillance and Society* 1.3: 240–253. http://www.surveillance-and-society.org.

Escobar, A. 1995. *Encountering Development: The Making and Unmaking of the Third World*. Princeton: Princeton University Press.

Fischer, B., and B. Poland. 1998. Exclusion, "Risk," and Social Control – Reflections on Community Policing and Public Health. *Geoforum* 29.2: 187–197.

Foucault, Michel. 1991. *Discipline and Punish: The Birth of the Prison*. London: Penguin.

Galea, Sandro, Nicholas Freudenberg, and David Vlahov. 2005. Cities and Population Health. *Social Science and Medicine* 60: 1017–1033.

Garfinkel, Harold. 1967. *Studies in Ethnomethodology*. Englewood Cliffs, NJ: Prentice-Hall.

Garrett, Laurie. 2000. *Betrayal of Trust: The Collapse of Global Public Health*. New York: Oxford University Press.

Goffman, Erving. 1963. *Stigma: Notes on the Management of Spoiled Identity*. Englewood Cliffs, NJ: Prentice-Hall.

Goode, E., and N. Ben-Yehuda. 1994. Moral Panics: Culture, Politics, and Social Construction. *Annual Review of Sociology* 20: 149–71.

Gostin, L. O., R. Bayer, and A. L. Fairchild. 2003. Ethical and Legal Challenges Posed by Severe Acute Respiratory Syndrome. *JAMA* 290.24: 3229–37.

Greenberg, Joshua, and Sean Hier. 2001. Crisis, Mobilization, and Collective Problematization: Illegal Migrants and the Canadian News Media. *Journalism Studies* 2.4: 563–83.

Hancock, Trevor. 1986. Lalonde and Beyond: Looking Back at "A New Perspective on the Health of Canadians." *Health Promotion* 1.1: 93–100.

Hannigan, John. 1995. *Environmental Sociology: A Social Constructionist Perspective*. New York: Routledge.

Hier, Sean. 2003. Risk and Panic in Late Modernity: Implications of the Converging Sites of Social Anxiety. *British Journal of Sociology* 54.1: 3–20.

Hier, Sean, and Joshua Greenberg. 2002. Constructing a Discursive Crisis: Risk Problematization and *Illegal* Chinese in Canada. *Ethnic and Racial Studies* 25: 490–513.

Hier, Sean, and Kevin Walby. 2006. Competing Analytical Paradigms in the Sociological Study of Racism in Canada. *Canadian Ethnic Studies* 38.1: 83–104.

Hung, Ho-fung. 2004. The Politics of SARS: Containing the Perils of Globalization by More Globalization. *Asian Perspectives* 28.1: 19–44.

Ibarra, P. R., and J. I. Kitsuse. 1993. Vernacular Constituents of Moral Discourse: An Interactionist Proposal for the Study of Social Problems. In *Reconsidering Social Constructionism: Debates in Social Problems Theory*, ed. J. A. Holstein and G. Miller, 25–58. New York: Aldine de Gruyter.

IOM (Institute of Medicine Board on International Health). 1997. *America's Vital Interest in Global Health: Protecting Our People, Enhancing Our Economy, and Advancing Our International Interests*. Washington: National Academy Press.

Keil, Roger, and S. Harris Ali. 2006a. The Avian Flu: Some Lessons Learned from the 2003 SARS Outbreak in Toronto. *Area* 38.1: 107–109.

———. 2006b. Multiculturalism, Racism, and Infectious Disease in the Global City: The Experience of the 2003 SARS Outbreak in Toronto. *Topia* 16: 23–49.

King, Nicholas B. 2002. Security, Disease, Commerce: Ideologies of Postcolonial Global Health. *Social Studies of Science* 32.5–6: 763–789.

Labonte, Ronald, and Susan Penfold. 1981. Canadian Perspectives in Health Promotion: A Critique. *Health Education* (April): 4–9.

Lalonde, Marc. 1981. *A New Perspective on the Health of Canadians*. Ottawa: Ministry of Supply and Services Canada.

Leung, C., and J. Guan. 2004. *Yellow Peril Revisited: Impact of SARS on the Chinese and Southeast Asian Canadian Communities*. Toronto: Chinese Canadian National Council.

Li, P. S. 1998. *The Chinese in Canada*. Toronto: Oxford University Press.

———. 1994. Unneighbourly Houses or Unwelcome Chinese: The Social Construction of Race in the Battle over "Monster Homes" in Vancouver, Canada. *International Journal of Comparative Race and Ethnic Studies* 1: 14–33.

Link, Bruce G., and Jo C. Phelan. 2001. Conceptualizing Stigma. *Annual Review of Sociology* 27: 363–385.

Lupton, Deborah. 1995. *The Imperative of Health: Public Health and the Regulated Body*. London: Sage.

———. 1999. *Risk and Sociocultural Theory: New Directions and Perspectives*. Cambridge: Cambridge University Press.

Mak, Winnie W. S., Phoenix K. H. Mo, Rebecca Y. M. Cheung, Jean Woo, Fanny M. Cheung, and Dominic Lee. 2006. Comparative Stigma of HIV/AIDS, SARS, and Tuberculosis in Hong Kong. *Social Science and Medicine* 63: 1912–1922.

McMichael, A. J. 2001. Human Culture, Ecological Change, and Infectious Disease: Are We Experiencing History's Fourth Great Transition? *Ecosystem Health* 7.2: 107–115.

Naylor, David. 2003. *Learning from SARS – Renewal of Public Health in Canada. A Report of the National Advisory Committee on SARS and Public Health*. Ottawa: Public Health Agency of Canada.

Packard, R. M. 1989. *White Plague, Black Labor: Tuberculosis and the Political Economy of Health and Disease in South Africa*. Berkeley: University of California Press.

Palmer, K. 2003. Doctors slam flawed SARS battle. *Toronto Star*, Sept. 17.

Parker, P., and P. Aggleton. 2003. HIV and AIDS-related Stigma and Discrimination: A Conceptual Framework and Implications for Action. *Social Science and Medicine* 57: 13–24.

Petersen, A. R. 1996. The "Healthy" City, Expertise, and the Regulation of Space. *Health and Place: An International Journal* 2.3: 157–65.

Petersen, Alan, and Deborah Lupton. 1996. *The New Public Health: Health and Self in the Age of Risk*. Thousand Oaks, CA: Sage.

Rider, David. 2003. Fear of virus fuels racism: Ontario must do more to stop return to days of "Yellow Peril," Asian leaders say. CanWest News Service, April 4.

Robinson, J. 2006. *Ordinary Cities: Between Modernity and Development*. New York: Routledge.

Rosenthal, E. 2003. From China's provinces, a crafty germ breaks out. *New York Times*, April 27, 1, 18.

Sanford, Sarah, and S. Harris Ali. 2005. The New Public Health Hegemony: Response to Severe Acute Respiratory Syndrome (SARS) in Toronto. *Social Theory and Health* 3: 105–125.

Sarasin, Philip. 2006. *Anthrax: Bioterror as Fact and Fantasy*. Trans. Giselle Weiss. Cambridge: Harvard University Press.

———. 2008. Vapors, Viruses, Resistance(s): The Trace of Infection in the Work of Michel Foucault. In *Networked Disease: Emerging Infections in the Global City*, ed. S. Harris Ali and Roger Keil, 267–280. Oxford: Wiley-Blackwell.

Shah, Nayan. 2001. *Contagious Divides: Epidemics and Race in San Francisco's Chinatown*. Berkeley: University of California Press.

Shamir, Roven. 2005. Without Borders? Notes on Globalization as a Mobility Regime. *Sociological Theory* 23.2: 197–217.

Sorensen, Chris. 2003. Chinese Canadians feeling backlash. *Toronto Star*, April 4.

Stallings, Robert. 2002. Weberian Political Sociology and Sociological Disaster Studies. *Sociological Forum* 17.2: 281–305.

Statistics Canada. 2005. Population by Mother Tongue, by Census Metropolitan Areas, Toronto. http://www40.statcan.ca/101/cst01/demo12c.htm.

Szasz, Andrew. 1994. *Ecopopulism: Toxic Waste and Movement for Environmental Justice*. Minneapolis: University of Minnesota Press.

Tator, C., F. Henry, and W. Mattis. 1997. *Challenging Racism in the Arts: Case Studies of Controversy and Conflict*. Toronto: University of Toronto Press.

Ungar, Sheldon. 2001. Moral Panic versus the Risk Society: The Implications of the Changing Sites of Social Anxiety. *British Journal of Sociology* 52.2: 271–291.

Van Wagner, Estair. 2008. The Practice of Biosecurity in Canada: Public Health, Legal Preparedness, and Toronto's SARS Crisis. *Environment and Planning* 40.7: 1647–1663.

Walker, David. 2003. *For the Public's Health: Initial Report of the Ontario Expert Panel on SARS and Infectious Disease Control*. Toronto: Ministry of Health and Long Term Care.

Walker-Renshaw, Barbara. 2003. Legal Issues in Outbreak Management. Paper presented at the Medical Emergencies and the Law Conference on SARS, Global Epidemics, and Other Disasters. Joint Program of Osgoode Hall Law School of York University and the Faculty of Medicine, University of Toronto, Toronto, Sept. 16.

Yang, Lawrence Hsin, Arthur Kleinman, Bruce G. Link, Jo C. Phelan, Sing Lee, and Byron Good. 2007. Culture and Stigma: Adding Moral Experience to Stigma Theory. *Social Science and Medicine* 64: 1524–1535.

Zamudio, Margaret M., and Francisco Rios. 2006. From Traditional to Liberal Racism: Living Racism in the Everyday. *Sociological Perspectives* 49.4: 483–501.

Zheng, Da. 2005. Encountering the Other: SARS, Public Health, and Race Relations. *Americana: The Journal of American Popular Culture* 4.1. http://www.americanpopular culture.com/journal/articles/spring_2005/zheng.htm.

8

Mental Disorders, Health Inequalities and Ethics: A Global Perspective

*Emmanuel M. Ngui, Lincoln Khasakhala,
David Ndetei and Laura Weiss Roberts*

Introduction

The burden and inequalities in mental healthcare throughout the world are critically important health issues, and taken together present immense ethical challenges. In this paper we examine mental health issues globally, with special emphasis given to the developing nations, because of the limited research and increased burden of mental disorders in these nations. We also provide an overview of the ethical considerations in international mental health and renewed interest in approaches that integrate mental health into primary care services (Table 1).

Mental disorders account for an enormous global burden of disease that is largely underestimated and underappreciated. In a given year, about 30% of the population worldwide is affected by a mental disorder and over two thirds of those affected do not receive the care they need (Chisholm et al., 2007; Kessler et al., 2005b; Wittchen, Jonsson, & Olesen, 2005). About 14% of the global disease burden is attributed to neuropsychiatric disorders, mostly depression, alcohol-substance abuse and psychoses (Murray & Lopez, 1996; Prince et al., 2007). In the USA, about 57.7 million adults experience a mental disorder annually, and 1 in 17 people have a serious mental health condition (Kessler, Chiu, Demler, & E.E., 2005a). These figures translate into hundreds of millions of people suffering from mental disorders

Source: *International Review of Psychiatry*, 22(3) (2010): 235–244.

Table 1: Summary of key issues in global mental health

Issue	Key points
Burden of mental disorders	• The global burden of mental disorders is enormous, under-appreciated, and largely unmet. • Annually, about 30% of the population worldwide is affected by a mental disorder and over two thirds of those affected do not receive the care they need. Depression, alcohol and substance abuse and psychoses are among the most prevalent conditions. • Mental health problems have major economic and social cost. • Many nations have limited capacity (e.g. infrastructure, workforce, resources) needed to assess, identify and treat mental health disorders.
Ethics, human rights and social justice	• Human rights and social justice frameworks are critical in understanding and addressing mental health inequalities. • In many nations, limited or no policies exist to address basic needs and human rights of people with mental illness and standards of ethical conduct of research and treatment of mental disorders are inadequate or lacking. • Ethical principles of beneficence, autonomy, respect for persons and non-malfeasance for people living with mental disorders foster human dignity, and promote human rights and social justice.
Mental health inequalities and unmet needs	• Mental disorders are associated and embedded within the broader social and economic context. • Poverty increases the risk of developing mental disorders, which in turn increase the risk of living in poverty due disability or loss of gainful employment. • Mental disorders are determined by multiple and interacting social, psychological and biological factors. The underlying social determinants of mental disorders (e.g. low levels of education, unemployment) also are key determinants of living in poverty. • Unmet mental health needs contribute to profound suffering and deaths largely because people cannot access needed treatment. • Shortage of mental health providers and resources result in unnecessary institutionalization of people with mental illness even though these conditions can be managed effectively in the community if services were available. • In most developing nations, the burden of caring for people with mental illness disproportionately falls on women and children.
Stigma and discrimination	• Mental health stigma and discrimination are major barriers to effective management of mental disorders. • Stigma, myths, and misconceptions of mental illness contribute to much of the discrimination and human rights violation experienced by people with mental disorders. • Stigma and discrimination increase social isolation and unmet needs for mental health services, negatively influence choice of mental health careers, and limit development of policies and human rights protections for people living with mental disorders and their families. • Limited access to modern psychiatric services increase beliefs that mental disorders are untreatable which further increases stigma and discrimination and patient's reliance on traditional healers who may not have adequate skills and training to help people with serious mental disorders.

Integration of mental health into primary health care services	• In many developing nations, mental health services are provided at the tertiary level with limited or no integration to primary care interventions.
	• The majority of individuals with mental disorders and their families live in overt poverty and cannot access, afford, appropriate and available specialized mental health services provided at tertiary levels health facilities serviced by psychiatrists.
	• Extreme and growing shortage of mental health workers further compounds the problem of access to mental health services resulting in limited access to services and reliance on traditional healers in some nations.
	• In the absence of integrated proper functioning health systems it is impossible to provide mental health services for most individuals with mental disorders and their families in developing nations.
	• Integrating mental health into primary care services is a critical, affordable, and cost-effective approach to delivering services for people living with mental disorders.
	• Such integrated systems of primary care can reduce unmet needs and social stigma and discrimination by decreasing social isolation, neglect, and institutionalization of people with mental disorder.
Impact on economic development	• The economic burden of mental disorders is great. Mental disorders significantly impair economic growth through their effects on labour supply, earnings, participation, and productivity.
	• Unmet mental health care needs are associated with increased risk of social problems (e.g. school drop out, alcohol and drug use, disability, unemployment, unsafe sexual behaviours, crime and poverty) that may influence economic growth.
	• In many developing nations, limited efforts have been made to address or modify the social determinants of health, including actions that allow people to adopt and maintain healthy life styles and those that create living conditions and environments that support health.
	• Mental health promotion is an integral part of health promotion theory and practice where persons with mental illness need affordable, available, accessible, and appropriate sustainable mental health services for them to continue education (children and youth) or remain in an economic sustaining livelihoods (employment).
	• These associations play a major role in risk behaviours, such as unsafe sexual behaviour, road trauma and physical inactivity resulting in lack of meaningful, or dismissal from, employment, and in turn becomes an associated cause for depression and alcohol and drug use among people with mental disorders and their families.
Mental health data	• Lack of reliable mental health data within and across nations is pervasive and a critical barrier in addressing unmet mental health needs.
	• Limited data hinder better understanding of mental health needs, limit policy, interventions, and resources needed to address mental disorders.
	• Data limitations put mental health needs on the back burner of policy development and resource allocation.
	• Better collection of mental health data are needed in the developing nations and among rural and racial groups in developed nations.

globally including depression (154 million), schizophrenia (25 million), and alcohol use disorders (91 million) (Schmidt, Norman, & Boshuizen, 1990; WHO/WONCA, 2008). Suicides account for about 1 million deaths annually (WHO, 2002).

The projected burden of mental health disorders is expected to reach 15% by the year 2020, where common mental disorders (depression, anxiety, and substance-related disorders including alcohol) will disable more people than complications arising from AIDS, heart disease, traffic accidents and wars combined. Almost one third (28%) of disability adjusted life-years in 2005 were attributed to neuropsychiatric disorders (e.g. unipolar affective disorder (10%) (Murray & Lopez, 1996).

The burden of unmet mental health needs is especially high among children and youths (Flisher et al., 1997; Kataoka, Zhang, & Wells, 2002; Ngui & Flores, 2007). About 10% to 20% of all children are affected by one or more mental or behavioural problems (Murthy et al., 2001). In the USA, 1 in 5 children suffer from a mental disorder, with 1 in 10 affected by a serious mental or emotional disorder (US Department of Health and Human Services, 1999). Only 15% to 30% of these children, however, receive the treatment they need (WHO/WONCA, 2008). Kataoka found that 79% of children 6 to 7 years of age with mental disorders do not receive the care they need (Kataoka et al., 2002). Data from the developing nations is less reliable, but estimates from the Western Cape region of South Africa suggest that 17% of children and adolescents suffer from mental disorders (Kleintjes et al., 2006), whereas in conflict areas such as Mosul, Iraq it is as high as 35% of children and youths (Al-Jawadi & Abdul-Rhman, 2007).

In light of this evidence, the ethical concerns associated with international mental health disparities are profound. Human rights and social justice frameworks are arguably central ethical tenets of public health (Beauchamp, 1999). According to the International Covenant of Economic, Social and Cultural Rights, 'everyone has a right to the highest attainable standard of physical and mental health' (Earle, 2006, p. 327). As such, addressing global mental health inequalities and the underlying determinants of mental disorders promotes human rights and social justice in any society. These frameworks call for the ethical care of people living with mental illness and global advocacy of beneficence, autonomy, respect for individuals, non-malfeasance and empowerment of all people, and particularly those who are marginalized, stigmatized and discriminated against (Roberts & Dyer, 2004; Sheppard, 2002). Paul Farmer observes that the needs of the world's poor are often not recognized and the underlying structural inequalities that contribute to these conditions are frequently neglected or ignored by the international public health and foreign policy communities (Farmer, 2003). He calls for the inclusion and integration of human rights agenda, including resource equity and social justice in health diplomacy and international health assistance (Mann, 1996).

Initiatives and strategies to address health must systematically incorporate mental health as a key part of overall health. Application of the human rights and social justice frameworks in mental health require concerted effort and commitment to address the underlying determinants of mental health problems including fair, equitable, and ethical distribution of resource distribution (e.g. treatment), inclusive mental health and primary care policies, and strengthened legal and human rights protection for people living with mental disorders and their families.

In many developing nations, standards for ethical conduct of research and treatment are inadequate or lacking. Addressing inequalities and unmet mental health needs, especially in the developing nations, will require establishment and strengthened ethical standards in research and treatment of people with mental health problems. Goodman (2008) observes that 'ethics is essential for building trust in developing world . . . ethics and trust are required for a successful research programme . . . health of communities depends on more and better research; and that such research is necessary for reducing disparities' (Goodman, 2008, p. 89). Indeed, repeated demonstrations of integrity by economically established countries towards all people affected by mental illness and its burden throughout the world is a precondition for ethically sound and humane healthcare. Moreover, intensive and more appropriately attuned ethics education is critically important in understanding and addressing mental health inequalities and in the preparation of clinicians caring for people living with mental illness (Chipp et al., in press; Hoop, DiPasquale, Hernandez, & Roberts, 2008; Jain, Dunn, & Roberts, in press; Lehrmann, Hoop, Hammond, & Roberts, 2009; Roberts, Johnson, Brems, & Warner, 2008) in rural and underserved areas.

Inequalities in Mental Health

Inequalities in mental health exist, are pervasive and often ignored as illustrated by the neglect of a mental health focus in the Millennium Development Goals (United Nations, 2000). According to the World Health Organization (WHO), health inequalities can be defined as 'differences in health status or in the distribution of health determinants between different population groups' (WHO, 2007). Health inequity is those inequalities in health considered unfair and unjust (Kawachi, Subramanian, & Almeida-Filho, 2002). Inequity in mental health exists in access to care, use and outcomes of care (e.g. morbidity and mortality) and can occur by geographical region (rural/urban), gender, socio-economic status, racial or ethnic background and sexual orientation among other things.

Mental health inequalities are strongly associated and embedded within the broader social and economic context. An inverse relationship between socio-economic status and mental disorders has been documented (Dalgard, 2008; Hunt, McEwen, & McKenna, 1979; Kessler et al., 1994). In almost all nations the poor are at a higher risk of developing mental disorders compared to the non-poor. Poverty, is both a 'determinant and a consequence of poor mental health' (Murali & Oyebode, 2004, p. 217). Mental disorders increase the likelihood of living in poverty, perhaps because of their influence on functionality and ability to get or sustain employment. Conversely, poverty increases the likelihood of developing mental disorders (Bostock, 2004; Das et al., 2007; Murali & Oyebode, 2004).

The consequences of mental health inequalities include continued unnecessary suffering and premature deaths, increased stigma and marginalization, lack of investment in mental health workforce and infrastructure, and limited or lack of treatment for people suffering from these conditions. In many developing nations with mental health policies, scarce resources and infrastructure, ineffective advocacy and the lack of political will limits effective mental health legislations and interventions (WHO, 2005b). These nations often lack effective mental health champions who can galvanize communities and policy makers to address mental health needs. Families of people with mental health problems are often marginalized and are limited in their ability to champion for mental health issues due to the stigma associated with these disorders. Some progress, however, is being made redress the challenges posed by mental health problems (Eaton, 2009), but these efforts are few and need to be scaled up to adequately meet mental health needs.

Unmet Mental Health Needs

Mental health inequalities have contributed to profound suffering and death worldwide largely because people cannot access the treatment they need. Estimates for untreated serious mental disorders in developing countries range from 75% to 85% (WHO, 2004). Over 80% of people suffering from mental disorders (e.g. epilepsy, schizophrenia, depression, intellectual disability, alcohol use disorders and those committing suicide), live in developing countries (Bertolote, Fleischmann, De Leo & Wasserman, 2004). Untreated cases range from 32.2% for schizophrenia (including other non-affective psychosis) to 56.3% for depression, to 78.1% for alcohol and drug use disorders (Kohn, Saxena, & Levav, 2004). In Kenya, for example, the number of unidentified cases of mental illness attending a National Hospital was 40% (Makanyengo, Othieno, & Okech, 2005); with unidentified cases of depression between 53% and 66.2% at the sub-district and district hospitals, respectively. Almost a quarter of patients attending general health facilities in Kenya have undiagnosed alcohol abuse problems (Ndetei et al., 2009). Rural areas in developing nations, as in economically established countries (Roberts, 2007), are especially affected by mental health disparities.

Many developing nations have no policies to address the basic needs and rights of individuals with mental illness, which contributes to limited prioritization of mental health in health planning, resource allocation, and workforce development, further increasing unmet mental health needs. Research shows that in developing nations patients often leave hospitals without knowing their diagnosis or what medications they are taking (Gerteis, Edgman-Levitan, Daley, & Delbanco, 1993; Ndetei, Mutiso, Khasakhala, & Kokonya, 2007b), wait too long for referrals, appointments, and treatment (Murray & Tantau, 1998) and are not respected or given adequate emotional support (Botelho, Lue, & Fiscella, 1996; Sobel, 1995). In many communities, the burden of caring for the sick is placed on women and increasingly children because of the high adult morbidity and mortality due to HIV/AIDS and other infectious diseases. This has resulted in age and gender inequities in primary caregiver's responsibilities for people living with mental illness. Moreover, increased international migration of health workers from developing to the developed nations (Connell, Zurn, Stilwell, Awases, & Braichet, 2007; Kirigia, Gbary, Muthuri, Nyoni, & Seddoh, 2006; Stilwell et al., 2003; Stilwell et al., 2004) and internal migration from rural poorer communities to more wealthier urban communities in the developing nations has further worsened the shortage of mental healthcare workers (Mwaniki & Dulo, 2008; Ndetei, Khasakhala, & Omolo, 2008; Ndetei et al., 2007b). As a result, the majority of people with mental illness in developing nations go untreated despite the availability of effective treatment. These large treatment gaps are not surprising given that in many developing countries there is no budget for mental health services. Not only are mental health services scarce, but individuals who have mental disorders attending public medical services are required to meet the cost of their treatment (psycho-active drugs), while treatment for physical health problems is freely provided (Ndetei, Khasakhala, Kingori, Oginga, & Raja, 2007a; WHO, 2005b). This disproportionately affects poorer people who are at greater risk of having mental disorders.

Stigma and Discrimination

The burden of mental disorders in developing countries is compounded by high rates of stigma and discrimination, which are major obstacles in the provision and utilization of mental health services (Horwitz, Roberts, & Warner, 2008; Okasha, 2002; Onyut et al., 2009; Ssebunnya, Kigozi, Lund, Kizza, & Okello, 2009). Research documents increasing social distance and stigmatization of people living with mental disorders in sub-Saharan Africa (Adewuya & Makanjuola, 2005, 2008) even among mental health providers (Ndetei et al., 2009). The stigma, myths and misconceptions surrounding mental illness contribute to much of the discrimination and human rights violations experienced by people with mental disorders (Ndetei et al., 2007a). The laws, practices and social norms in many nations give extensive powers to guardians of people with mental disorders to decide where they live, their movements, their personal and financial affairs, and their care including their commitment to mental hospitals (Ndetei et al., 2007a). Research, however, shows that clinicians and others, including family members, inaccurately judge what patients value (Gerhart, Koziao-McLain, Lowenstein, & Whiteneck, 1994; Laine et al., 1996; Roberts et al., 2003; Roberts, Warner, Anderson, Smithpeter, & Rogers, 2004a; Roberts, Green Hammond, Warner, & Lewis, 2004b), resulting in unnecessary restrictions in the rights to work, education, marriage and participation in community or family functions.

Stigma associated with mental disorders can also influence career choices resulting in fewer people choosing to work in the mental healthcare field. Studies involving medical students in Colombia ($n = 375$) (Pailhez, Bulbena, López, & Balon, 2010), Saudi Arabia ($n = 54$) (El-Gilany, Amr, & Iqbal, 2010), and Spain ($n = 207$) (Pailhez et al., 2010), and medical residents in Romania ($n = 112$) (Voinescu, Szentagotai, & Coogan, 2010), published in a special collection recently demonstrated the negative attitudes that exist towards the medical specialism of psychiatry. For example, 82% of the Saudi Arabian students and 52% of the Romanian students in these survey projects endorsed the statement that 'if a student expresses interest in psychiatry, he or she risks being . . . seen by others as odd, peculiar, or neurotic'. Large proportions of students had been actively discouraged by their medical school teachers, family members, friends, and fellow students from going into psychiatry (El-Gilany, Amr, & Iqbal, 2010; Voinescu, Szentagotai, & Coogan, 2010).

Limited knowledge of the causes, symptoms and treatment of mental illness often leads to common but erroneous beliefs that these conditions are caused by individuals themselves or by supernatural forces, possession by evil spirits, curse or punishment following the individual's family or is part of family lineage (Mohit, 2001). Disturbingly, physicians in training in some developing or economically disadvantaged countries hold these same beliefs, even after undergoing psychiatric training (Roberts, 2010). For example, 23–40% of Nigerian medical students in one study endorsed supernatural causes of mental illness, such as charms, evil spirits, and witchcraft (Aghukwa, 2010). These beliefs increase stigma, discrimination, and social isolation of individuals living with mental illness and limits resources for their care. Without effective diagnosis and treatment options, mental disorders are seen as untreatable, resulting in patients being undervalued and perceived as not able to contribute to society. In developing nations and in some communities in developed nations, the limited availability of modern mental health services and providers is offset by reliance on traditional and faith healers (Beals et al., 2005; Hewson, 1998; Ngoma, Prince, & Mann, 2003; Ovuga, Boardman, & Oluka, 1999; Sorsdahl et al., 2009). Although these alternative healers play a critical role, they often lack the necessary training and skills to provide effective care for people with serious mental illness.

Mental Disorders and Economic Development

The economic burden of mental disorder is great. In the USA the indirect costs associated with these disorders is estimated to be over $79 billion, with about $63 billion reflecting the loss of productivity because of illnesses (Manderscheid, Druss, & Freeman, 2007). In Canada, the economic burden of mental illness in 2003 was about $34 billion ($1,056 per capita), with depression and schizophrenia accounting for about $5 billion and $2.7 billion annually, respectively (Patra et al., 2007). Mental health conditions cost between 3% and 4% of the gross national product in the European Union member countries (Gabriel & Liimatainen, 2000; WHO, 2005c).

The effect of mental disorders extends beyond individual and family suffering to national economic development. Mental health well-being is strongly related to many economic development sectors (e.g. education, employment, law enforcement and incarceration) (Gureje & Jenkins, 2007) and several Millennium Development Goals (United Nations, 2000), (e.g. eradicating extreme poverty and hunger, reduce child mortality and improve maternal health) (Gureje & Jenkins, 2007; Miranda & Patel, 2005; WHO, 2002). These conditions can influence economic growth through their effects on labour supply, earnings, participation and productivity (Dewa & Lin, 2000; WHO, 2002). Depression, for instance, can negatively affect education, employment, and productivity (Berndt et al., 1998; Kessler & Frank, 1997), but productivity gains after effective treatment exceed direct treatment costs (Simon et al., 2001).

Economic loss associated with decreased labour force participation and institutionalization of people with mental disorders is great. In the USA, about 3% of men and 4.5% of women cannot work or engage in regular activities because of mental or emotional problems. Men with mental disorders earn 21% less than men without mental disorders (Robins & Regier, 1991). In Kenya, the economic loss associated with institutionalization of mental and behavioural disorders is about $13 million (Kirigia & Sambo, 2003), a large amount in a country where over half of the population live on less than a dollar per day and have no safe drinking water (UNDP, 2000). Unmet mental health needs can create social problems (e.g. unemployment, substance abuse, poverty) that may increase crime and political instability. Sen observes 'there is plenty of evidence that unemployment has many far-reaching effects other than loss of income, including psychological harm, loss of work motivation, skill and self-confidence, increase in ailments and morbidity (and even mortality rates), disruption of family relations and social life, hardening of social exclusion and accentuation of racial tensions and gender asymmetries' (Sen, 1999, p. 94). In many developing nations these social problems are further compounded by poor governance, corruption and social morbidity due to natural and manmade disasters (e.g. wars) which increase mental health problems, erode social cohesion and capital, and limit economic growth (Dewa & Lin, 2000; Njenga, 2002; WHO, 2002).

Integrating Mental Health into Primary Health Care Services

The mismatch between the global burden of mental disorders and availability of mental health resources is alarming. According to WHO, there is less than one psychiatrist for every 100,000 people in much of south-east Asia, and less than one psychiatrist for every 1 million people in sub-Saharan Africa (Jacob et al., 2007; WHO, 2005b). Nigeria, for example, has 100 psychiatrists for its population of 114 million (Gureje & Lasebikan, 2006). Globally, only 2% of national budgets are devoted to mental health (WHO, 2005b). About 70% of African and 50% of south-east Asian countries devoted less than 1% of their health budget on mental health (Jacob et al., 2007).

Given the scarcity of mental health providers in developing nations, the few psychiatric hospitals that exist are often understaffed, crowded, and may not provide the quality of care needed. Most psychiatric hospitals are located in urban settings and away from family members, which further increases the social isolation and cost for families. In some countries, these hospitals are simply 'warehouses' where patients are kept from the rest of the society because of limited resources and capacity to manage effectively their conditions. In developed nations (e.g. USA), deinstitutionalization of people with mental illness results in many patients, mostly racial/ethnic minorities, being incarcerated because of limited access and availability of basic mental health services in the community.

One key strategy for addressing inequalities in mental health care is to ensure the integration of mental health with other primary care services. Ongoing efforts to implement and enhance primary care in developing countries (Rohde et al., 2008; Tejada de Rivero, 2003; Walley et al., 2008) must include mental healthcare, as a critical component of overall population health and wellbeing. Chan and Van Weel observe that:

> For too long, mental disorders have been largely overlooked as part of strengthening primary care. This is despite the fact that mental disorders are found in all countries, in women and men, at all stages of life, among the rich and poor, and in both rural and urban settings. It is also despite the fact that integrating mental health into primary care facilitates person-centred and holistic services, and as such, is central to the values and principles of the Alma Ata Declaration. (WHO/ WONCA, 2008, p. vii)

Reasons for integrating mental health into primary care include the enormous social and economic burden, the interwoven nature of physical and mental problems, and the significant treatment gaps of mental health problems (WHO/WONCA, 2008). Moreover, since primary care for mental health problems is affordable and cost-effective, such integration would generate good outcomes; promote access to care, and respect for human rights (WHO/WONCA, 2008).

Community mental health services can help reduce social stigma and discrimination by reducing the social isolation, neglect, and institutionalization of people living with mental health problems. Effective community management of mental disorders also helps people realize that people with mental illness can live productive lives, contribute to society, and be integrated with society.

Data and the Global Burden of Mental Health Problems

Efforts to address mental health problems must also address the pervasive lack of reliable data within and across nations (WHO, 2005a). Health systems in many developing nations do not routinely collect mental health data (Ndetei et al., 2007b), which can limit the ability of nations to accurately determine the burden of these conditions and develop plans to address them. Data limitations put mental health needs on the backburner for most policy makers and make it difficult for governments and international agencies to devote more resources to address mental disorders. Strategies, such as adding reliable mental health measures to ongoing population surveys (e.g. The Demographic Health Survey) can significantly improve availability of data for advocacy, programme planning and policy formation in many countries.

In conclusion, health reform agendas in the developed and developing nations need to provide legal protection, services, and human rights to people living with mental disorders. These policies must protect people with mental disorders from abuse, neglect, and discrimination, and afford them the care they need. Justice requires that people with mental illness receive the same societal and legal protection given to other people with physical health conditions. Ethical and human rights challenges in caring for people living with mental illness and their families exist. These include: (1) justification to provide mental health services to communities when primary health-care services are inaccessible, unavailable, and unaffordable and therefore unsustainable in rural and hard-to-reach areas; (2) lack of public awareness on mental health and limited knowledge about the causes of mental illness which have resulted in mental health being given low priority by the policy makers and health providers, (3) the vicious circle between mental ill-health and poverty, (4) the role played by stigma towards individuals who have mental illness and their families, and (5) inadequate developed mental health policies, resulting in limitations to bring about major reforms in the implementation of mental health policies and service delivery needed by mental health systems.

Although the idea of health without mental health sounds absurd, mental health is perhaps the most neglected aspect of health in developed and developing nations. Addressing mental disorders often appears to be an afterthought in health and social policy development, added to existing 'more important health issues' rather than a part of individual and population overall health and wellbeing. In defining health, the WHO clearly articulated the importance of mental health by including it with overall physical and social well-being. By putting it in between the state of 'physical' and 'social' well-being, this definition symbolically shows how mental health ties physical health and social wellbeing together. Neglect of mental health needs in health policies often translates to neglect in research, funding, services, and infrastructure (e.g. the development of competent mental health workforce) especially in poor and underserved communities (WHO, 2001a, 2001b). Mental health is vital to our understanding of health and economic development and must be prioritized in health planning, resource allocation and fully integrated with other primary care services.

References

Adewuya, A.O., & Makanjuola, R.O. (2005). Social distance towards people with mental illness amongst Nigerian university students. *Social Psychiatry & Psychiatric Epidemiology*, *40*(11), 865–868.

Adewuya, A.O., & Makanjuola, R.O. (2008). Social distance towards people with mental illness in southwestern Nigeria. *Australian & New Zealand Journal of Psychiatry*, *42*(5), 389–395.

Aghukwa, C. (2010). Medical students' beliefs and attitudes toward mental illness: Effects of a psychiatric education. *Academic Psychiatry*, *34*, 67–70.

Al-Jawadi, A.A., & Abdul-Rhman, S. (2007). Prevalence of childhood and early adolescence mental disorders among children attending primary health care centers in Mosul, Iraq: A cross-sectional study. *BMC Public Health*, *7*, 274.

Beals, J., Novins, D.K., Whitesell, N.R., Spicer, P., Mitchell, C.M., & Manson, S.M. (2005). Prevalence of mental disorders and utilization of mental health services in two American Indian reservation populations: Mental health disparities in a national context. *American Journal of Psychiatry*, *162*(9), 1723–1732.

Beauchamp, D.E. (1999). Public health as social justice. In D.E. Beauchamp & B. Steinbock (Eds.), *New Ethics for the Public's Health* (pp. 105–114). New York: Oxford University Press.

Berndt, E.R., Finkelstein, S.N., Greenberg, P.E., Howland, R.H., Keith, A., Rush, A.J. et al. (1998). Workplace performance effects from chronic depression and its treatment. *Journal of Health Economics*, *17*(5), 511–535.

Bertolote, J.M., Fleischmann, A., De Leo, D., & Wasserman, D. (2004). Psychiatric diagnoses and suicide: Revisiting the evidence. *Crisis: Journal of Crisis Intervention & Suicide*, *25*(4), 147–155.

Bostock, J. (2004, November). The high price of poverty. *Poverty and debt are major risk factors for mental ill health in deprived communities and groups. Mental Health Today*, 27–29.

Botelho, R., Lue, B., & Fiscella, K. (1996). Family involvement in routine health care: A survey of patients' behaviors and preferences. *Journal of Family Practice*, *42*, 572–576.

Chipp, C., Dewane, S., Brems, C., Johnson, M.E., Warner, T.D. & Roberts, L.W. (in press). 'If only someone had told me . . .': Lessons from rural providers. *Journal of Rural Health* *7*(27), doi:10.1186/1475-9276-1187-1127.

Chisholm, D., Flisher, A.J., Lund, C., Patel, V., Saxena, S., Thornicroft, G. et al. (2007). Scale up services for mental disorders: A call for action. *Lancet*, *370*(9594), 1241–1252.

Connell, J., Zurn, P., Stilwell, B., Awases, M., & Braichet, J. -M. (2007). Sub-Saharan Africa: Beyond the health worker migration crisis. *Social Science and Medicine*, *64*, 1876–1891.

Dalgard, O.S. (2008). Social inequalities in mental health in Norway: possible explanatory factors. *International Journal of Equity in Health*, *7*(27), doi:10.1186.1475-9276-1187-1127.

Das, J., Do, Q.T., Friedman, J., McKenzie, D., & Scott, K. (2007). Mental health and poverty in developing countries: Revisiting the relationship. *Social Science and Medicine, 65*(3), 467–480.

Dewa, C., & Lin, E. (2000). Chronic physical illness, psychiatric disorder and disability in the workplace. *Social Science and Medicine, 51*, 41–50.

Earle, S. (2006). Promoting public health: Part 1. *Nursing Management, 13*(7), 32–35.

Eaton, J. (2009). A new movement for global mental health and its possible impact in Nigeria. *Nigerian Journal of Psychiatry, 7*(1), 14–15.

El-Gilany, A., Amr, M., & Iqbal, R. (2010). Students' attitudes toward psychiatry at Al-Hassa Medical College, Saudi Arabia. *Academic Psychiatry, 34*, 71–74.

Farmer, P. (2003). *Pathologies of power: Health, human rights, and the new war on the poor.* Berkeley, CA: University of California Press.

Flisher, A.J., Kramer, R.A., Grosser, R.C., Alegria, M., Bird, H.R., Bourdon, K.H. et al. (1997). Correlates of unmet need for mental health services by children and adolescents. *Psychologic Medicine, 27*(5), 1145–1154.

Gabriel, P. & Liimatainen, M. (2000). *Mental Health in the Workplace.* Geneva: International Labour Office. Available at http://www.ilo.org/public/english/employment/skills/disability/download/execsums.pdf (accessed 17 February 2010).

Gerhart, K., Koziao-McLain, J., Lowenstein, S., & Whiteneck, G. (1994). Quality of life following spinal cord injury: Knowledge and attitudes of emergency care providers. *Annals of Emergency Medicine, 23*(4), 807–812.

Gerteis, M., Edgman-Levitan, S., Daley, J., & Delbanco, T. (1993). *Through the Patient's Eyes: Understanding and Promoting Patient-Centered Care.* San Francisco: Jossey-Bass.

Goodman, K. (2008). Ethics, evidence and innovation. In *Global Forum Update on Research for Health.* M. Gehner, S. Jupp & S.A. Matlin (Eds.), *Volume 5: Fostering innovation for global health* (pp. 88–90). Woodbridge: Pro-Brook.

Gureje, O., & Jenkins, R. (2007). Mental health in development: Re-emphasising the link. *The Lancet, 369*(9560), 447–449.

Gureje, O., & Lasebikan, V.O. (2006). Use of mental health services in a developing country. *Results from the Nigerian survey of mental health and well-being. Social Psychiatry & Psychiatric Epidemiology, 41*(1), 44–49.

Hewson, M.G. (1998). Traditional healers in Southern Africa. *Annals of Internal Medicine, 128*(12/1), 1029–1034.

Hoop, J., DiPasquale, T., Hernandez, J., & Roberts, L.W. (2008). Ethics and Culture in mental health care. *Ethics and Behavior, 18*, 353–372.

Horwitz, R., Roberts, L.W., & Warner, T.D. (2008). Mexican immigrant women's perceptions of health care access for stigmatizing illnesses: A focus group study in Albuquerque, New Mexico. *Journal of Health Care for the Poor and Underserved, 19*(3), 857–873.

Hunt, S., McEwen, J., & McKenna, S. (1979). Social inequalities and perceived health. *Effective Health Care, 2*(4), 151–160.

Jacob, K.S., Sharan, P., Mirza, I., Garrido-Cumbrera, M., Seedat, S., Mari, J.J. et al. (2007). Mental health systems in countries: where are we now? *Lancet, 370*(9592), 1061–1077.

Jain, S., Dunn, L. & Roberts, L. (in press). Psychiatric residents' needs for education about informed consent, principles of ethics and professionalism, and caring for vulnerable populations: Results of a multisite survey. *Academic Psychiatry.*

Kataoka, S.H., Zhang, L., & Wells, K.B. (2002). Unmet need for mental health care among US children: variation by ethnicity and insurance status. *American Journal of Psychiatry, 159*(9), 1548–1555.

Kawachi, I., Subramanian, S.V., & Almeida-Filho, N. (2002). A glossary for health inequalities. *Journal of Epidemiology and Community Health, 56*(9), 647–652.

Kessler, R., Chiu, W., Demler, O., & Walters, E. (2005a). Prevalence, severity, and comorbidity of 12-month DSM-IV disorders in the National Comorbidity Survey Replication(NCS-R). *Archives of General Psychiatry, 62*(6), 617–627.

Kessler, R., Demler, O., Frank, R., Olfson, M., Pincus, H., Walters, E. et al. (2005b). Prevalence and treatment of mental disorders, 1990 to 2003. *New England Journal of Medicine*, *352*(24), 2515–2523.

Kessler, R.C., & Frank, R.G. (1997). The impact of psychiatric disorders on work loss days. *Psychological Medicine*, *27*(4), 861–873.

Kessler, R.C., McGonagle, K.A., Zhao, S., Nelson, C.B., Hughes, M., Eshleman, S. et al. (1994). Lifetime and 12-month prevalence of DSM-III-R psychiatric disorders in the United States. *Results from the National Comorbidity Survey. Archieves of General Psychiatry*, *51*(1), 8–19.

Kirigia, J., Gbary, A., Muthuri, L., Nyoni, J., & Seddoh, A. (2006). The cost of health professionals' brain drain in Kenya. *BMC Health Services Research*, *6*, 89.

Kirigia, J., & Sambo, L. (2003). Cost of mental and behavioural disorders in Kenya. *Annals of General Hospital Psychiatry*, *2*(1), 7.

Kleintjes, S., Flisher, A., Fick, M., Railoun, A., Lund, C., Molteno, C. et al. (2006). The prevalence of mental disorders among children, adolescents and adults in the Western Cape, South Africa. *South African Psychiatry Review*, *9*, 157–160.

Kohn, R., Saxena, S., & Levav, I. (2004). The treatment gap in mental health care. *Bulletin of the World Health Organization*, *82*, 858–866.

Laine, C., Davidoff, F., Lewis, C., Nelson, E., Kessler, R., & Delbanco, T. (1996). Important elements of outpatient care: A comparison of patients' and physicians' opinions. *Annals of Internal Medicine*, *125*, 640–645.

Lehrmann, J., Hoop, J.G., Hammond, K., & Roberts, L. (2009). Medical students' affirmation of ethics education. *Academic Psychiatry*, *33*, 470–477.

Makanyengo, M.A., Othieno, C.J., & Okech, M.L. (2005). Consultation liaison psychiatry at Kenyatta National Hospital, Nairobi. *East Africa Medical Journal*, *82*(2), 79–84.

Manderscheid, R., Druss, B., & Freeman, E. (2007). *Data to manage the mortality crisis: Recommendations to the substance abuse and mental health services administration.* Washington DC: SAMHSA, Accessed March 3, 2010 at: http://www.promoteac ceptance.samhsa.gov/10by10/summit_presentations.aspx.

Mann, J. (1996). Health and human rights. *British Medical Journal*, *312*(7036), 924–925.

Miranda, J.J., & Patel, V. (2005). Achieving the Millennium Development Goals: Does mental health play a role? *Public Library of Science (PLoS) Medicine*, *2*(10), e291.

Mohit, A. (2001). Mental health and psychiatry in the Middle East: Historical development. *Eastern Mediterranean Health Journal*, *7*, 336–347.

Murali, V., & Oyebode, F. (2004). Poverty, social inequality and mental health. *Advances in Psychiatric Treatment*, *10*, 216–224.

Murray, C. & Lopez, A. (Eds). (1996). *The global burden of disease: A comprehensive assessment of mortality and disability from diseases, injuries and risk factors in 1990 and projected to 2020. Global Burden of Disease and Injury Series.* Cambridge, MA: Harvard School of Public Health on behalf of the World Health Organization and the World Bank.

Murray, M., & Tantau, C. (1998). Must patients wait? *Journal of Quality Improvement*, *24*, 423–425.

Murthy, R., Bertolote, J., Epping-Jordan, J.A., Funk, M., Prentice, T., Saraceno, B. et al. (2001). *The World Health Report Mental Health: New Understanding New Hope.* Geneva: World Health Organization.

Mwaniki, D. & Dulo, C. (2008). *Migration of Health Workers in Kenya: The Impact on Health Service Delivery*, EQUINET Discussion paper 55. Harare: EQUINET, ECSA HC and IOM. Available at http://www.equinetafrica.org/bibl/docs/DIS55HRndetei.pdf (accessed 22 January 2010).

Ndetei, D., Khasakhala, L., Kingori, J., Oginga, A. & Raja, S. (2007a). Baseline study: The mental health situation in Kangemi informal settlement Nairobi, Kenya. Available at http://www.basicneeds.org.uk (accessed February 7, 2010).

Ndetei, D., Khasakhala, L., Kuria, M., Mutiso, V., Ongecha, F. & Kokonya, D. (2009). The prevalence of mental disorders in adults in different level general medical facilities in Kenya: A cross-sectional study. *Annals of General Psychiatry*, 8(1), Available at http://www.annals-general-psychiatry.com/content/2018/2011/2011 (accessed 22 January 2010).

Ndetei, D., Khasakhala, L. & Omolo, J. (2008). *Incentives for Health Worker Retention in Kenya: An Assessment of Current Practice*. EQUINET Discussion Paper 62. Harare: AMHF, EQUINET, ECSA-HC. Available at http://www.equinetafrica.org/bibl/docs/DIS62HRndetei.pdf (accessed 22 January 2010).

Ndetei, D., Mutiso, V., Khasakhala, L., & Kokonya, D. (2007b). The challenges of human resources in mental health in Kenya. *South African Psychiatric Review*, *10*, 33–36.

Ngoma, M.C., Prince, M., & Mann, A. (2003). Common mental disorders among those attending primary health clinics and traditional healers in urban Tanzania. *The British Journal of Psychiatry*, *183*(4), 349–355.

Ngui, E.M., & Flores, G. (2007). Unmet needs for specialty, dental, mental, and allied health care among children with special health care needs: Are there racial/ethnic disparities? *Journal of Health Care for the Poor & Underserved*, *18(4)*, 931–949.

Njenga, F.G. (2002). Challenges of balanced care in Africa. *World Psychiatry*, *1*(2), 96–98.

Okasha, A. (2002). Mental health in Africa: The role of the WPA. *World Psychiatry*, *1*(1), 32–35.

Onyut, L.P., Neuner, F., Ertl, V., Schauer, E., Odenwald, M., & Elbert, T. (2009). Trauma, poverty and mental health among Somali and Rwandese refugees living in an African refugee settlement – An epidemiological study. *Conflict Health*, *3*(6), doi:10.1186/1752-1505-1183-1186.

Ovuga, E., Boardman, J., & Oluka, E.G.A.O. (1999). Traditional healers and mental illness in Uganda. *Psychiatric Bulletin*, *23*, 276–279.

Pailhez, G., Bulbena, A., López, C., & Balon, R. (2010). Views of psychiatry: A comparison between medical students from Barcelona and Medellín. *Academic Psychiatry*, *34*, 61–66.

Patra, J., Popova, S., Rehm, J., Bondy, S., Flint, R., & Giesbrecht, N. (2007). *Economic Cost of Chronic Disease in Canada 1995–2003*. Ontario: Ontario Chronic Disease Prevention Alliance and the Ontario Public Health Association.

Prince, M., Patel, V., Saxena, S., Maj, M., Maselko, J., Phillips, M.R. et al. (2007). No health without mental health. *Lancet*, *370*(9590), 859–877.

Roberts, L., Warner, T., Anderson, C., Smithpeter, M., & Rogers, M. (2004a). Schizophrenia research participants' responses to protocol safeguards: Recruitment, consent, and debriefing. *Schizophrenia Research*, *67*, 283–291.

Roberts, L., Warner, T., Nguyen, K., Geppert, C., Rogers, M., & Roberts, B. (2003). Schizophrenia patients' and psychiatrists' perspectives on ethical aspects of symptom re-emergence during psychopharmacological research participation. *Psychopharmacology*, *171*, 58–67.

Roberts, L.W. (2010). Stigma, hope, and challenge in psychiatry: Trainee perspectives from five countries on four continents. *Academic Psychiatry*, *34*, 1–4.

Roberts, L.W., & Dyer, A.R. (2004). *Concise Guide to Ethics in Mental Health Care*. Washington DC: American Psychiatric Publishing.

Roberts, L.W., Green Hammond, K., Warner, T., & Lewis, R. (2004b). Influence of ethical safeguards on research participation: Comparison of perspectives of people with schizophrenia and psychiatrists. *American Journal of Psychiatry*, *161*, 2309–2311.

Roberts, L.W., Johnson, M.E., Brems, C., & Warner, T.D. (2007). Ethical disparities: Challenges encountered by multidisciplinary providers in fulfilling ethical standards in the care of rural and minority people. *Journal of Rural Health*, *23*, 89–97.

Roberts, L.W., Johnson, M.E., Brems, C., & Warner, T.D. (2008). When providers and patients come from different backgrounds: perceived value of additional training on ethical care practices. *Transcultural Psychiatry*, *45*, 553–565.

Robins, L., & Regier, D. (1991). *Psychiatric Disorders in America: The Epidemiological Catchment Area Study*. New York: Free Press.

Rohde, J., Cousens, S., Chopra, M., Tangcharoensathien, V., Black, R., Bhutta, Z.A. et al. (2008). 30 years after Alma-Ata: Has primary health care worked in countries? *Lancet, 372*(9642), 950–961.

Schmidt, H.G., Norman, G.R., & Boshuizen, H.P. (1990). A cognitive perspective on medical expertise: Theory and implication. *Academic Medicine, 65*(10), 611–621.

Sen, A. (1999). *Development as Freedom*. New York: Anchor.

Sheppard, M. (2002). Mental health and social justice: Gender, race and psychological consequences of unfairness. *British Journal of Social Work, 32*(6), 779–797.

Simon, G.E., Barber, C., Birnbaum, H.G., Frank, R.G., Greenberg, P.E., Rose, R.M. et al. (2001). Depression and work productivity: The comparative costs of treatment versus nontreatment. *Journal of Occupational & Environmental Medicine, 43*(1), 2–9.

Sobel, D.S. (1995). Rethinking medicine: improving health outcomes with cost-effective psychosocial interventions. *Psychosomatic Medicine, 57*(3), 234–244.

Sorsdahl, K., Stein, D.J., Grimsrud, A., Seedat, S., Flisher, A.J., Williams, D.R. et al. (2009). Traditional healers in the treatment of common mental disorders in South Africa. *Journal of Nervous & Mental Disease, 197*(6), 434–441.

Ssebunnya, J., Kigozi, F., Lund, C., Kizza, D., & Okello, E. (2009). Stakeholder perceptions of mental health stigma and poverty in Uganda. *BMC International Health and Human Rights, 9*, 5.

Stilwell, B., Diallo, K., Zurn, P., Dal Poz, M., Adams, O., & Buchan, J. (2003). Developing evidence-based ethical policies on the migration of health workers: Conceptual and practical challenges. *Human Resources for Health, 1*, 8.

Stilwell, B., Diallo, K., Zurn, P., Vujicic, M., Orvill, A., & Dal Poz, M. (2004). Migration of health-care workers from developing countries: Strategic approaches to its management. *Bulletin of the World Health Organization, 82*(8), 595–600.

Tejada de Rivero, D.A. (2003). Alma-Ata revisited. *Perspectives in Health, 18*(2), 1–6.

UNDP (2000). *Human Development Report 2000*. Oxford: Oxford University Press.

United Nations. (2000). United Nations Millennium Declaration: Resolution adopted by the general assembly (No. A/RES/55/2 (8th plenary meeting). Accessed February 20, 2010). New York: United Nations General Assembly.

US Department of Health and Human Services (1999). *Mental Health: A Report of the Surgeon General*. Washington, DC: Department of Health and Human Services. Available at http://www.surgeongeneral.gov/library/mentalhealth/home.html (accessed January 29, 2010).

Voinescu, B., Szentagotai, A., & Coogan, A. (2010). Attitudes towards psychiatry – A survey of Romanian medical residents. *Academic Psychiatry, 34*, 75–78.

Walley, J., Lawn, J.E., Tinker, A., de Francisco, A., Chopra, M., Rudan, I. et al. (2008). Primary health care: making Alma-Ata a reality. *Lancet, 372*(9642), 1001–1007.

WHO (1998). *Mental Disorders in Primary Care*. Geneva: World Health Organization.

WHO (2001a). *Atlas: Mental Health Resources in the World*. Geneva: World Health Organization.

WHO (2001b). *The World Health Report 2001. Mental Health: New Understanding, New Hope*. Geneva: World Health Organization.

WHO (2002). *Health, Economic Growth, and Poverty Reduction: The Report of Working Group 1 of the Commission on Macroeconomics and Health*. Geneva: World Health Organization, Commission on Macroeconomics and Health.

WHO (2005a). *Atlas of Child and Adolescent Mental Health Resources, Global Concerns, Implications for the Future*. Geneva: World Health Organization.

WHO (2005b). *Atlas: Mental Health Resources in the World*. Geneva: World Health Organization.

WHO (2005c). *The Economics of Mental Health in Europe*. Helsinki: WHO European Ministerial Conference on Mental Health. Available at http://euro.who.int/document/mnh/ebrief09.pdf (accessed 12 January 2010).

WHO World Mental Health Survey Consortium (2004). Prevalence, severity, and unmet need for treatment of mental disorders in the World Health Organization World Mental Health Surveys. *Journal of the American Medical Association*, *291*, 2581–2590.

WHO (2007). Ten Statistical highlights in global public health (Part1): World Health Statistics. Geneva: World Health Organization.

WHO/WONCA (2008). *Integrating Mental Health into Primary Care: A Global Perspective*. Geneva: World Health Organization and World Organization of Family Doctors (WONCA).

Wittchen, H.U., Jonsson, B., & Olesen, J. (2005). Towards a better understanding of the size and burden and cost of brain disorders in Europe. *European Neuropsychopharmacology*, *15*(4), 355–356.

9

The 2006 Hugh Rodman Leavell Lecture "Globalization, Poverty, and Health"

Paulo Marchiori Buss

The first words I say should express my deep gratitude to the World Federation of Public Health Associations for the privilege of being the Leavell Lecturer at this XI World Congress and the VIII Brazilian Congress on Collective Health.

This award means a lot to me. First, I received it from the largest and most important public health association in the world. The Federation brings together more than 70 national associations from all around the world – institutions critical to the health of their countries' populations – and health professional members working at national health services, universities, public health schools, academies, and institutes. I pay special homage to Margaret Hilson, a dear friend of mine and an extraordinary supporter of the global public health, who introduced me to the Federation.

Second, I'm honored to receive an award named after Hugh Rodman Leavell, the Professor of Public Health and Preventive Medicine who has been influencing my thinking, from the start of my career until today. Leavell was Professor Emeritus at the Harvard School of Public Health and co-wrote with E. Gurney Clark, a seminal book for the doctors of my generation: *Preventive Medicine for the Doctor in His Community*. In the book, they described the natural history of disease, putting forward creative explanatory models, now widely known, of the health–disease process. They helped us organize our thinking and understand and distinguish different levels of applicability of health promotion and disease prevention measures – the principal objectives

Source: *Journal of Public Health Policy*, 28(1) (2007): 2–25.

of public health and its professionals. Leavell served the World Federation as Executive Director for many years and is thus rightfully honored here and immortalized by his very own Federation with this award.

Public health – both a field of knowledge and a social practice – has faced gigantic challenges throughout its history. The late twentieth century and the beginning of this millennium challenge us with two defying processes: globalization and poverty. Daily, these two phenomena deeply influence the health of the population which is the first and foremost concern of public health and public health professionals. So to tackle this better, we must try to comprehend these phenomena better.

Globalization and Poverty

The planetary or global economy, strictly speaking, has existed since the end of the sixteenth century, a time when Europeans explored Africa, Asia, and the Americas. The European colonialist expansion brought both positive and negative social and economic consequences to these territories, as well as to the European population.

Most authors, however, view *globalization* as a social, economic, and cultural process of the past two or three decades, characterized, not explicitly, by

- an increase in the international trade of goods, products, and services;
- trans-nationalization of mega corporations;
- free circulation of capital;
- privatization of the economy with a decrease in importance of governments and nation-states;
- reduction of protectionist commercial barriers and the regulation of international trade as manifest in the rules of the World Trade Organization (WTO);
- facilitated transit of people and goods among countries; and
- expansion of communication, an information society aided by technology, importantly the internet.

Critics of globalization include many authors and several United Nations (UN) agencies. The "short twentieth century", an expression coined by the historian Eric Hobsbawm (1) to describe the recently completed century, brought an extraordinary "revolution in transport and communication that has practically annihilated time and distance". In bringing together unequal cultures and economies, the globe becomes the "basic operational unit, and older units such as 'national economies', which are defined by the policies of territorial states, change into [mere] impediments to transnational activities".

The World Commission on the Social Dimension of Globalization, established by the International Labor Organization (2), insists that the current process of globalization is generating unbalanced outcomes, both between and within countries. Wealth is being created, but too many countries and people are not sharing in its benefits. Many of them live in the limbo of informal economy, without formal rights and in a swathe of poor countries that subsist precariously on the margins of the global economy. Even in economically successful countries some workers and communities have been adversely affected by globalization.

The Commission alerts us that "these global imbalances are morally unacceptable and politically unsustainable", noting "the unfairness of key global rules on trade and finance and their asymmetric effects on rich and poor countries", as well as "the failure of current international policies to respond adequately to the challenges posed by globalization".

We see "market opening measures and financial and economic considerations predominate over social ones". Overseas Development Assistance falls far short of the minimum amounts required, even for achieving the Millennium Development Goals (MDG), and tackling growing global problems. The multilateral system responsible for designing and implementing international policies is also under-performing. It lacks policy coherence as a whole and is not sufficiently democratic, transparent, and accountable. These rules and policies are the outcome of a system of global governance largely shaped by powerful countries and powerful players. There is a serious democratic deficit at the heart of the system. Most developing countries still have very limited influence in global negotiations on rules and in determining the policies of key financial and economic institutions" (2). The failure of the Doha Round at the WTO illustrates the problem.

According to 2001 Nobel Prize winner in Economics, Joseph Stiglitz, it was developed countries that profited from globalization. Countries whose internal savings and technological development, together with strong protectionism applied only to others – despite the golden rule of trade liberalization – made them the privileged addresses of the world's wealth.

More recently, even the World Bank, in its World Development Report of 2006 (3), finally admitted that market forces and free trade will not solve the problem of poverty in the world or even reduce it to bearable levels. The report itself affirms that "only equity is capable of increasing our capacity to reduce poverty."

Internal and foreign debts, trade barriers, etc. and the protectionism for industry and agriculture in richer countries (which hinder the developing countries' primary and industrial goods) are the roots of the enormous fiscal crisis presently faced by developing countries. They contribute to the increasing social debt they have with their people. Almost all taxes collected in these countries, as well as international loans granted by the International Monetary Fund under strict conditions, are used to postpone paying off debts acquired in the past in less good times, often under non-democratic and corrupt governments. Debt increases because of abusive interest rates imposed unilaterally by the international financial community. Consequently, programs destined to fight poverty and other social programs end up under-financed and ineffective (4).

One of the most harmful aspects of globalization is the brutal attacks promoted by international speculative capital on fragile national economies of poor or middle-income countries. The so-called *hot money* has injured social budgets, including health, in poor countries. Approximately US$1.8 trillion of speculative, nonproductive capital is currently in circulation in the world (5). This capital has no nation and, therefore, no responsibility for people or to countries. The world must control it both nationally and internationally to diminish its pernicious effects, both globally and locally. John Williamson, an economist, coined the expression "Washington Consensus" to name the set of recommendations for Latin America regarding economic reforms. These define the conceptual basis of the globalization process as we understand it today. Despite his previous positions Williamson now recognizes, in his recent book, the imperative to control capital flow in the so-called emerging markets (6).

Besides bad economic results, the international division of production and labor that takes place with globalization, also led to important social, environmental, and sanitary consequences. In the labor domain, unemployment moved from developed to developing countries (due, in great part, to protectionist policies and agricultural subsidies in richer countries). Economic activities with higher risks to workers and environment or the ones that produce dangerous waste (the so-called "dirty industries") have been transferred to poorer countries, whose legislation protecting workers and the environment are more tolerant.

Unsustainable patterns of urbanization, industrialization, waste generation, and energy consumption in more developed, rich, and industrialized countries pose destructive threats to the environment, including progressive global warming. This results in losses in food production, desertification, pollution of air, soil, rivers, aquifers, and oceans, the depletion of woods and forests, and unrecoverable damage to biodiversity.

Recently, United Nations University specialists warned that within 5 years, the world is going to have at least 50 million so-called "environmental refugees" (6). These refugees are people who have had to leave their houses and/or lands because of tornadoes, tsunamis, earthquakes, long-lasting droughts, deforestation, desertification, and other natural disasters – phenomena resulting from uncontrolled economic activities affecting the environment. Many are "refugees in their own countries".

Following disasters, diseases appear in survivors favored by weaker health, social, and economic conditions. These "new refugees" present and constitute new public health problems, the responsibility of national and local governments (and in the event of global disasters, the United Nations). We – public health specialists and health professionals and administrators – should provide the care these people deserve.

Responsibility for the terrible social and economic results of globalization should be attributed not only to developed countries and to international financial corporations and organizations, but also to the political and economic elites and governments of many developing countries, those manifesting a low level of social commitment and often corruption.

The politics and governance of many developing countries wastes resources and results in ineffective environmental protection, health promotion, disease prevention, and health assistance initiatives – if they exist at all. Social, environmental, and sanitary programs in these countries are often vertical, unarticulated, and drained by corruption.

Although necessary, foreign aid and eased export rules aimed at improving trade balances for poor countries, remain insufficient to launch development. What countries gain through foreign trade is not distributed to the poor population, remaining heavily concentrated in the hands of few, generally national or trans-national export corporations.

The result is that many are cast-aways from the benefits of globalization and yet vulnerable to its costs. Unhappily, the benefits they receive from public policies in health are very limited.

Globalization and Poverty

Poverty is a multidimensional concept (as well as a multidimensional real-life situation). In the past, poverty referred exclusively to the income of the individual: such as those who live on less than US$1 per day, adjusted to its purchasing power in the country or region. Although the wealth of the world – presently estimated at US$20 trillion per year – continues to grow, approximately 1.2 billion people live on less than US$1 per day (a situation categorized as of "extreme poverty") and half of the world's population lives on less than US$2 per day (7). In Sub-Saharan Africa, almost half the population lives on less than US$1 per day, compared to 37% of the population in South Asia (or 448 million people). In Latin America and the Caribbean, 222 million people are poor, of whom 96 million, or 18% of the population, are indigent (8).

Thanks to the critical work of Amartya Sen (Nobel Prize in Economics, 1998), we learned that a universal poverty line could not be established and applied to everyone in the same way – without taking personal characteristics and circumstances into account. Sen pointed out that the analysis of poverty should also concentrate on the capacity of the individual to take advantage of his/her opportunities (9). The analysis must consider health, nutrition, and education, which reflect the individual's basic working capacity in a society. The power of health promotion directed at the poor and strategies for individual and collective "empowerment" rest on observations such as Sen's.

The poor, on the other hand, are precisely those living in the worst social, environmental, and sanitary conditions. They face the greatest difficulty accessing public services – specifically, health services. Studies conducted around the world show that lower income people are precisely the ones who, despite being underprivileged, have poorest access to adequate housing, potable water, sanitation, food, education, transportation, leisure, stable and risk-free jobs, as well as to health services.

Poverty and Health

Disparities in wealth exist between countries and regions, as well as between the rich and the poor within each country. Table 1 shows the differences in health between the countries, grouped by level of development. Health indicators are worse for poorer and less developed countries.

Life expectancy at birth is 27 years longer in high income countries than in least developed countries; the infant mortality rate is 100 per 1,000 live births in less developed countries and only 6‰ in high-income countries; the difference in the under 5 mortality is even higher: 159 per 1,000 live births in least developed countries and 6 in high-income countries. Health inequalities between rich and poor people within poor countries remain larger – health and nutrition levels (morbidity, disabilities and mortality) and for access to social and health services.

Studying selected health indicators in the poorest countries in the world, Gwatikin *et al.* in Carr (10) showed (see Figure 1) that under 5 mortality was 2.2 times higher among the poorest fifth of the world compared to the richest fifth; malnutrition among women was 1.9 times higher; and the prevalence of stunted growth in children was 3.2 times higher.

In Brazil, my country, studies show that, as in many parts of the world, infant mortality is related to family income, the mother's level of education, housing conditions, where the child and family dwell, and their social conditions (11) (Figure 2).

Among black people (as skin color is a proxy in Brazil for social situation), the average infant mortality rate is 34 per 1,000 live births vs. 23 per 1,000 in the white population; 35 per 1,000 among the poor and 16 per 1,000 among the rich; 40 per 1,000 among mothers with less than 3 years of education while only 17 per 1,000 among mothers with 8 or more years of education; 35 per 1,000 among the rural population and 27 per 1,000 in the urban population; and 67 per 1,000 in a poorer northeastern state vs. 16 per 1,000 in a richer southern state.

The rich and the poor also differ in use of health services. Studies conducted in 50 poor countries between 1992 and 2002 show that the use of oral rehydration therapy is 1.3 times higher among the rich when compared to the poor; vaccination in children is 2.3 times more frequent; three or more antenatal care visits is 3.1 times more common among the richest fifth; and use of modern oral contraceptives is 4.4 times higher. The rich births are 4.8 times more likely to be attended by skilled health personnel (12) (Figure 3).

The differences in *per capita* total expenditures on health are also impressive, as shown in Figure 4. Less developed countries spend an average of

Table 1: Life expectancy and mortality rates, by country development category (1995–2000)

Development category	Population (1,999 millions)	Annual average income (US dollars)	Life expectancy at birth (years)	Infant mortality (deaths by age 1 per 1,000 live births)	Under five mortality (deaths before age 5 per 1,000 live births)
Least-developed countries	643	296	51	100	159
Other low-income countries	1,777	538	59	80	120
Lower-middle-income countries	2,094	1,200	70	35	39
Upper-middle-income countries	573	4,900	71	26	35
High-income countries	891	25,730	78	6	6
Memo: sub-Saharan Africa	642	500	51	2	151

Source: Human Development Report 2001, Table 8, and CMH calculations using World Development Indicators of the World Bank, 2001.

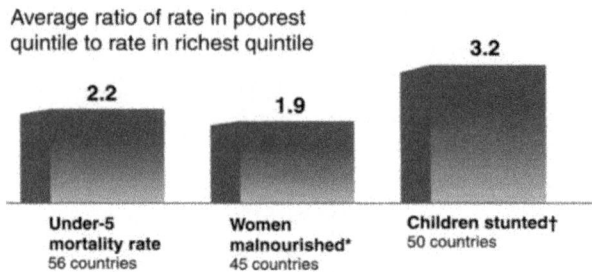

Note: Averages are not weighted for population size.
* BMI < 18.5, defined as weight in kilograms divided by the square of height in meters.
†Low height for age in relation to an international reference population of well-nourished children.
Source: D. Gwatkin et al., *Initial Country-Level Information About Socio-Economic Differences In Health, Nutrition, and Population,* Volumes I and II (November 2003).

Figure 1: Health inequalities in less developed countries, 1990–2002

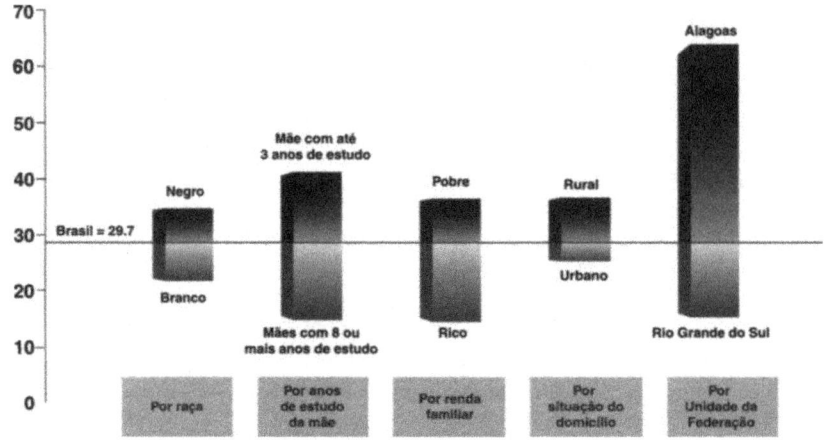

Source: UNICEF Brasil. *Relatório da Situação da Infância e Adolescência Brasileiras.* Brasília: UNICEF Brasil; 2005.

Figure 2: Infant mortality, Brazil, 2000

11 dollars *per capita* per year, against 241 dollars in middle-income countries and approximately 2,000 dollars in high-income countries.

These data show that globalization has made countries poorer and increased poverty, exclusion, and social and economic inequalities. These inequalities are heavily echoed in the health of individuals and the population as a whole.

Globalization and Disease

Globalization affects health when it helps spread transmissible diseases – particularly new or re-emerging infections – more widely. Since international travel has been facilitated and trade intensified, microorganisms can be easily transported by people, animals, insects, and food from country to country,

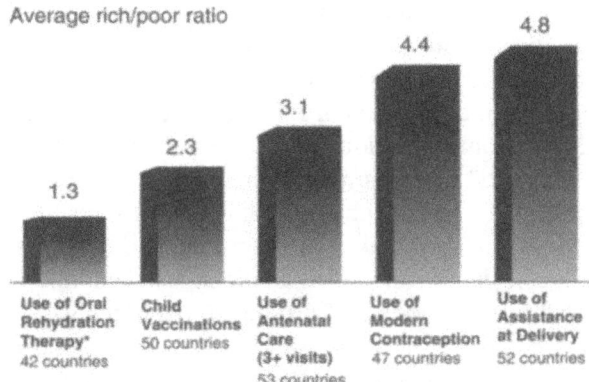

Average rich/poor ratio

| | Use of Oral Rehydration Therapy* 42 countries | Child Vaccinations 50 countries | Use of Antenatal Care (3+ visits) 53 countries | Use of Modern Contraception 47 countries | Use of Assistance at Delivery 52 countries |

Note: Represents the average of the ratios of the richest fifth to poorest fifth, not weighted for population size and excluding countires with use than 1 percent.
*Percent of children with diarrhea in the two weeks preceding the survey who had received oral rehydration salts, other recommended home fluids, or increased liquids.
Among key health services, the gap in use between rich and poor is greatest for modern contraception and skilled delivery assistance. The wealthiest women are four to five times more likely than the poorest women to use these services.

Figure 3: Inequalities in the use of health services, latest surveys, 1999–2002. PRB (Population Reference Bureau). *The wealth gap in health,* pp. 5, May 2004. Available at http://www.prb. org, accessed 24 July 2005. Fonte

Income group	Total spending on health per person, US$
Least developed countries	11
Other low-income countries	
(per capita GNP < US$760 in 1998)	23
Lower-middle-income countries	
(US$761 <per capita GNP < US$3,030 in 1998)	93
Upper-middle-income countries	
(US$3,031 <per capita GNP < US$9,360 in 1998)	241
High-income countries	
(per capita GNP < US$9,360 in 1998)	1,907

Note: Income ranges are those specified in the Organisation for Economic Co-Operation and Development, *Development Co-Operation Report, 2000.*
Source: WHO, *Macroeconomics and Health: Investing in Health for Economic Development* (2001): 56.

Figure 4: *Per capita* health expenditures, by countries' level of income, 1997. http://www.prb. org/pdf04/ImprovingtheHealthWorld_eng.pdf

from any point of the globe to another. The spread of SARS and of the Dengue and Bird Flu viruses are recent examples. Person-to-person transmission of viral hemorrhagic fevers, as in the recent outbreaks of Marburg and Ebola hemorrhagic fevers in Africa, is a major path for epidemics (now facilitated by fast international air travel). Such outbreaks reinforce the importance and need to strengthen global networks of surveillance and diagnosis managed by the World Health Organization (WHO) and partners around the world.

HIV illustrates the problem, as it probably originated in a remote part of Africa and spread throughout the world in the last 20 years. Migrating birds may also spread of infectious diseases globally, Bird Flu and the West-Nile Virus, for example. Salmonellosis and *Eschericbia coli* infections have often been associated with contamination of fresh or industrialized food shipped between countries. "Old" diseases may re-emerge in one region and spread throughout the world. *Polio,* for example, has been found in many distant places following a recent outbreak in African and Middle Eastern countries due to flaws in vaccine coverage. Ongoing epidemics of cholera have affected 75 countries in the last 40 years and produced, over the last 2 years, more than 50 thousand cases and two thousand deaths in Angola alone. Yellow fever reappeared in African countries. We also find new versions of old diseases, such as the drug-resistant tuberculosis. The increase of antibiotic resistance in some microorganisms facilitates their global spread and must be taken into account.

I cannot ignore sexual tourism and its consequences. Many developing countries depend economically on international tourism. The globalized tourism industry tolerates the sexual trade of children, adolescents, and adults of both sexes. Certain tourist choose many destinations in the world today for opportunities related to sexual tourism, including Brazil, and many Caribbean, Asian, and African countries. The globalization of sexual trade results in the spread of sexually transmissible diseases and the psychological and emotional damage that results from the sexual abuse of children, adolescents, and even adults.

Increases in war and conflicts caused by economic and territorial disputes between countries and between groups or ethnicities within nation-states also reflect globalization. They cause thousands of deaths, injuries, and post-conflict physical, emotional and psychological disabilities, afflicting principally youth, the major victims of conflicts. Mutilations caused by injuries, landmines, or by deliberate injury of prisoners, exploitation and abuse of women for revenge, and genocidal attacks on children and the elderly count among recent war crimes. State terrorism and terrorism by particular groups contribute to these tragic statistics.

Devastation of infrastructure results from wars and conflicts. Destruction of health and sanitary services and harm to the environment affect the health of the people. Many are affected directly, but indirectly and strongly, entire populations suffer. Governments redirect money from social programs, such as education and health, to finance the military, thus hindering people's access to essential services and worsening health conditions.

The 20th century was one of the most violent periods in human history: conflicts directly or indirectly caused the loss of approximately 191 million people, half of whom were civilians (13). The last years of the 20th century and the first years of the 21st century sadly indicate an increasing trend for these harmful events.

Globalized violence has created refugees, forcing thousands to leave conflict areas, becoming political refugees. Groups of people taken by force from their original homes, often face worse physical and mental health conditions in their new places (14).

Drug trafficking (cocaine, heroin, marijuana, and synthetic chemical drugs) has expanded immensely with globalization, with the use of drugs in almost all societies, causing dreadful consequences for the health of drug users. International drug trade, moreover, remains associated with international gun trafficking; an explosive combination with astonishing consequences, as described in the World Report on Violence and Health (13).

One of the paradoxes of the current process of globalization is that, despite the fact that the history of mankind has reached a stage where agricultural technology can produce an abundance of food products, hunger remains very prevalent in the world, and in parts of the planet incites true genocide. The Food and Agriculture Organization (FAO) warns that no fewer than 852 million people suffer from chronic hunger and malnutrition, causing the death of 5 million children every year and costing billions of dollars in productivity losses and decreases in national incomes (15). Every year, moreover, 20 million babies are born underweight, usually because of malnourished mothers.

In Sub-Saharan Africa – currently the world region most affected by poverty and its consequences – FAO (15) estimates that no less than 33% of the population are considered malnourished – a rate that reaches 55% in Central Africa and approximately 40% in Southern and Eastern Africa. In addition to urgent outside aid to tackle the cruel situation in countries such as Niger and Malawi today, specialists agree that only technical and financial cooperation, plus investments in water, the sustainability of ecosystems, and enhancing people's own capacities can overcome malnutrition.

Market-oriented sectoral reforms, extolled in the recent past by international organizations, constitute another consequence of globalization and cause further health inequities (16). These reforms leave no place for public health or for health promotion, as the focus is exclusively on the medical care of individuals and how to finance it. The same applies to the imported models for training human resources – which may be ill-suited to a country's cultural patterns and national health systems. Thus, it is imperative that we abandon this kind of reform in favor of a course of action that would implement egalitarian and solidarity-based public health systems. They should take the *health of the population* into account and not simply *do business with disease*.

The Opportunities of Globalization

Globalization presents, however, positive aspects. If we remember the last half of the 20th century, for example, right after the trauma of World War II, we see that the creation of the UN, including the WHO, represented an important step towards international dialogue, peaceful coexistence of nations, and cooperation for the progress of all people and countries in the world. Deception and subsequent loss of trust has, nonetheless, caused many member States, organizations, and individuals to demand a broad reform of the UN system.

In the 1990s, the UN offered guidance to sectoral organizations to carry out a set of large thematic conferences "in order to prepare the world for the 21st century". The major conferences are listed below.

These conferences have generated important reports with substantial recommendations, which had they been heeded and implemented by countries and even by the UN itself, could already have caused impressive political, social, economic, and environmental development for the world as a whole. But the conference recommendations are seen to express contradictory political interests, hence have not been implemented, and have been relegated to mere internationalist rhetoric.

In the year 2000, closing the series of conferences that took place in the previous decade, the UN organized the World Summit, in which all Member-States made a new global commitment to development, with an all-encompassing perspective derived from agreements already reached. The policy thrust was reflected in the Millennium Declaration (17). The MDGs are presented below.

The UN Millenium Development Goals

- **Goal 1: Eradicate extreme poverty and hunger**
- Reduce by half the proportion of people living on less than a dollar a day
- Reduce by half the proportion of people who suffer from hunger

- **Goal 2: Achieve universal primary education**
- Ensure that all boys and girls complete a full course of primary education

- **Goal 3: Promote gender equality and empower women**
- Eliminate gender disparity in primary and secondary education preferably by 2005, and at all levels by 2015

1990's United Nations Conferences

1990 – World Summit for Children
1990 – World Conference on Education for All
1992 – United Nations Conference on Environment and Development
1993 – World Conference on Human Rights
1994 – International Conference on Population and Development
1995 – United Nations Fourth World Conference on Women
1995 – World Summit for Social Development
1996 – United Nations Second Conference on Human Settlements (Habitat II)
1996 – World Food Summit
2000 – Millennium Summit: Millennium Declaration and MDG
2002 – World Summit on Sustainable Development
2005 – World Summit: 2005 Outcomes

- **Goal 4: Reduce child mortality**
- Reduce by two-thirds the mortality rate among children under five

- **Goal 5: Improve maternal health**
- Reduce by three-quarters the maternal mortality ratio

- **Goal 6: Combat HIV/AIDS, malaria and other diseases**
- Halt and begin to reverse the spread of HIV/AIDS
- Halt and begin to reverse the incidence of malaria and other major diseases

- **Goal 7: Ensure environmental sustainability**
- Integrate the principles of sustainable development into country policies and programmes; reverse loss of environmental resources
- Reduce by half the proportion of people without sustainable access to safe drinking water
- Achieve significant improvement in lives of at least 100 million slum dwellers, by 2020

- **Goal 8: Develop a global partnership for development**
- Develop further an open trading and financial system that is rule-based, predictable and non-discriminatory, includes a commitment to good governance, development and poverty reduction – nationally and internationally
- Address the least developed countries special needs. This include tariff and quota-free access for their exports; enhanced debt relief for heavily indebted poor countries; cancellation of official bilateral debt; and more generous official development assistance for countries committed to poverty reduction
- Address the special needs of landlocked and small island developing States
- Deal comprehensively with developing countries debt problems through national and international measures to make debt sustainable in the long term
- In cooperation with the developing countries, develop decentand productive work for youth
- In cooperation with pharmaceutical companies, provide access to affordable essential drugs in developing countries
- In cooperation with the private sector, make available the benefits of new technologies – especially information and communications technologies

The goals were subdivided into 18 measures and 48 indicators. Health relates directly to at least 18 of these indicators, whose 1990 values should have improved by now (18). The conclusions of WHO's recently conducted evaluation should inspire us to reflection and action:

- If the state of affairs observed in the last 5 years continue, most poor countries of the world won't be able to meet the modest goals established for reducing infant mortality and under-five mortality. Moreover, the goals for the measles vaccine coverage of children under one year of age will not be met.
- Maternal mortality is being reduced only in countries that already have low rates. In high-rate countries, rates have either stabilized or increased.
- A few indicators related to health services have improved: the proportion of women receiving care by trained professionals during labor; the use of insecticide-impregnated mosquito-nets in areas with a high prevalence of malaria; and coverage of assisted treatment of tuberculosis.

The first seven goals incorporate commitments to be met primarily by the developing countries in order to gradually provide universal access to minimum levels of well-being. Goal 8, which reads to "develop a global partnership for development", encompasses both developed countries commitments to support the efforts of developing countries and elements intended to redress international asymmetries and thus benefit developing nations. It would target official development assistance and a trade/financial system capable of providing viable workouts for debt overhangs.

Richer countries agreed that they needed to invest 0.7% of their national income in aid in order to attain the Millennium Goals. However, the percentage of internal wealth that richer countries send to poorer countries has been halved in the last 40 years, falling from 0.48% in the period 1960–1965 to 0.24% today (19) (Figure 5).

Without doubt, any contemporary national or international struggle to increase external help from developed countries, will seek at least the amount agreed to in the MDGs. These modest goals would be achieved if donor countries invested US$80 per person per year in aid programs. I note that this aid is equivalent to about 1/5 of the rich countries' defense budgets or half of what they spend on agricultural subsidies.

Economists for Peace and Security have compared military expenses and official aid expenses, with shocking results (20). The world's military expenses in 2003 amounted to US$956 billion. The United States alone spent US$417 billion. To attain the Millennium Goals, the world would have to invest not more than US$760 billion over the next 10 years – less than the amount the world spends on arms in just one year (Figure 6).

The United States' *per capita* expenditures for its military amounts to US$1,217. On the other hand, foreign aid *per capita* amounted to only US$46 (Figure 7). Only 23% of this amount, moreover, was sent to the most underprivileged. Thus, for every 25 dollars that the United States spends on the military, it spends only one dollar on foreign aid and just 23 cents on those who need it most. European Union *per capita* military expenditures are US$358.00 and the *per capita* expenditures on foreign aid US$61. Stiglitz and Bilmes,

Net ODA as percentage of GNI 1960-2003
OECD countries

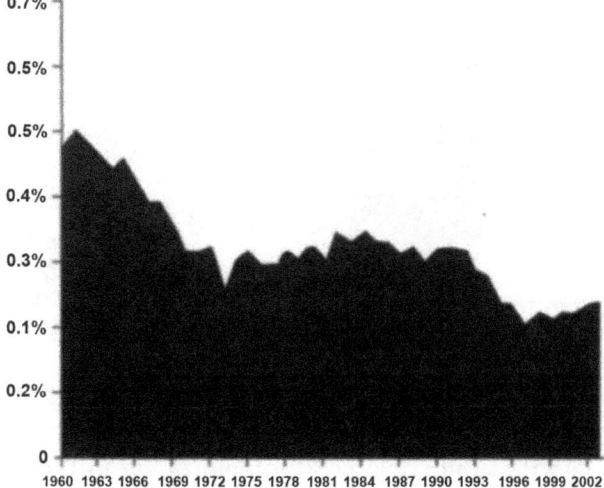

Source: Compiled by Oxfam from DAC data.

Figure 5: Governments spend less than ever on aid. OXFAM. *Paying the price: Why rich countries must invest now in a war against povery.* Available at: http://www.oxfam.org.uk/what_we_do/issues/debt_aid/downloads/mdgs_price_summ.pdf

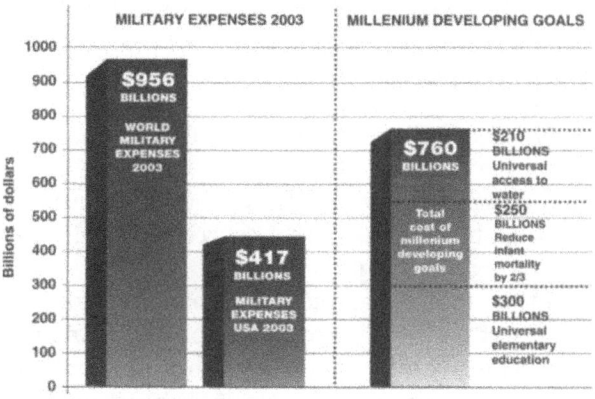

Figure 6: Military expenses vs MDG expenses

economics professors at Columbia and Harvard universities, respectively, estimate expenditures of one trillion dollars for the Iraq War alone (21).

The Commission on Macroeconomics and Health, created by the WHO in 2000, emphasizes its conclusion that investments in health that expand the coverage of essential health services for the world's poor, using a relatively small number of specific interventions, are fundamental for promoting economic development, reducing poverty, and promoting world security (22).

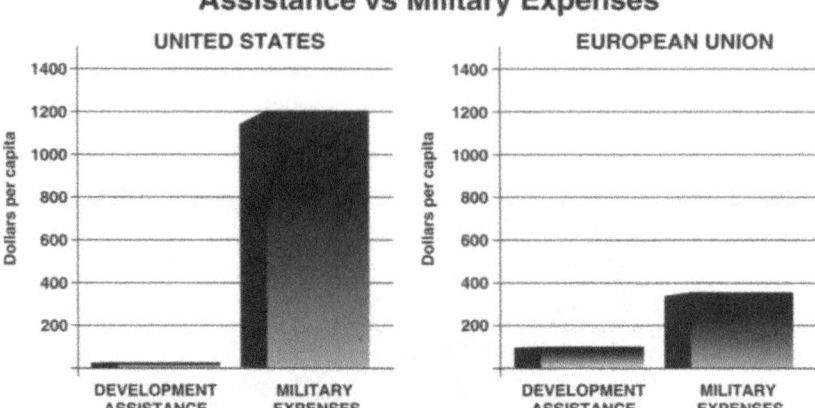

Figure 7: Assistance vs Military expenses in US and EU

Organized efforts to improve child immunization in the poorest countries of the world constitute an important example of good opportunities
enhanced by globalization. The Global Alliance for Vaccines and Immunization (GAVI), a collaboration between the World Bank, WHO, UNICEF, developed countries, private foundations (including the largest of them, the Bill
and Melinda Gates Foundation) and other partners, has created a Vaccine
Fund that supports basic immunization (DTP + polio) plus vaccines against
hepatitis type B and Hib in 70 countries with a *per capita* GDP less than
US$1,000. Six million children have already received the DTP and polio
vaccines (23).

At this point, I must mention Ilona Kickbusch's protest, in her 2004 Leavell
Lecture. It was outrageous, she said, that global health governance in the
world's national governments would allow charitable institutions, such as
the Gates Foundation, to allocate more resources for health than the United
Nations' own health organization, the WHO.

One recent and successful example of international mobilization, notable
for its potential impact on health promotion concerning non-communicable
diseases and risk factors, is the Framework Convention on Tobacco Control,
adopted in May 2005 by the 56th World Health Assembly. In September
2005, the New York Presidential Summit analyzed and adopted 32 proposals
for international treatises (24).

One of the most daring propositions regarding equity and eradication of
poverty would assure a minimum income to all people in a given country, a
measure which is today being called *citizen's income* or *existence income* (25).
Renowned economists, politicians, and institutions, such as Keynes, Tobin,
Friedman, Galbraith, and Moynihan and the Basic Income European Network,
under the leadership of Van Parijs, have defended various versions of this

proposition (25). Alas, the idea does not seem merely a theoretical formulation or Utopia. We recognize that at various moments in the 20th century, after the global depression of the 1930s, countries such as Denmark, the United Kingdom, Germany, Netherlands, Belgium, Ireland, Luxembourg, France, Portugal, and several provinces in Spain established similar comprehensive citizen's income programs, with very positive results (25).

A similar proposition is being developed in Brazil and could stimulate both Brazil and other countries to take concrete actions against poverty. The strategy, as Amartya Sen said, may be used as a means of "overcoming economic freedom deprivation, which leads to the loss of social liberty" (9).

An effective way to control the circulation of speculative capital already exists: taxing non-productive short-term international financial transactions (the so-called hot money) to create a world fund to finance global priorities (basic human and environmental needs) such as global warming, poverty, hunger, and health. This fund could get between US$100 and 300 billion. A global initiative seeks support from all citizens of the world to put the tax in place. It could be implemented and operated by multilateral cooperation or by the UN once approved by national parliaments. Called "Tobin Tax," it is named after the Nobel Prize winner James Tobin of Yale University, who first introduced the idea (5).

The 59th World Health Assembly analyzed the Report of the Commission on Intellectual Property Rights, Innovation, and Public Health (26) and, after exhausting discussions, approved a proposal from Brazil and Kenya to prepare a mid and long-term plan to increase resources for research on health problems that affect the poorer disproportionately and to analyze intellectual property rights for drugs and other products used to tackle these health problems.

In 2005, the WHO created the Global Commission on Social Determinants of Health seeking to devise evidence-based recommendations to inform decision-making on policies plus global and national instruments to act upon fundamental health determinants, as these are essentially social (27). In Brazil, the president has created a counterpart which I'm honored to coordinate (28). I have great expectations for this Commission, whose report will be analyzed in one of the next World Health Assemblies. We hope it will lead to a pact among countries to tackle health determinants both globally and within each country.

A multitude of initiatives thrive around the world, not unlike the ones I have mentioned here. These initiatives have different attributes and focus on different ways to reduce poverty in the world, specific regions or countries, among particular groups – women, children, elderly, etc. They confront particular health-related situations or problems – hunger, malaria, AIDS, etc. As public heath professionals, we must identify these initiatives and give them our support both globally and locally.

There is not, however, only one way to change the equation

Globalization + poverty and exclusion = worse health conditions
into
Globalization + equity and inclusion = health

The only thing we can be sure of is that global solutions should be interconnected with national and local initiatives specifically intended to confront concrete expressions of globalization, poverty, and the health-disease situation. For this commitment and this struggle, I'm sure the world can count on the global community of public health workers and on the resolute action of both the World Federation and Abrasco (Brazil's public health association).

As the elected president of the World Federation of Public Health Associations, I'm committed to the fight. I invite all public health workers in the world to join in the cause against unfair globalization, poverty, exclusion, the arms race, and violence. I invite you, moreover to struggle on behalf of a sustainable environment, equity in health, peace, and solidarity between populations of the world, so that we can attain better health conditions and a better quality of life, not in a distant future, but today, here, and now!

Thank you very much!

References

1. Hobsbawm E. *Age of Extremes – The Short Twentieth Century: 1914–1991*. New York: Pantheon Books; 1995.
2. ILO/World Commission on the Social Dimension of Globalization. *A Fair Globalization: Creating Opportunities for All*. Geneva: International Labour Organization; 2004.
3. World Bank. *World Development Report 2006: equity and development*. Available at http://web. worldbank.org/WBSITE/EXTERNAL/EXTDEC/EXTRESEARCH/EXTWDRS/EXTWDR 2006/0,menuPK:477658~pagePK:64167702~piPK:64167676~theSitePK:477642,00. html, accessed 24 July 2006.
4. Buss PM. Globalization and disease. *Cadernos Saúde Pública*. 2002;18(6):1783–8.
5. Tobin Tax Initiative. *Fact sheet on Tobin Tax*. Available at http://www.ceedweb.org/ iirp/, accessed 18 July 2005.
6. Globo O, (Brazilian newspaper, July 23 2005 edition, p. 24), referring John Williamson' new book *"Curbing the boom-bust cycle: stabilizing capital flows to emerging markets"*.
7. World Bank. *World Development Report 2000/2001: Attacking Poverty*. New York: Oxford University Press; 2000.
8. ECLAC. *Statistical Yearbook for Latin America and the Caribbean 2005*. Santiago de Chile: Economic Commission for Latin America and the Caribbean; 2006.
9. Sen A. *Development as Freedom*. New York: Anchor Books; 2000.
10. Carr D. *Improving the health of the world's poorest people*. PRB (Population Reference Bureau). Health Bulletin 1, February 2004, 39pp. Available at http://www.prb.org/ pdfo4/ImprovingtheHealthbrief_Eng. pdf, accessed 24 July 2005.
11. UNICEF Brasil. *Relatório da Situação da Infância e Adolescência Brasileiras*. Brasília: UNICEF Brasil; 2005.

12. PRB (Population Reference Bureau). *The wealth gap in health* p. 5, May 2004. Available at http://www.prb.org/Template.cfm?Section=PRB&template=/ContentMan agement/ContentDisplay.cfm&ContentID=11007, accessed 24 July 2005.
13. WHO. *World Report on Violence and Health 2002*. Geneva: World Health Organization; 2002.
14. Buss PM. International health: approaches to concepts and practices. in: PAHO, *International Health: A North–South Debate*. Washington: Pan American Health Organization, Human Resources Development Series 9; 1992, pp. 239–48.
15. Food and Agriculture Organization. *The State of Food Insecurity in the World 2004*. Available at http://www.fao.org/docrep/007/y5650e/y5650e00.htm, accessed 15 August 2005.
16. World Bank. *World Development Report 1993: Investing in Health*. Washington: World Bank; 1993.
17. United Nations. *55 General Assembly, Resolucion 55/2*. New York: United Nations; 2000.
18. WHO. *Health and the Millennium Development Goals*. Geneva: World Health Organization; 2005.
19. OXFAM. Paying the price: why rich countries must invest now in a war against poverty. Available at http://www.oxfam.org.uk/what_we_do/issues/debt_aid/down loads/mdgs_price_summ.pdf, accessed 25 October 2005.
20. ECAAR (Economists Allied for Arms Reduction). *Gastos militates vs. gastos socials*. Available at http://www.eumed.net/paz/tepys/gm-gs.htm, accessed 22 June 2006.
21. Folha On Line (Brazilian e-news), accessed 24 July 2006.
22. WHO. *Macroeconomics and Health: Investing in Health for Economic Development – Report of the Commission on Macroeconomics and Health*. Geneva: World Health Organization; 2002.
23. GAVI & Vaccine Fund. *Fact sheet*. Available at www.vaccinealliance.org, accessed 18 July 2005.
24. WHO. *Framework Convention on Tobacco Control*. 55th World Health Assembly Resolutions and Decisions. Available at http://policy, who.int/cgi-bin/om_isapi.dll?infobase= WHA&softpage=Browse_Frame_Pg42, accessed 27 July 2005.
25. Suplicy EM. *Renda de cidadania: A saída é pela porta*. São Paulo: Cortez Editora; 2004.
26. WHO. *Report of the Commission on Intellectual Property Rights, Innovation and Public Health*. Available at http://www.who.int/intellectualproperty/documents/there port/en/, accessed 20 July 2006.
27. WHO. *Commission on Social Determinants of Health*. Available at www.who.int/ social_determinants, accessed 20 April 2006.
28. CNDSS (National Commission on Social Determinants of Health). Available at http://www.determinants.fiocruz.br, accessed 20 June 2006.

I

10

Placing Gender at the Centre of Health Programming: Challenges and Limitations

Carol Vlassoff and Claudia Garcia Moreno

Introduction

Gender refers to those characteristics of men and women that are socially constructed. Sex refers to biologically determined characteristics. While biologically determined difference are universal, social differences between women and men are learned, changeable over time and vary both within and between cultures. Gender interacts closely with biological differences, and other social varibale such as class, in determining health exposures and outcomes. Unfortunately, "gender" is increasingly used inappropriately as a substitute for "sex", particularly in the biomedical literature, a tendency which has created confusion (Fischman et al., 1999).

Gender roles change over time and over an individual's life stages. They also vary according to culture, but in practically all cultures women have a lower status than men and female roles are valued less than those of men. Exceptions are found in a small number of societies which are matrilineal and matrifocal such as the Akan-speaking people of Ghana which emphasize the powerful social, political and economic roles of women, even of unschooled market women (Terborg-Penn & Benton Rushing, 1996). Even in these societies, however, matrilineality does not necessarily give more independent power to women; entitlements usually devolve through men – often

Source: *Social Science & Medicine*, 54(11), International Health in the 21st Century: Trends and Challenges (2002): 1713–1723.

uncles – of the wife or mother. Moreover, matrilineal inheritance systems are rapidly giving way to patrilineal inheritance patterns, particularly with respect to land, as a result of the formalization of customary rights such as land registration that often discriminate against women (IFPRI, 1995).

Because it pertains to the roles performed by men and women and the power relationships between them, gender affects most areas of human existence, including health. Gender interacts with both biological differences and social factors. It affects access to health care (Kutzin, 1993; Velez, Hendrickx, Roman, & Agudelo, 1997), health seeking behaviour (Vlassoff & Bonilla, 1994; Tanner & Vlassoff, 1998), health status (Santow, 1996) and the way health policies and programmes are developed and implemented (Hatcher Roberts & Vlassoff, 1995; AbouZahr, Vlassoff, & Kumar, 1996; Hatcher Roberts & Kitts, 1996). In this paper we argue that a gender analysis is fundamental to health and health planning and discuss the challenges and limitations involved in incorporating gender into the new global health policy of Health for All in the 21st century.

Gender and Its Relationship to Health

Primary health care (PHC), while concerned with socio-economic inequalities, was promoted without adequate consideration of gender. While PHC advocated for "community participation" as one of its central tenets, it tended to assume a homogenous grouping of society members with shared interests, which is usually not the case (Garcia-Moreno & Piza, 1991). Furthermore, gender roles, in particular women's roles, although central to virtually all components of PHC, were largely ignored (Leslie, 1989).

PHC activities were promoted without thought as to how and by whom the activities proposed would be carried out, and how those responsible for these activities could be helped to perform these roles. Moreover, these roles were expected to be undertaken in a more effective way, thus requiring an increased time commitment from women, while not necessarily providing tangible benefits to them (Rosenfield & Maine, 1985; Leslie, 1989, Wong, Li, Burris, & Xiang, 1995). For example, the time needed for making and adequately using oral rehydration solutions for a small child with diarrhea turns out to be substantial.

PHC programmes have also perpetuated the notion that child care is almost exclusively a woman's role. Immunization is a case in point. It is usually assumed that women are responsible for getting their children immunized, and most immunization messages are targeted at women. However, a study by Brugha, Kevani and Swan (1996) in Ghana found that the father's participation in the decision to immunize his child not only increased the uptake, but also led to earlier uptake and timely completion (that is, by 12 months) of the immunization schedule. This example illustrates that the effectiveness of a health programme can be improved by using appropriate health education messages which promote a more equal sharing of child care responsibilities.

It has not been easy for the health community to incorporate the concept of gender into its terminology and programming. This may be because, until recently, sex differences in health and illness have been poorly explored and understood. Traditionally, a male model in medical science was taken as the norm, and sex differences were acknowledged mainly in relation to reproduction and pregnancy where differences are particularly obvious. Moreover, the concept of gender was developed by the social sciences which have played a relatively minor role in medical research. They also tend to be given less credibility than the so-called hard sciences. Nonetheless, it is increasingly recognized that the prevention and control of disease and the maintenance of good health require a holistic perspective in which people's perceptions, needs and priorities are central. People's access to and control over resources that promote and protect health is another area in which gender clearly has an impact.

While the concepts of gender and equity are related they are also distinct concepts. Whitehead notes that equity refers to differences that are unnecessary and avoidable – in access to services, socially ascribed characteristics, educational attainment, economic or other areas. In addition, these differences "are considered unfair and unjust" (Whitehead, 1992). While men and women are physically and physiologically different the differences between them are natural and unavoidable and should not translate into status and power inequalities. Nonetheless, social justice may require that men and women be treated differently (e.g. in terms of the kinds of services provided for men and women) in order to achieve equality in the opportunity for an outcome such as health.

The social differences between men and women that interact with biological variables do fall under considerations of equity, such as differences in how girls and boys are treated with respect to the provision of basic needs including education, nutritious food and health care. Equality of result must also be taken into account, and this may mean that different inputs are necessary to achieve health. Just as individuals differ in size and nutritional intake requirements, so men, because they are usually larger in stature, may require a larger quantity of food than women. Similarly, because of their different biological functions, including menstruation, pregnancy and childbirth, women may require additional supplements to make up for the depletion of micronutrients such as iron, zinc or Vitamin A. However, in order to achieve an equal result in which both males and females in a household are fed and nourished adequately and equitably, the quality of the food should not differ according to whether it is served to girls or boys and both should have access to the quantity required for their physical well-being and growth. Thus, equality applied to gender goes beyond equity: what is required is both gender equality and equity.

How Gender Contributes to Health
Policies and Programmes

Inclusion of a gender analysis can improve health planning and programming in at least five ways. In this section we briefly illustrate these contributions with examples from recent research. The examples are not exhaustive of all health problems or all situations: we focus on gender issues in stable health care settings. However, many of these issues are even more pronounced in emergency and humanitarian crises (Byrne, 1996; Palmer & Zwi, 1998). For example, reproductive health services are poorly addressed within the context of humanitarian relief to people displaced by armed conflict and violence, even though the provision of such services is firmly on the policy agendas of international relief organizations (Palmer, Lush, & Zwi, 1999).

A Gender Analysis Improves Detection and Treatment
of Health Problems in Underreported Groups

Several recent studies have shown that health statistics that are based upon official data from health centres or hospitals may underestimate female morbidity and mortality because women often do not go to these centres for detection and treatment of their health problems (Vlassoff & Bonilla, 1994; Santow, 1996). Women are more likely than men to self-treat or to seek care from other sources such as traditional healers (Beljaev, Brohult, & Haque, 1986; Ettling, Krongthong, Krachaklin, & Bualombai, 1989; Okonofua, Feyisetan, Davies-Adetugbo, & Sanusi, 1992; Agyepong, 1992; Mwenesi, 1993; Kikwawila Study Group, 1995; Santow, 1996). In addition, women's use of services is affected by cost, time, mobility and distance in different ways than men's (Velez et al., 1997; Kutzin, 1990). Because women who do not report to health services are not included in registry data, it is often assumed that they suffer less from various diseases when this may not be the case. This gender bias in health statistics has been demonstrated in India with regard to leprosy (Vlassoff et al., in press; Rao, Garole, Walawalkar, Khot, & Karandikar, 1996), in Colombia with regard to leishmaniasis (Velez et al., 1997), in Thailand (Ettling et al., 1989), India (Beljaev, Sharma, Brohult, & Haque, 1986) and Colombia (Bonilla & Rodriguez, 1993) with regard to malaria, and in chronic respiratory disease in India (Mishra, Malhotra, & Gupta, 1990).

Another reason for the under detection of women with respect to certain diseases is the fact that, in areas of high fertility, women may be pregnant for a great deal of their adult lives, and hence may be excluded from mass screening and treatment campaigns. A study of the impact of the exclusion of pregnant and breastfeeding women in Sierra Leone from treatment for onchocerciasis with Ivermectin (*Mectizan*) (Yumkella, 1996) found that many women were repeatedly excluded from the regular treatment rounds because they were either pregnant or breastfeeding. Of 427 pregnant or lactating women at the beginning of the study, only five were treated according to the recommended treatment guidelines (i.e. 1 month after they finished breast-feeding) due to lack of resources, lack of information or the fact that they had decided to wait for a subsequent treatment campaign. One year later, 265 of these women were followed up, and only 22% had received treatment, 38% had been excluded because they were again either pregnant or lactating, and the rest said they were awaiting the next treatment round a year later. These results are important not only because a sizeable proportion of women remained repeatedly untreated but also because they represented a reservoir for the transmission of onchocerciasis in the larger population. Fortunately, pregnant women are no longer excluded from Ivermectin campaigns.

In the case of tuberculosis (TB), recent research suggests that women may have more difficulty in receiving adequate treatment than men, especially where there are restrictions on women's movement and an unwillingness to pay for their treatment (Khan, Walley, Newell, & Imdad, 2000). More research is required in different cultural settings on gender issues affecting TB treatment-seeking behaviour (Hudelson, 1996; Khan et al., 2000). For example, there is some evidence that women do not receive as prompt and comprehensive treatment for TB as men (Long, Joansson, Diwan, & Winkvist, 1999). The reasons for this remain to be investigated. Whatever the reasons, this is a cause for concern since incomplete treatment contributes to an increase in resistant strains of the disease which has major cost and public health implications.

Although the above constraints refer mainly to women, gender considerations apply equally to men. In a study of gender differences in knowledge and perceptions of tuberculosis in Vietnam beliefs about the disease were closely related to different gender roles in the society. Men were believed to be more at risk due to work and life-style factors, influencing how doctors responded to male and female patients (Long et al., 1999). Gender roles, especially the expectation that men should be strong and capable of suffering without complaint, may determine how quickly people report illnesses that they consider embarrassing or something to be endured without complaint. The cultural expectation of stoicism with respect to illness also applies to women. In Kenya, Nigeria and Sierra Leone rural women are expected to work until they are virtually unable to stand up, and they are expected to go about their duties uncomplainingly, even when they are ill (Vlassoff, Hardy, & Rathgeber, 1995).

A Gender Analysis Improves Understanding of the Epidemiology of Health Problems

Several recent studies have shown that a gender analysis provides a more comprehensive understanding of the epidemiology of disease and other health problems than is otherwise available. Several examples are presented below.

Injuries and violence are increasingly receiving attention as major public health problems. There are important differences in the nature and experience of violence between women and men. There also are important differences in the types of injury which affect women and men globally, with road traffic accidents being the main injury-related cause of death in males, while for females it is self-inflicted injuries (WHO, 1999). Mortality rates for all forms of injury, except those caused by fire, is higher in men (WHO, 1999). In contrast, domestic burns, mostly derived from kitchen work, are mostly in women and children. It is increasingly recognized that violence-related injuries in women are greatly underreported. In addition, the way these issues have been conceptualized and defined has tended to exclude the experiences of women. For example, accidents in the home are rarely reported, yet new evidence shows that exposure to toxic chemicals may be more frequent than previously assumed (Pronczuk, personal communication). Research suggests that generally males are greater risk takers than females in the road traffic environment (Dora and Philips, in press). However, risk taking varies not only on the basis of gender but also with age. For example, research in the United Kingdom found that older female pedestrians are two and a half times more at risk of injury than male pedestrians of the same age for the same distance walked (Ward et al., 1994).

Violence is also experienced differently by men and women. Although men have higher mortality and injury from violence, a growing number of studies have identified violence against women as a major public health concern which impacts significantly on women's mental and physical health (WHO, 1997). The nature and consequences of violence are different for women and girls than they are for males. Women are most at risk from men they know and they experience violence primarily in the family. While street violence among men is most likely to result in injury or death, violence against women is associated with a wide range of both acute and long-term health consequences, including slight and severe injuries, depression, post-traumatic stress and other mental health problems, unwanted pregnancies and sexually transmitted diseases (Heise, 1993; Heise, Ellsberg, & Gottemuller, 1999). Domestic violence can go on for many years and increase in severity over time. Most violence is perpetrated by men, regardless of the sex of the victim. It is important to understand how gender contributes to increasing men's and women's risk of violence, both as perpetrators and as victims.

A Gender Analysis Elucidates Psycho-Social Dimensions of Disease for Men and Women

Very little attention has been devoted to gender differences in the psychological and social dimensions of disease, although there is increasing evidence of differences in the stigma associated with leprosy (Rao et al., 1996), TB (Liefooghe, Moran, Habib, & De Muynck, 1997; Khan et al., 2000) and

HIV/AIDS (Ingram & Hutchinson, 1999; UNAIDS, 1998; Sowell et al., 1997; Bunting, 1996).

A recently completed study in five African sites of the psycho-social dimensions of onchocercal skin disease investigated gender differences in the stigma associated with the condition. Stigma was experienced differently by men and women, and gender differences were consistent across the geographically and culturally varied study sites (Pan-African Study Group on Onchocercal Skin Disease, 1995; Vlassoff et al., 2000). The results indicated that felt stigma, expressed as the tendency to deny or minimize the impact of stigma as a means of coping with the disease, was greater among women than men. Women were more concerned than men about the impact of onchodermatitis on their appearance and life chances, whereas men worried more about its impact on sexual and reproductive capacity, fearing that the disease caused infertility and sexual weakness. An understanding of such issues provides opportunities for positive interventions in areas seemingly unrelated to onchocerciasis. For example, onchodermatitis could be an entry point for opening up a discussion about fears related to infertility and sexual difficulties, and for discussing sexually transmitted diseases and its accompanying risks more broadly. This example of how the same disease may be perceived differently by men and women indicates the potential for making public health services and information more relevant to the concerns of the affected population.

A Gender Analysis Improves Relevance of Public Health Services

It is now increasingly acknowledged that the application of a male model in medicine is neither adequate nor correct. For example, the male model in the development of health technology is a growing concern in industrialized countries, as evidence accumulates that illnesses such as heart disease and cancer affect men and women differently, and that technology for their diagnosis and treatment may need to be adapted to these differences. Pharmaceutical companies are under considerable pressure to include women in clinical trials of new drugs and to consider the relative effectiveness of existing drugs for women, as well as men. This pressure began in the 1980s as the perceived risks of clinical research decreased and its perceived benefits increased (McCarthy, 1994). Women's advocacy groups also played a role, as well as the growing realization that women, while excluded from clinical trials, were not excluded from the clinical applications of research (Wright & Chew, 1996). The establishment of the Office of Women's Health in the United States National Institutes of Health in 1990, as well as the Office of Women's Health in the Food and Drug Administration in 1993, stimulated greater attention to this issue as a public health priority (NIH, 1999).

It is also becoming increasingly clear that the standard male model in medical science is often incorrect when applied to men themselves. By assuming, for example, that men should be strong and suffer their pain in silence, health professionals may miss opportunities to help male patients cope with their disease and their inner suffering. This seems to be especially true of disfiguring and stigmatizing diseases such as leprosy, TB and lymphatic filariasis. Research with men in Recife, Brazil, by a female urologist, Gerusa Dreyer, has revealed tremendous pain and suffering on the part of male patients from hydrocele, the swelling of lymphatic tissue in the testicles as a result of the build-up of lymphatic fluid. While the pathology associated with hydrocele has been studied by medical scientists for many years, it is only recently that its psycho-social dimensions have received attention. Patient support groups have been introduced which have offered psychological relief through the sharing of common problems, as well as the sharing of simple therapeutic methods such as bathing the swollen area with soap and water. Group sharing of experiences and success with this treatment, which depends on patient motivation rather than medical intervention, is helping to promote this cost-effective and patient-controlled methods of disease alleviation (Dreyer, Noroes, & Addis, 1996).

A Gender Analysis Increases Potential for Greater Public Participation in Health

An understanding of gender, and how it influences the behaviour of individuals, families and communities, greatly increases the potential for enhanced public participation in the promotion of healthy lifestyles and related preventive actions and disease control activities. Many development projects targeted at women have failed to recognize the context in which they will ultimately be applied (Boserup, 1989). It has been shown repeatedly that health education or other interventions undertaken without an understanding of gender issues are often misguided and fail.

For example, women's authority within the domestic arena may be negatively affected by malaria control activities because of the intrusion of vector control personnel, the need to reorganize the household environment to eliminate breeding grounds, and the creation of a perception that the disease originates within the household for which women are responsible (Winch, Lloyd, Hoemeke, & Leontsini, 1994). Similarly, social marketing to promote insecticide treated mosquito nets have not sufficiently addressed gender relations within the household that determine the use and maintenance of the nets (Tanner & Vlassoff, 1998).

On the other hand, there is growing recognition that people who are given relevant, accurate and complete information about the purpose of health interventions and the need for behavioural change are more likely to adopt and sustain the activities themselves. An example is a South African health education initiative, "Soul City," a multi-media project involving television, radio and booklets that provides socially relevant health information to its audience. It is rooted in a township (shanty-town) setting and is based on the real life experiences of people living in a poor Johannesburg slum community. Gender power relations are intricately woven into the programmes that address a variety of health problems such as HIV and STDs, child abuse, violence against women, and discrimination in the labour force. The scripts are developed by a professional team, but before they are finalized they are discussed with township community members and revised according to the feedback received. The first external evaluation of the project showed not only that it reached a wide and ethnically pluralistic audience, but also that it had affected the behaviour of up to 87% of those interviewed (CASE, 1995). Among those who watched television for instance, 21% of those who watched Soul City said they always used condoms as opposed to 15% of those who did not. The second evaluation, completed in 1999, also found that Soul City had a positive impact on attitudes: for example, the understanding that domestic violence is a form of violence increased from 17% to 31% among men and from 12% to 20% among women (Samuels et al., 1999).

Challenges of Incorporating Gender in Health Planning and Programming

Despite the many arguments in favour of incorporating gender issues into health policies and programmes, many obstacles remain. Some of these are briefly discussed below, as well as opportunities for placing gender at the centre of health programming.

Creating Awareness and Understanding

The concepts of gender and gender analysis evolved from feminist thinking which emphasized the social and cultural nature of many of the differences between women and men, particularly the unequal power and status attributed to male and female roles. This thinking slowly filtered into development debates, including those on health and development. However, gender issues have not been mainstreamed into the training of health professionals who are therefore not exposed to these concepts and their importance to public health. Although there has been some progress in recent years in building a theoretical and practical basis for the incorporation of gender into health issues, empirical and other data demonstrating the importance of gender is still scarce and is often overlooked.

In addition, the gender and health community has at times simplified or over-generalized gender differences in health. There is still a need for more complex analyses that look at the interactions between biology and gender and between gender and other social variables, particularly socio-economic status. Furthermore, there is a need to move from the diagnosis of gender-related problems to identifying responses and interventions to address them. Health providers and policy makers often feel that gender inequalities are beyond their control and that they can only be addressed by profound societal change. Clearly, broad societal change is required in the long run, but there are also ways in which the health systems can begin to incorporate gender issues into their policies and programmes. At a minimum, they should not exacerbate existing inequalities between women and men.

Often changes may be small and simply require an increased level of awareness and sensitivity to gender concerns and disparities. Examples include: ensuring that female health providers are available to examine female clients in areas where women are not comfortable with male providers, providing separate lines for women and men to collect TB treatment and making contraceptives available during immunization days. Gender and other differences in the experience of TB by affected persons need to be studied further to identify appropriate strategies in different contexts.

Linking the Social and Biomedical Sciences

Multi-disciplinary research methods which integrate social, and particularly gender, dimensions into epidemiological research are increasingly required to address the complex health problems presented in our globalizing world. A combination of qualitative and quantitative approaches are needed to enable a better understanding of the gender differences in patterns of disease and health seeking behaviour, as well as their determinants and consequences (Bird and Rieker, 1995). In recent years examples of interdisciplinary research with a gender focus have been supported by the Special Programme for Research and Training in Tropical Diseases (TDR) of WHO, and their impact on both health and social science research has been significant (Vlassoff, 1997; Vlassoff, 1992a, b). A number of the examples cited in this paper are derived from such interdisciplinary research.

Recognizing that Gender Inequalities Is Everyone's Concern

The view that gender is principally a Northern feminist issue continues to be used by some as an argument against integrating gender into the health system. However, a growing number of women and men in developing countries are calling for equality between women and men through the promotion of gendersensitive policies. In the recent world conferences on human rights in Vienna, on population and development in Cairo, and on women in Beijing, developing countries played a key role in international agreements that promote gender equity and equality and women's empowerment, as a basis for better health and sustainable development. Both the Programme of Action of the International Conference on Population and Development and the Beijing Platform for Action argue strongly for integrating a gender analysis into all health and development activities. This was reinforced in 1997 by the agreed conclusion on gender mainstreaming into all activities of the United Nations system by the United Nations Economic and Social Council (ECOSOC). All of these international agreements have been endorsed by governments from both North and South and by World Health Organization (WHO).

The global health policy, "Health for All in the 21st Century", adopted by the 51st World Health Assembly, also underscores gender as one of the cornerstones for achieving Health for All, along with human rights, equity and ethics.

Medicine and the Health System Are Gendered Institutions

The health system, and the medical hierarchy within it, replicate many of society's power relationships: between doctors and nurses, doctors and clients and nurses and clients. These relationships are discussed in detail in Marks (1994), with examples of domination and subordination in the nursing field in England, the United States, South Africa and India, resulting from class and racially defined distinctions and gender inequalities. In these hierarchical environments, being female does not necessarily make a health provider sensitive to the needs of female clients, particularly when these clients are of a lower social class. Female health providers are themselves in disadvantaged situations where they often occupy the lower ranks of the professional ladder and frequently experience gender discrimination at home (Fonn & Xaba, 1995).

Incorporating gender into existing policies and programmes entails changes in the structure of the health care system to allow for equal opportunities for women at all levels, but especially at higher levels (Hartigan, 2001). In situations where the imbalance is pronounced, affirmative action policies such as targetting a certain percentage of posts at higher levels for women, as well as recruiting more men to positions that have been almost exclusively female, could be introduced. In Bangladesh, where many roles tend to be rigidly defined by gender, UNICEF and several NGOs have deliberately recruited female drivers and guards in order to change such gender-based stereotypes in occupations. Similarly, efforts are currently being made to recruit males to traditionally female occupations such as village health workers and health volunteers.

Because of the traditional association of medicine with male doctors (female nurses and assistants occupying the lower ranks), there is a tendency to assume that men will occupy the top positions in any health system, an assumption which is often true. In international health today few females are in the high professional categories, while they predominate in the lower ranks. Changing this unequal hierarchy will require changes in the training and socialization of health professionals at all levels, as well as changes in the opportunities provided to females to advance to the highest levels of the health profession. Women also need to be involved in policy development, so that their perspective is incorporated into policies and programmes from their inception, rather than, as is often the case, added on after policies have already been formulated, if at all.

Another example is provided in a study of editors, authors and reviewers of four U.S. epidemiology journals (Dickersin, Fredman, Flegal, Scott, & Crawley, 1998) which showed that one in seven editors in chief was a woman, a position she shared with a man. For all journals the proportion of editors who were women ranged from five (6.5%) of 77 in 1982 to 42 (16.3%) of 258 in 1994. The study concluded that fewer women in public health held editorial positions (12.8%) than were authors or reviewers (26.7%). Given that these positions are influential in shaping science, they recommend that the reasons for this, including the possibility of selection bias favouring men, should be further investigated.

In most environments, a change is also required in the operations of the health system as a whole. A commitment to an enabling working environment for both sexes is needed. Often this involves a deliberate change in language, attitudes and behaviour. All too often high level professionals are referred to as "he", a habit that reinforces traditional norms and attitudes, and stands in the way of promoting the opposite perception that females could also attain these positions. Working hours should be more flexible to allow women to accommodate breast-feeding or other child-care needs. It should be recognized that reproduction is essential to human survival, and women should not be penalized because they play the predominant role in the fulfilment of this role. Similarly, health care facilities should be more sensitive to the problem of conflicting hours for working women who need to bring their children for services, or to attend themselves. Messages on child health should address both men and women, rather than perpetuating a model that puts the responsibility for child care only on mothers.

Linking Gender to Training and Performance of Health Professionals

An understanding and appreciation of gender differentials and their implications for health and health seeking behaviour should be incorporated into the training of health professionals and into the development of health sector responses. Linking gender sensitivity to performance appraisals assures that the rhetoric is taken seriously and translated into practice. Indicators of gender sensitivity in work performance could include whether the employee has participated in a gender training course, performance in the course, use of gender sensitive language, efforts to bring minority employees (e.g. women, people of colour) into the mainstream of office activities and advancement opportunities, and application of gender analysis to the work. While gender sensitivity will usually mean providing greater opportunities to women in lower positions, it may also mean providing opportunities for men to work in traditionally female positions such as nursing or secretarial work.

Translating Political Will into Resources for Change

Addressing inequalities between women and men is a long-term process requiring profound changes of bureaucratic structures and organizations at all levels. It requires changes in work and management cultures, in staffing and budgeting, in agenda setting and training programmes. If the integration of gender is to become reality, and not just rhetoric, clear commitment is needed, particularly from senior management in health and other sectors, as well as resource allocations to undertake the research, capacity building and other necessary changes at all levels.

Another fundamental requirement is that the necessary budget allocations be provided to allow planning, implementation and training in this area to be seriously undertaken. The funding of programmes and projects should also include consideration of their gender sensitivity, and their implementation should be monitored from this perspective. Personnel sensitive to gender issues should be assigned to planning, implementation and evaluation activities, and their participation given equal weight to those of other evaluation personnel. An important component of the success of research, programmes and projects will be the degree to which they promote gender equality.

Developing and Using Practical Tools

The lack of practical tools to implement a gender approach in health systems is another limitation. As awareness increases, there is an emerging demand for concrete examples of how to apply a gender analysis to health programming and policy making. Knowledge in this area is slowly beginning to be built up and a number of tools have been developed. Examples include: WHO (1995), IPPF (1995), SIDA (1997), Harding and Sills (1999), and Liverpool School of Tropical Medicine and DFID (1999). The number of such tools also needs to be expanded in order to assure widespread applicability and use. Further, these resources need to be validated by operations research to assess their impact and effectiveness.

An illustration of such a tool which has been validated for impact and effectiveness is a methodology developed in South Africa to sensitize health workers to gender issues, especially the problem of lack of respectful and understanding attitudes to poor female clients in peripheral health services. The methodology is participatory, taking health workers through a series of workshops in which they explore issues such as reasons for becoming health workers, how their clients see them, the status of women in society and possible solutions to problems at work. TDR supported research on the testing and evaluation of this methodology using a combination of qualitative and quantitative research methods (Vlassoff and Fonn, in press), which demonstrated the importance of listening to health workers and allowing them to express their frustrations and problems, as well as the positive impact of the methodology on the performance of the health workers and the health system as a whole. It was also found to be a useful way of eliciting, in a rapid and cost-effective way, a plethora of problems in the health system and identifying ways to address them. As a result, it has been used as a change management and health system development tool in various settings (Fonn, Xaba, Tint, Conco, & Varkey, 1998).

What mechanisms can be used to monitor gender sensitivity in the health system? The collection of sex-disaggregated data is a necessary but not a sufficient first step. It should be a standard practice to disaggregate all epidemiological data (on mortality, morbidity, disability, etc.) by sex and age for all diseases and health conditions. This allows for gender and other analyses of data and to monitor the sex-specific burden of disease over the lifetime. It also provides the basis for understanding the broader gender dimensions of health and illness, including how illness is experienced by women and men, how services are provided and how people use them. Qualitative research, including rapid assessment methods (Vlassoff, 1992b) will assist in elucidating these dimensions.

Conclusion

The key to placing gender values firmly in place in Health for All renewal is a change in philosophy at all levels of the health sector. This change in philosophy requires, firstly, a recognition that "gender" is not synonymous with "women" or with sex – it is a concept that sees men and women within the context of their culturally defined roles, constraints and potentialities. This context itself often needs to be challenged, such as when gender stereotypes place women in an inferior position or condone harmful practices such as female genital cutting or violence against women. Secondly, this change in philosophy requires the acknowledgement that men and women are different but equal, and that both men and women have a fundamental right to adequate health care and health information that responds to their specific needs. Similarly, it must be realized that the male model in health is not adequate. For example, although research on women is complicated by possible reproductive risks, women should not be automatically excluded from clinical trials and other research. Rather, the rationale for exclusion needs to be carefully considered, whether exclusion is truly justified by the potential risks involved, and if so, what other options may be considered to permit an understanding of the female dimension. Finally, it must be recognized that both gender equity and equality are required in attaining health for all. As the factors affecting the sexual attribution of roles are capable of being changed, inequalities in the division of these roles must be addressed in the process of renewal of the Health for All policy.

Integrating a gender analysis into health policies and programmes is necessary for the achievement of social justice. Furthermore, it can improve the coverage and effectiveness of health programmes. The agreements made by most countries at the recent international conferences, such as the International Conference on Population and Development in 1994 and the Fourth World Conference on Women in 1995, offer a unique opportunity for this, as does the growing interest and concern for gender equality in the international development assistance community. The five year follow up review of the Cairo Conference, recently completed, indicated that most developing countries have taken the commitments made half a decade ago seriously and a number of key actions were identified at the Special Session of the General Assembly of the United Nations in June–July, 1999, aimed to further advance women's equality and empowerment (UNFPA, 1999).

Positive aspects of the Cairo +5 review process included the willingness of most countries to move forward and measure progress on the Programme of Action. It also permitted an open discussion of sensitive issues such as female genital cutting, rape and incest. Gender issues are critically important in the "Key Actions" called for, such as the equality and empowerment of women, women's rights, ending of discrimination against the girl child, measures to prevent the spread of HIV/AIDS and respond to the needs of youth. The review of the Beijing Platform for Action, recently completed, provided an ongoing opportunity for advancing gender equality in areas beyond the Cairo agenda, including occupational health, mental health and substance abuse, as well as broader social and economic issues.

WHO has only recently begun to tackle gender issues in a concerted way and much remains to be done. Gender must be mainstreamed throughout the organization, including in its regional and country offices. Its draft gender policy aims to provide a framework for WHO's own work. More broadly, WHO is developing a framework to assist in integrating gender into the work of technical programmes. This can form the basis for conceptualizing and designing gender-sensitive policies, practical guidelines and monitoring and evaluation tools for mainstreaming gender into the health sector. Ideally, these tools should be complementary to or integrated into other tools for prioritizing and decision making, including cost-effectiveness studies and burden of disease estimates. WHO should also provide ongoing evidence, based on research findings, regarding sex and gender differences in the determinants and consequences of specific diseases. The basis for evidence of this kind has been established by locating responsibility for gender mainstreaming within the Evidence and Information for Health Policy cluster. This location has the potential to give gender and health issues a strong scientific basis and ensure they are integrated across all WHO technical units. This function needs to be supported and strengthened with the necessary resources to undertake and disseminate research of this kind.

The integration of gender into health programming and policy making should ultimately contribute both to the achievement of social justice and to more effective health policies and programmes which will benefit both women and men.

References

AbouZahr, C., Vlassoff, C., & Kumar, A. (1996). Quality health care for women: *A global challenge. Health Care for Women International, 17,* 449–467.

Agyepong, I. A. (1992). Malaria: Ethnomedical perceptions and practice in an Adangbe farming community and implications for control. *Social Science and Medicine, 35,* 131–137.

Beljaev, A. E., Sharma, G. K., Brohult, J. A., & Haque, M. A. (1986). Studies on the detection of malaria at Primary Health Centres. Part II. Age and sex composition of patients subjected to blood examination in passive case detection. *Indian Journal of Malariology, 23,* 19–25.

Bonilla, E., & Rodriguez, A. (1993). Determining malaria effects in rural Colombia. Social *Science and Medicine, 37,* 1109.

Boserup, E. (1989). Population, the status of women and rural development. *Population and Development Review, 15*(Suppl), 45–60.

Brugha, R. F., Kevani, J. P., & Swan, V. (1996). An investigation of the role of fathers in immunization uptake. *International Journal of Epidemiology, 25,* 840–845.

Bunting, S. M. (1996). Sources of stigma associated with women with HIV. *ANS Advances in Nursing Science, 19,* 64–73.

Byrne, B. (1996). *Gender, conflict and development.* Brighton: University of Sussex.

CASE (Community Agency for Social Enquiry) (1995). *Let the sky to be the limit.* Soul City Evaluation Report, Janaca Education, Johannesburg.

Dickersin, K., Fredman, L., Flegal, K. M., Scott, J. D., & Crawley, B. (1998). Is there a sex bias in choosing editors? Epi journals as an example. *Journal of the American Medical Association, 280,* 260–264.

Dreyer, G., Noroes, J., & Addis, D. (1996). The silent burden of sexual disability associated with lymphatic filariasis. *Acta Tropica, 63,* 57–60.

Ettling, M. B., Krongthong, T., Krachaklin, S., & Bualombai, P. (1989). Evaluation of malaria clinics in Maesot, Thailand: Use of serology to assess coverage. *Transactions of the Royal Society of Tropical Medicine and Hygiene,* 325–330.

Fischman, R. J., Wick, J. G., & Koenig, B. A. (1999). *The use of "sex" and "gender" to define and characterize meaningful differences between men and women in National Institutes of Health 1999, Agenda for Research on Women's Health in the 21st Century,* vol. 1. Washington: Office of Research on Women's Health.

Fonn, S., & Xaba, M. (1995). *Health workers for change.* Geneva: WHO/TDR.

Fonn, S., Xaba, K., Tint, K. S., Conco, D., & Varkey, S. (1998). Reproductive health services from rhetoric to implementation: South African experience. *Reproductive Health Matters, 6,* 22–32.

Garcia-Moreno, C., & Piza, L. E. (1991). *Gender and primary health care.* Oxfam Gender and Development Unit, Newspack No. 15.

Hartigan, P. (2001). The importance of gender in defining and improving quality of care: Some conceptual issues. *Health Policy and Planning, 16,* 7–12.

Hatcher Roberts, J., & Vlassoff, C. (Eds.) (1995). *The female client and the health provider.* Ottawa: International Development Research Centre.

Heise, L. (1993). Violence against women: The hidden health burden. *World Health Statistics Quarterly, 46*(1), 78–85.

Hudelson, P. (1996). Gender differentials in tuberculosis: The role of socio-economic and cultural factors. *Tubercle and Lung Disease, 77,* 391–400.

Ingram, D., & Hutchinson, S. A. (1999). HIV-positive mothers and stigma. *Health Care for Women International, 20,* 93–103.

International Food Policy Research Institute. (1995). Land inheritance: Are matrilineal systems giving way to patrilineal systems and what are the implications? *Gender CG Newsletter, 1*(2).

International Planned Parenthood Federation. (1995). *Gender awareness for population and development*. An IPPF English Language Bibliography. IPPF, London.

Khan, A., Walley, J., Newell, J., & Imdad, N. (2000). Tuberculosis in Pakistan: Socio-cultural constraints and opportunities in treatment. *Social Science and Medicine, 50*, 247–254.

Kikwawila Study Group. (1995). *WHO/TDR workshop on qualitative research methods*. Report on the field work, Social and Economic Research Project Reports, 14, Special Programme for Research and Training in Tropical Diseases, World Health Organization, Geneva.

Kutzin, J. (1993). *Obstacles to women's access: Issues and options for more effective interventions to improve women's health*. HRO Working Papers No. 13, The World Bank, Washington.

Leslie, J. (1989). Women's time: A factor in the use of child survival technologies? *Health Policy and Planning, 4*, 1–16.

Liefooghe, R. G., Moran, M. B., Habib, S., & De Muynck, A. O. (1997). Treatment adherence of tuberculosis patients in Bethonia Hospital, Sialkot. *Journal of the College of Physicians and Surgeons, Pakistan, 7*, 140–144.

Liverpool School of Tropical Medicine, Department for International Development (DFID) (1999). Guidelines for the analysis of gender and health, Liverpool.

Long, N. H., Joansson, E., Diwan, V. K., & Winkvist, A. (1999). Different tuberculosis in men and women: Beliefs from focus groups in Vietnam. *Social Science and Medicine, 49*, 815–822.

Marks, S. (1994). *Divided sisterhood: Race, class and gender in the South African nursing profession*. New York: St. Martin's Press.

McCarthy, C. R. (1994). Historical background of clinical trials. Involving women and minorities. *Academic Medicine, 69*(9), 695–698.

Mishra, V. N., Malhotra, M., & Gupta, S. (1990). Chronic respiratory disorders in females of Delhi. *Journal of the Indian Medical Association, 88*, 77.

Mwenesi, H. A. (1993). *Mothers definition and treatment of childhood malaria on the Kenyan Coast*. Social and Economic Project Reports No. 13, Special Programme for Research and Training in Tropical Diseases (TDR), WHO, Geneva.

National Institutes of Health. (1999). *Agenda for research on women's health for the 21st century*. Office of Research on Women's Health, Washington.

Okonofua, F. E., Feyisetan, B. J., Davies-Adetugbo, A., & Sanusi, Y. O. (1992). Influence of socioeconomic factors on the treatment and prevention of malaria in pregnant and non-pregnant adolescent girls in Nigeria. *Journal of Tropical Medicine and Hygiene, 95*, 309.

Palmer, C. A., Lush, L., & Zwi, A. B. (1999). The emerging international policy agenda for reproductive health services in conflict settings. *Social Science and Medicine, 49*, 1689–1703.

Palmer, C. A., & Zwi, A. B. (1998). Women, health and humanitarian aid in conflict. *Disasters, 22*, 236–249.

Pan-African Study Group on Onchocercal Skin Disease. (1995). *The importance of onchocercal skin disease*. Applied Field Research Reports No. 1, WHO/TDR, Geneva.

Terborg-Penn, R., & Benton Rushing, A. (Eds.) (1996). *Women in Africa and the African diaspora*. Washington, DC: Howard Universtiy Press.

Rao, S., Garole, V., Walawalkar, S., Khot, S., & Karandikar, N. (1996). Gender differentials in the social impact of leprosy. *Leprosy Review, 67*, 1–10.

Rosenfield, A., & Maine, D. (1985). Maternal mortality – a neglected tragedy: Where is the M in MCH? *The Lancet, 2*, 83–85.

Samuels, T., Stevens, L., Kimmie, Z., Kola, S., Schneider, M., & Barker, K. (1999). *Learning the easy way II: Evaluation of the Soul City III TV and print series*. Community Agency for Social Enquiry (CASE).

Santow, G. (1996). Gender differences in health risks and use of services. In Population Division. (Ed.), *Women and population. Proceedings of the United Nations expert group meeting on population and women*, Gaborone, Botswana, 1992 (pp. 125–140). New York: Department for Economic and Social Information and Policy Analysis, United Nations.

Sowell, R. L., Lowensstein, A., Moneyham, L., Demi, A., Mizuno, Y., & Seals, B. F. (1997). Resources, stigma, and patterns of disclosure in rural women with HIV infection. *Public Health and Nursing, 14*, 302–312.

Swedish International Development Corporation Agency (SIDA). 1997. *Handbook for mainstreaming a gender perspective in the health sector*. Department for Democracy and Social Development, Heath Division: Stockholm.

Tanner, M., & Vlassoff, C. (1998). Treatment seeking behaviour for malaria: A framework based on endemicity, age and gender. *Social Science and Medicine, 46*, 523–532.

UNAIDS. (1998). Gender and HIV/AIDS. UNAIDS Technical Update, UNAIDS, Geneva.

UNFPA. (1999). *Report of the ad hoc committee of the whole of the 21st special sesssion of the general assembly*. Key actions for the further implementation of the Programme of Action of the International Conference on Population and Development, New York.

Velez, D. I., Hendrickx, E., Roman, O., & Agudelo, S. (1997). *Gender and leishmaniasis in Colombia: A redefinition of existing concepts*. Gender and Tropical Diseases Resource Papers No. 3, WHO/TDR, Geneva.

Vlassoff, C. (1992a). Listening to the people: Improving disease control using social science approaches (leading article). *Transactions of the Royal Society of Tropical Medicine and Hygiene, 86*, 465–466.

Vlassoff, C. (1992b). Special Issue: Rapid assessment methods for the control of tropical diseases. *Health Policy and Planning, 7*(1).

Vlassoff, C. (1997). The gender and tropical diseases task force of TDR: Achievements and challenges. *Acta Tropica, 67*, 173–180.

Vlassoff, C., & Bonilla, E. (1994). Gender-related differences in the impact of tropical diseases in women: What do we know? *Journal of Biosocial Science, 26*, 37–53.

Vlassoff, C., & Fonn, S. (in press). Health workers for change: A health planning tool.

Vlassoff, C., Hardy, R., & Rathgeber, E. (1995). *Towards the healthy women counselling guide*. Geneva: WHO/TDR.

Vlassoff, C., Khot, S., & Rao, S. (in press). Double jeopardy: Women and leprosy in India. In M.E. Khan (Ed.) *Work, health and contraception from women's perspectives*. Baroda: Centre for Operations Research and Training.

Vlassoff, C., Weiss, M., Ovuga, E. B. L., Eneanya, C., Titi Nwel, P., Babalola, S. S., Awedoba, A. K., Theophilus, B., Cofie, P., & Shetabi, P. (2000). Gender and the stigma of onchocercal skin disease in Africa. *Social Science and Medicine, 50*, 1353–1368.

Whitehead, M. (1992). The concepts and principles of equity and health. *International Journal of Health Services, 22*, 429–445.

Winch, P. J., Lloyd, L. S., Hoemeke, L., & Leontsini, E. (1994). Vector control at the household level: An analysis of its impact on women. *Acta Tropica, 56*, 327–339.

Wong, G. C., Li, V. C., Burris, M. A., & Xiang, Y. (1995). Seeking women's voices: Setting the context for women's health interventions in two rural counties in Yunnan, China. *Social Science and Medicine, 41*(8), 1147–1157.

World Health Organization, Women's Health Project. (1995). *Health workers for change – a manual*. World Health Organization, Geneva.

World Health Organization. (1997). *Violence against women, WHO/FRH/WHD/97.8*. World Health Organization, Geneva.

Wright, D., & Chew, N. (1996). Women as subjects in clinical research. Applied Clinical Trials, September (pp. 44–52).

Yumkella, F. (1996). *Women, onchocerciasis and Ivermectin in Sierra Leone*. Gender and Tropical Diseases Resource Papers No. 2, WHO/TDR, Geneva.

Equity and Inequality

11

Deprivation and Health

Douglas Black

Above all, a nation cannot last as a money-making mob: it cannot with impunity, – it cannot with existence, – go on despising literature, despising science, despising art, despising nature, despising compassion, and concentrating its soul on Pence.[1]

These words were wrung from John Ruskin as he contemplated the monetarism of Victorian England, with its great gulf fixed between Disraeli's "two nations." In the end, Ruskin went mad. A little over a century later, even though the clock has been determinedly put back in the past decade, we are more fortunate in one respect. Surveying a similar problem with equal compassion, but also with a wealth of scientifically based understanding, Peter Townsend has remained eminently sane. This distinguished sociologist has recently retired from his chair in Bristol; and in September a celebration of his work took place at Dartington Hall. I was pleased and honoured to take part in this tribute to a man who has contributed more than any other to the recognition of socioeconomic deprivation as a great and growing evil with grim effects on health.

Before I turn to the report whose preparation brought the celebrants together, I would like to lay open the roots of my own concern with these matters. As a medical student and young doctor I saw barefoot children in Dundee, and after doing a clinic in Manchester I would bring home a flea several times each year. Even in the lean years after the war, the welfare state improved matters considerably, but by no means completely; but the recurrence of monetarism over the past dozen years has again created evils that

Source: *British Medical Journal*, 307(6919) (1993): 1630–1631.

should have been relegated to history. In Britain we have homelessness and massive unemployment, potent causes of illness, and we have a health service which is being covertly denationalised, at great expense occasioned by the unnecessary creation of an artificial "internal market." (There has been even more rapid regress in the United States – under Ronald Reagan's brand of monetarism it took only eight years for tuberculosis to be re-established as a health hazard on the streets of New York.) In 1974 Peter Townsend was already aware of the dangers inherent in hierarchical corporate management, regarding it as "the largest single threat to free access to health care and the aim of a healthy society."[2] After a lifetime of professional detachment, I am both sad and angry to see the attempted destruction of a system of health care which was as comprehensive as any in the world, and at a lower cost than any other comparable system.

The Black Report . . .

What was so special about the 1980 so called "Black Report" on inequalities in health related to social class? It was certainly not the discovery of a new phenomenon – the association between social deprivation and ill health had been stressed by John Brotherston, chief medical officer in the Scottish Home and Health Department 1964–77, in a Galton lecture[3] and by Peter Townsend himself in his Geoffrey Vickers lecture in 1974.[2] And it was in full awareness of the problem that David Ennals, secretary of state for social services 1976–9 commissioned the study which led to the report. Each member of the group might pick on different reasons to explain the impact of the report. I would like to suggest a couple.

Firstly, the group, having met at fairly frequent intervals over a three year period and enjoying the services of two qualified research assistants and the back up of a large Department of State, was able to assemble a mass of relevant statistical information which compelled attention. In this, we were greatly assisted by the comprehensive information on health status compiled by the Office of Population Censuses and Surveys. It is also much to the credit of the departmental civil servants that we enjoyed the same support, even after a change of government to one which did not find our activities or conclusions especially palatable.

Secondly, we were able to identify a number of questions that were both important and open to research. We drew attention to several imperfections in the then available indices of health and of social status; much better indices have now been developed, and the search for causal mechanisms (which must surely be multiple) has been made easier by the study, using the newer indices, of subnational populations defined by postcode or municipal wards. Such studies confirm the constancy of the association between deprivation and ill health; and they may also bring to light potentially causal factors which might not appear so clearly in national statistics. In all this work, conveniently brought together in the 1992 edition of *The Health Divide*,[4] Peter Townsend has been a major worker, and a stimulant of work in others.

. . . and After

The differences in mortality associated with social deprivation are considerable, and they are at their greatest in the very early stages of life. One of our prime recommendations was that children be given a better start in life. We based this on the long term duration of any improvement in health which could be achieved by better maternity or infant care; but that is not the whole story. More recent studies by Marmot in relation to heart disease[5] and by Barker more generally[6] support the generalisation in the Court report that childhood illness casts long shadows ahead, producing impairment of health years in the future.[7]

In seeking for an explanation of the association between social deprivation and ill health we recognised that on occasion the onset of ill health could lead to impoverishment, and also that shifts of individuals between classes could be statistically confounding. We did, however, adopt the position that in the main it was social deprivation which was the independent variable, damaging health in a variety of ways, acting differently at different stages of life. For children, important factors were lack of safe play areas, domestic overcrowding, and lack appropriate stimulation. For workers, inappropriate living conditions and deleterious lifestyles (some of which are shamefully stimulated by advertising) may have their effects made worse by specific hazards of particular occupations, which may be both poorly paid and intrinsically hazardous. The accumulated burden of a socially deprived life continues to oppress the diminished band of those who survive into old age. Many details have since been added to the picture, but they have in general strengthened its outlines, rather than demanded any radical revision.

Another matter on which we had some discussion within the group was the relative importance, in alleviating the health effects of social deprivation, of measures which could be broadly called social, and of measures directly related to health services. Our view, that poverty and its effects was the root cause of the ill health associated with it, naturally led us to advocate a wider strategy of social measures. After all, the surest way to alleviate the effects of poverty must be to alleviate poverty itself. But since the millenium is not at hand, except perhaps in the most formal sense, it is also worth considering the value of improving health care, even if only as a palliative. At the time of the report there was little quantitative evidence to suggest that specifically medical measures might appreciably reduce mortality at the population level, though their potential to help in individual episodes of illness was well recognised. It has now, however, been shown in several countries that mortality due to those diseases for which curative measures are available is falling more rapidly than mortality from those diseases which are not yet open to medical intervention.[8] Common sense might suggest this would be so, but it is nice now to have some figures.

I think that Peter Townsend might agree with me that if we are serious in seeking to diminish morbidity and mortality due to social deprivation, we must return to the values of the welfare state and pursue them with greater determination.

References

1. Ruskin J. *Sesame and lilies*. London: Smith, Elder and Co, 1865.
2. Townsend P. Inequality and the health service. *Lancet* 1974;i:1179–90.
3. Brotherson J. Inequality: is it inevitable? In: Peel J, Carter CO, eds. *Equalities and Inequalities in Health*. London: Academic Press, 1976.
4. Whitehead M. *The health divide*. 2nd ed. London: Penguin, 1992.
5. Marmot MG, Shipley MJ, Rose G. Inequalities in death – specific explanations of a general pattern? *Lancet* 1984;i:1003–6.
6. Barker DJP. The fetal and infant origins of adult disease. *BMJ* 1990;301:1111.
7. Department of Health and Social Security. *Health Services development: Court report on child health services*. London: HMSO, 1978.
8. Boys RJ, Forster DP, Jojan P. Mortality from causes amenable and non-amenable to medical care: the experience of eastern Europe. *BMJ* 1991;303:879–83.

12

Development as Capability Expansion

Amartya Sen

Introduction

In his *Grundlegung zur Metaphysik de Sitten*, Immanuel Kant argues for the necessity of seeing human beings as ends in themselves, rather than as means to other ends: "So act as to treat humanity, whether in thine own person or in that of any other, in every case as an end withal, never as means only."[1] This principle has importance in many contexts – even in analysing poverty, progress and planning. Human beings are the agents, beneficiaries and adjudicators of progress, but they also happen to be – directly or indirectly – the primary means of all production. This dual role of human beings provides a rich ground for confusion of ends and means in planning and policy-making. Indeed, it can – and frequently does – take the form of focusing on production and prosperity as the essence of progress, treating people as the means through which that productive progress is brought about (rather than seeing the lives of people as the ultimate concern and treating production and prosperity merely as means to those lives).

Indeed, the widely prevalent concentration on the expansion of real income and on economic growth as the characteristics of successful development can be precisely an aspect of the mistake against which Kant had warned. This problem is particularly pivotal in the assessment and planning of economic development. The problem does not, of course, lie in the fact that the pursuit of economic prosperity is typically taken to be a major goal of planning and policy-making. This need not be, in itself, unreasonable. The problem relates to the level at which this aim should be taken as a goal. Is it just an intermediate goal, the importance of which is contingent on what it ultimately contributes to human lives? Or is it the object of the entire

Source: *Journal of Development Planning*, 19 (1989): 41–58.

Table 1: Economic prosperity and life expectancy, 1985

Country	GNP per capita	Life expectancy at birth
China	310	69
Sri Lanka	380	70
Brazil	1640	65
South Africa	2010	55
Mexico	2080	67
Oman	6730	54

Source: World Development Report 1987 (New York, Oxford University Press, 1988), table 1.

exercise? It is in the acceptance – usually implicitly – of the latter view that the ends – means confusion becomes significant – indeed blatant.

The problem might have been of no great practical interest if the achievement of economic prosperity were tightly linked – in something like a one-to-one correspondence – with that of enriching the lives of the people. If that were the case, then the pursuit of economic prosperity as an end in itself, while wrong in principle, might have been, in effect, indistinguishable from pursuing it only as a means to the end of enriching human lives. But that tight relation does not obtain. Countries with high GNP per capita can nevertheless have astonishingly low achievements in the quality of life, with the bulk of the population being subject to premature mortality, escapable morbidity, overwhelming illiteracy and so on.

Just to illustrate an aspect of the problem, the GNP per capita of six countries is given in table 1, along with each country's respective level of life expectancy at birth.

A country can be very rich in conventional economic terms (i.e., in terms of the value of commodities produced per capita) and still be very poor in the achieved quality of human fife. South Africa, with five or six times the GNP per capita of Sri Lanka or China, has a much lower longevity rate, and the same applies in different ways to Brazil, Mexico, Oman, and indeed to many other countries not included in this table.

There are, therefore, really two distinct issues here. First, economic prosperity is no more than one of the means to enriching lives of people. It is a foundational conclusion to give it the status of an end. Secondly, even as a means, merely enhancing average economic opulence can be quite inefficient in the pursuit of the really valuable ends. In making sure that development planning and general policy-making do not suffer from costly confusions of ends and means, we have to face the issue of identification of ends, in terms of which the effectiveness of the means can be systematically assessed.

This paper is concerned with discussing the nature and implications of that general task.

The Capability Approach: Conceptual Roots

The particular line of reasoning that will be pursued here is based on evaluating social change in terms of the richness of human life resulting from it. But the quality of human life is itself a matter of great complexity. The approach that will be used here, which is sometimes called the "capability approach", sees human life as a set of "doings and beings" – we may call them "functionings" – and it relates the evaluation of the quality of life to the assessment of the capability to function. It is an approach that I have tried to explore in some detail, both conceptually and in terms of its empirical implications.[2] The roots of the approach go back at least to Adam Smith and Karl Marx, and indeed to Aristotle.

In investigating the problem of "political distribution", Aristotle made extensive use of his analysis of "the good of human beings", and this he linked with his examination of "the functions of man" and his exploration of "life in the sense of activity".[3] The Aristotelian theory is, of course, highly ambitious and involves elements that go well beyond this particular issue (e.g., it takes a specific view of human nature and relates a notion of objective goodness to it). But the argument for seeing the quality of life in terms of valued activities and the capability to achieve these activities has much broader relevance and application.

Among the classical political economists, both Adam Smith and Karl Marx explicitly discussed the importance of functionings and the capability to function as determinants of well-being.[4] Marx's approach to the question was closely related to the Aristotelian analysis (and indeed was apparently directly influenced by it).[5] Indeed, an important part of Marx's programme of reformulation of the foundations of political economy is clearly related to seeing the success of human life in terms of fulfilling the needed human activities. Marx put it thus: "It will be seen how in place of the wealth and poverty of political economy come the rich human being and rich human need. The rich human being is simultaneously the human being in need of a totality of human life-activities – the man in whom his own realization exists as an inner necessity, as need."[6]

Commodities, Functionings and Capability

If life is seen as a set of "doings and beings" that are valuable, the exercise of assessing the quality of life takes the form of evaluating these functionings and the capability to function. This valuational exercise cannot be done by focusing simply on commodities or incomes that help those doings and beings, as in commodity-based accounting of the quality of life (involving a confusion of means and ends). "The life of money-making", as Aristotle put it, "is one undertaken under compulsion, and wealth is evidently not the good we are seeking; for it is merely useful and for the sake of something else."[7] The task is that of evaluating the importance of the various functionings in human life, going beyond what Marx called, in a different but related context, "commodity fetishism".[8] The functionings themselves have to be examined, and the capability of the person to achieve them has to be appropriately valued.

In the view that is being pursued here, the constituent elements of life are seen as a combination of various different functionings (a "functioning n-tuple"). This amounts to seeing a person in as it were, an "active" rather than a "passive" form (but neither the various states of being nor even the "doings" need necessarily be "athletic" ones). The included items may vary from such elementary functionings as escaping morbidity and mortality, being adequately nourished, undertaking usual movements etc., to many complex functionings such as achieving self-respect, taking part in the life of the community and appearing in public without shame (the last a functioning that was illuminatingly discussed by Adam Smith[9] as an achievement that is valued in all societies, but the precise commodity requirement of which, he pointed out, varies from society to society). The claim is that the functionings are constitutive of a person's being, and an evaluation of a person's well-being has to take the form of an assessment of these constituent elements.

The primitive notion in the approach is that of functionings – seen as constitutive elements of living. A functioning is an achievement of a person: what he or she manages to do or to be, and any such functioning reflects, as it were, a part of the state of that person. The capability of a person is a derived notion. It reflects the various combinations of functionings (doings and beings) he or she can achieve.[10] It takes a certain view of living as a combination of various "doings and beings". Capability reflects a person's freedom to choose between different ways of living. The underlying motivation – the focusing on freedom – is well captured by Marx's claim that what we need is "replacing the domination of circumstances and chance over individuals by the domination of individuals over chance and circumstances".[11]

Utilitarian Calculus versus Objective Deprivation

The capability approach can he contrasted not merely with commodity-based systems of evaluation, but also with the utility-based assessment. The utilitarian notion of value, which is invoked explicitly or by implication in much of welfare economics, sees value, ultimately, only in individual utility, which is defined in terms of some mental condition, such as pleasure, happiness, desire-fulfilment. This subjectivist perspective has been extensively used, but it can be very misleading, since it may fail to reflect a person's real deprivation.

A thoroughly deprived person leading a very reduced life, might not appear to be badly off in terms of the mental metric of utility, if the hardship is accepted with non-grumbling resignation. In situations of long-standing deprivation, the victims do not go on weeping all the time, and very often make great efforts to take pleasure in small mercies and to cut down personal desires to modest – "realistic" – proportions. The person's deprivation, then, may not at all show up in the metrics of pleasure, desire-fulfilment etc., even though he or she may be quite unable to be adequately nourished, decently clothed, minimally educated and so on.[12]

This issue, apart from its foundational relevance, may have some immediate bearing on practical public policy. Smugness about continued deprivation and vulnerability is often made to look justified on grounds of lack of strong public demand and forcefully expressed desire for removing these impediments.[13]

Ambiguities, Precision and Relevance

There are many ambiguities in the conceptual framework of the capability approach. Indeed, the nature of human life and the content of human freedom are themselves far from unproblematic concepts. It is not my purpose to brush these difficult questions under the carpet, In so far as there are genuine ambiguities in the underlying objects of value, these will be reflected in corresponding ambiguities in the characterization of capability. The need for this relates to a methodological point, which I have tried to defend elsewhere, that if an underlying idea has an essential ambiguity, a precise formulation of that idea must try to capture that ambiguity rather than attempt to lose it.[14] Even when precisely capturing an ambiguity proves to be a difficult exercise, that is not an argument for forgetting the complex nature of the concept and seeking a spuriously narrow exactness. In social investigation and measurement, it is undoubtedly more important to be vaguely right than to be precisely wrong.[15]

It should be noted also that there is always an element of real choice in the description of functionings, since the format of "doings" and "beings" permits additional "achievements" to be defined and included. Frequently, the same doings and beings can be seen from different perspectives, with varying emphases. Also, some functionings may be easy to describe, but of no great interest in the relevant context (e.g., using a particular washing powder in doing the washing).[16] There is no escape from the problem of evaluation in selecting a class of functionings as important and others not so. The evaluative exercise cannot be fully addressed without explicitly facing questions concerning what are the valuable achievements and freedoms, and which are not. The chosen focus has to be related to the underlying social concerns and values, in terms of which some definable functionings and capabilities may be important and others quite trivial and negligible. The need for selection and discrimination is neither an embarrassment nor a unique difficulty for the conceptualization of functioning and capability.[17]

In the context of some types of welfare analysis, for example, in dealing with extreme poverty in developing economies, we may be able to go a long distance in terms of a relatively small number of centrally important functionings and the corresponding capabilities, such as the ability to be well-nourished and well-sheltered, the capability of escaping avoidable morbidity and premature mortality and so forth.[18] In other contexts, including more general problems of assessing economic and social development, the list may have to be much longer and much more diverse.[19] The task of specification must relate to the underlying motivation of the exercise as well as dealing with the social values involved.

Quality of Life, Basic Needs and Capability

There is an extensive literature in development economics concerned with valuing the quality of life, the fulfilment of Basic needs and related matters.[20] That literature has been quite influential in recent years in drawing attention to neglected aspects of economic and social development. It is, however, fair to say that these writings have been typically comprehensively ignored in the theory of welfare economics, which has tended to treat these contributions as essentially *ad hoc* suggestions. This treatment is partly the result of the concern of welfare theory that proposals should not just appeal to intuitions but also be structured and founded. It also reflects the intellectual standing that such traditional approaches as utilitarian evaluation enjoy in welfare theory, and which serves as a barrier to accepting departures even when they seem attractive. The inability of utility-based evaluations to cope with persistent deprivations was discussed earlier, but in the welfare-economic literature the hold of this tradition has been hard to dislodge.

The charge of "*ad hoc*-ness" against the development literature relates to the different modes of arguing that are used in welfare theory and in development theory. As far as the normative structure is concerned, the latter tends to be rather immediate, appealing to strong intuitions that seem obvious enough. Welfare theory, on the other hand, tends to take a more circuitous route, with greater elaboration and defence of the foundations of the approach in question. To bridge the gap, we have to compare and contrast the foundational features underlying the concern with quality of life, Basic needs etc. with the informational foundations of the more traditional approaches used in welfare economics and moral philosophy, such as utilitarianism. It is precisely in this context that the advantages of the capability approach become perspicuous. The view of human life seen as a combination of various functionings and capabilities, and the analysis of human freedom as a central feature of living, provide a differently grounded foundational route to the evaluative exercise. This informational foundation contrasts with the evaluative bases incorporated in the more traditional foundations used in welfare economics.[21]

The "basic needs" literature has, in fact, tended to suffer a little from uncertainties about how basic needs should be specified. The original formulations often took the form of defining basic needs in terms of needs for certain minimal amounts of essential commodities such as food, clothing and shelter. If this type of formulation is used, then the literature remains imprisoned in the mould of commodity-centred evaluation, and tan in fact be accused of adopting a form of "commodity fetishism". The objects of value tan scarcely be the holdings of commodities. Judged even as means, the usefulness of the commodity-perspective is severely compromised by the variability of the conversion of commodities into capabilities. For example, the requirement of food and of nutrients for the capability of being well-nourished may greatly vary from person to person depending on metabolic rates, body size, gender, pregnancy, age, climatic conditions, parasitic ailments and so on.[22] The evaluation of commodity-holdings or of incomes (with which to purchase commodities) tan be at best a proxy for the things that really mutter, but unfortunately it does not seem to be a particularly good proxy in most cases.[23]

Rawls, Primary Goods and Freedoms

The concern with commodities and means of achievement, with which the motivation of the capability approach is being contrasted happens to be, in fact, influential in the literature of modern moral philosophy as well. For example, in John Rawls' outstanding book on justice (arguably the most important contribution to moral philosophy in recent decades), the concentration is on the holdings of "primary goods" of different people in making interpersonal comparisons. His theory of justice, particularly the "difference principle" is dependent on this procedure for interpersonal comparisons. This procedure has the feature of being partly commodity-based, since the list of primary goods includes "income and wealth", in addition to "the basic liberties", "powers and prerogatives of offices and positions of responsibility", "social bases of self-respect" and so on.[24]

Indeed, the entire list of "primary goods" of Rawls is concerned with means rather than ends; they deal with things that help to achieve what we want to achieve, rather than either with achievement as such or even with the freedom to achieve. Being nourished is not a part of the fist, but having the income to buy food certainly is. Similarly, the social bases of self-respect belong to the list in a way self-respect as such does not.

Rawls is much concerned that the fact that different people have different ends must not be lost in the evaluative process, and people should have the freedom to pursue their respective ends. This concern is indeed important, and the capability approach is also much involved with valuing freedom as such. In fact, it can be argued that the capability approach gives a better account of the freedoms actually enjoyed by different people than can be obtained from looking merely at the holdings of primary goods. Primary goods are means to freedoms, whereas capabilities are expressions of freedoms themselves.

The motivations underlying the Rawlsian theory and the capability approach are similar, but the accountings are different. The problem with the Rawlsian accounting lies in the fact that, even for the same ends, people's ability to convert primary goods into achievements differs, so that an interpersonal comparison based on the holdings of primary goods cannot, in general, also reflect the ranking of their respective real freedoms to pursue any given – or variable – ends. The variability in the conversion rates between persona for given ends is a problem that is embedded in the wider problem of variability of primary goods needed for different persona pursuing their respective ends.[25] Hence, a similar criticism applies to Rawlsian accounting procedure as applies to parts of the basic needs literature for their concentration on means (such as commodities) as opposed to achievements or the freedom to achieve.

Freedom, Capability and Data Limitations

The capability set represents a person's freedom to achieve various functioning combinations. If freedom is intrinsically important, then the alternative combinations available for choice are all relevant for judging a person's advantage, even though he or she will eventually choose only an alternative. In this view, the choice itself is a valuable feature of a person's life.

On the other hand, if freedom is seen as being only instrumentally important, then the interest in the capability set lies only in the fact that it offers the person opportunities to achieve various valuable states. Only the achieved states are in themselves valuable, not the opportunities, which are valued only as means to the end of reaching valuable states.

The contrast between the intrinsic and the instrumental views of freedom is quite a deep one, and I have discussed the importance of the distinction elsewhere.[26] Both views can be accommodated within the capability approach. With the instrumental view, the capability set is valued only for the sake of the best alternative available for choice (or the actual alternative chosen). This way of evaluating a capability set by the value of one distinguished element in it can be called "elementary evaluation".[27] If, on the other hand, freedom is intrinsically valued, then elementary evaluation will be inadequate since the opportunity to choose other alternatives is of significance of its own. To bring out the distinction, it may be noted that if all alternatives other than the chosen alternative were to become unavailable, then there would be a real loss in the case of the intrinsic view, but not in the instrumental, since the alternative chosen is still available.

In terms of practical application, the intrinsic view is much harder to reflect than the instrumental view, since our direct observations relate to what was chosen and achieved. The estimation of what could have been chosen is, by its very nature, more problematic (involving, in particular, assumptions about the constraints actually faced by the person). The limits of practical calculations are set by data restrictions, and this can be particularly hard on the representation of capability sets in full, as opposed to judging the capability sets by the observed functioning achievements.

There is no real loss involved in using the capability approach in this reduced form if the instrumental view of freedom is taken, but there is loss if the intrinsic view is accepted. For the latter, a representation of the capability set as such is important.

In fact, neither the instrumental view nor the intrinsic view is likely to be fully adequate. Certainly, freedom is a means to achievement, whether or not it is also intrinsically important, so that the instrumental view must be *inter alia* present in any use of the capability approach. Also, even if we find in general the instrumental view to be fairly adequate, there would clearly be cases in which it is extremely limited. For example, the person who fasts, that is, starves out of choice, can hardly be seen as being similarly deprived as a person who has no option but to starve because of penury. Even though their observed functionings may be the same, at least in the crude representation of functionings, their predicaments are not the same.

In practice, even if in general the capability approach is used in the reduced form of concentrating on the chosen functioning combination, some systematic supplementation would be needed to take care of cases a which the freedom enjoyed is of clear and immediate interest. There may be no great difficulty in doing this supplementation in many cases, once the problem is posed clearly enough and the data search is made purposive and precise. Sometimes it would be useful to redefine the functionings in what is called a "refined" way, to take note of some of the obviously relevant alternatives that were available, but not chosen. Indeed, fasting is an example of a "refined" functioning, and contrasts with the unrefined functioning of "starving", which does not specify whether or not this was by choice.[28] The important issue does not concern the existence or not of some actual word (such as fasting) that reflects the refined functioning (that is largely a matter of linguistic convention), but assessing whether or not such refining would be central to the exercise in question, and if central, deciding how this might be done.

As a matter of fact, the informational base of functionings is still a much finer basis of evaluation of the quality of life and economic progress than various alternatives more commonly recommended, such as individual utilities or commodity holdings. The commodity fetishism of the former and the subjectivist metric of the latter make them deeply problematic. Thus, the concentration on achieved functionings has merits over the feasible rivals (even though it may not be based on as much information as would be needed to attach intrinsic importance to freedom). And in terms of data availability, keeping track of functionings (including vital ones such as being well-nourished and avoiding escapable morbidity or premature mortality) is typically no harder – often much easier – than getting data on commodity use (especially divisions within the family), not to mention utilities.

The capability approach can, thus, be used at various levels of sophistica-tion, and how far we can go would depend much on the practical consideration of what data we can get and what we cannot. In so far as freedom is seen to be intrinsically important, the observation of the chosen functioning bundle cannot be in itself an adequate guide for the evaluative exercise, even though the freedom to choose a better bundle rather than a worse one can be seen to be, in some accounting, an advantage even from the perspective of freedom.[29]

The point can be illustrated with a particular example. An expansion of longevity is seen, by common agreement, as an enhancement of the quality of life (though, strictly speaking, I suppose one can think of it as an enhance-ment of the quantity of life). This is so partly because living longer is an achievement that is valued. It is also partly because other achievements, such as avoiding morbidity, tend to go with longevity (and thus longevity serves also as a proxy for some achievements that too are intrinsically valued). But greater longevity can also be seen as an enhancement of the freedom to live long. We often take this for granted on the solid ground that given the option, people value living longer, and thus the observed achievement of living longer reflects a greater freedom than was enjoyed.

The interpretative question arises at this precise point. Why is it evidence of greater freedom as such that a person ends up living longer rather than shorter? Why can it not be just a preferred achievement, but involving no difference in terms of freedom? One answer is to say that one always does have the option of killing, oneself, and thus an expansion of longevity expands one's options. But there is a further issue here. Consider a case in which, for some reason (either legal or psychological or whatever), one cannot really kill oneself (despite the presence in the world of poisons, knives, tall buildings and other useful objects). Would we then say that the person does not have more freedom by virtue of being free to live longer though not shorter? It can be argued that if the person values, prefers and wishes to choose living longer, then the change in question is in fact an expansion of the person's freedom, since the evaluation of freedom cannot be dissociated from the assessment of the actual options in terms of the person's evaluative judgments.[30]

The idea of Freedom takes us beyond achievements, but that does not entail that the assessment of freedom must be independent of that of achievements. The freedom to live the kind of life one would like to me has importance that the freedom to live the kind of life one would hate to have does not. Thus, the temptation to see more freedom in greater longevity is justifiable from several points of view, including noting the option of ending one's life and being sensitive to the evaluative structure of achievements which directly affect the metric of freedom. The bottom line of all this is to recognize that the use of the capability approach even in the reduced form of concentrating on the achieved functionings (longevity, absence of morbid-ity, avoidance of undernourishment etc.) may give more role to the value of freedom than might have been initially apparent.

Inequality, Class and Gender

The choice of an approach to the evaluation of well-being and advantage has bearings on many exercises. These include the assessment of efficiency as well as inequality. Efficiency, as it is formally defined, is concerned with noting overall improvements, and in standard economic theory, this takes the form of checking whether someone's position has improved without anyone's position having gone down. A situation is efficient if and only if there is no alternative feasible situation in which someone's position is better and no one's worse. Obviously, the content of this criterion depends crucially on the way individual advantage is defined. If it is defined in terms of utility, then this criterion of efficiency immediately becomes that of "Pareto optimality" (or "Pareto efficiency", as it is sometimes – more accurately – called), On the other hand, efficiency can be defined also in term of other metrics, including that of the quality of life based on the evaluation of functionings and capabilities.

Similarly, the assessment of inequality too depends on the chosen indicator of individual advantage. The usual inequality measures that can be found in empirical economic literatures tend to concentrate on inequalities of incomes or wealth.[31] These are valuable contributions. On the other hand, in so far as income and wealth do not give adequate account of quality of life, there is a case for baling the evaluation of inequality on information more closely related to living standards.

Indeed, the two informational bases are not alternatives. Inequality of wealth may tell us things about the generation and persistence of inequalities of other types, even when our ultimate concern may be with inequality of living standard and quality of life. Particularly in the context of the continuation and stubbornness of social divisions, information on inter-class inequalities in wealth and property ownership is especially crucial. But this recognition does not reduce the importance of bringing in indicators of quality of life to assess the actual inter-class inequalities of well-being and freedom.

One field in which inequalities are particularly hard to assess is that of gender differences. There is a great deal of general evidence to indicate that women often have a much worse deal than men do, and that girls are often much more deprived than boys. These differences may be reflected in many subtle as well as crude ways, and in various forms they can be observed in different parts of the world – among both rich and poor countries. However, it is not easy to determine what is the best indicator of advantage in terms of which these gender inequalities are to be examined. There is, to be sure, no need to look for one specific metric only, and the need for plurality of indicators is as strong here as in any other field. But there is still an issue of the choice of approach to well-being and advantage in the assessment of inequalities between women and men.

The approach of utility-based evaluation is particularly limiting in this context, since the unequal deals that obtain, particularly within the family,

are often made "acceptable" by certain social notions of "normal" arrangements, and this may affect the perceptions of women as well as men of the
comparative levels of well-being they respectively enjoy. For example, in the
context of some developing countries such as India, the point has been made
that rural women may have no clear perception of being deprived of things
that men have, and may not be in fact any more unhappy than men are.
This may or may not be the case, but even if it were so, it can be argued that
the mental metric of utility may be particularly inappropriate for judging
inequality in this context. The presence of objective deprivation in the form
of greater undernourishment, more frequent morbidity, lower literacy etc.
cannot be rendered irrelevant just by the quiet and ungrumbling acceptance
of women of their deprived conditions.[32]

In rejecting utility-based evaluations, it may be tempting to go in the
direction of actual commodities (enjoyed by women and men, respectively)
to check inequalities between them. There is here the problem, already discussed earlier in this paper, that commodity-based evaluations are inadequate
because commodities are merely means to well-being and freedom and do
not reflect the nature of the lives that the people involved can lead. But, in
addition, there is the further problem that it is hard – sometimes impossible –
to get information on how the commodities belonging to the family are
divided between men and women, and between boys and girls.

For example, studies on the division of food within the family tend to be
deeply problematic since the observation needed to see who is eating how
much is hard to carry out with any degree of reliability. On the other hand, it
is possible to compare signs of undernourishment of boys and girls, to check
their respective morbidity rates etc., and these functioning differences are
both easier to observe and of greater intrinsic relevance.[33]

There are indeed inequalities between men and women in terms of
functionings, and in the context of developing countries the contrast may
be sharp even in basic matters of life and death, health and illness, education and illiteracy. For example, despite the fact that when men and women
are treated reasonably equally in terms of food and health care (as they
tend to be in the richer countries, even though gender biases may remain
in other – less elementary – fields), women seem to have a greater ability
to survive than men, in the bulk of the developing economies, men out-
number women by large margins. While the ratio of females to males in
Europe and North America tends to be about 1.06 or so, that ratio is below
0.95 for the Middle East (including countries in Western Asia and North
Africa), South Asia (including India, Pakistan and Bangladesh) and China.[34]
This crude figure of the ratio of survived females to survived males already
tells a story that has much informational value in judging inter-gender inequalities. Sometimes there are sharp contrasts even within a country (e.g.,
the ratio of females to males varies within India all the way from 1.03 in
Kerala to 0.87 or 0.88 in Haryana and Punjab). From the point of view of

studying both the actual situations and the causal influences operating in the generation of inter-gender inequalities, these regional contrasts may be particularly important.

Being able to survive is of course only one capability (though undoubtedly a very Basic one), and other comparisons can be made with information on health, morbidity etc. The ability to read and write is also another important capability, and here it can be seen that the ratio of female to male literacy rates is often shockingly low in different parts of the world. The combined effects of low literacy rates in general (a deprivation of a basic capability across genders) and gender inequalities in literacy rates (unequal deprivation of this basic capability for women) tend to be quite disastrous denials for women. It appears that even leaving out many countries for which no reliable data exist, in a great many countries in the world, the female literacy rate is still below 50 per cent. In fact, it is below even 30 per cent for as many as 26 countries, below 20 per cent for 16 and below 10 per cent in at least five.[35]

In general, the perspective of functionings and capabilities provides a plausible approach to examining inter-gender inequalities. It does not suffer from the type of subjectivism that makes utility-based accounting particularly obtuse in dealing with entrenched inequalities. Nor does it suffer from the over-concentration on means that commodity-based accounting undoubtedly does, and in fact it has better informational sources in studying inequalities within the family than is provided by guesswork on commodity distributions (e.g., who is eating how much?). The case of inter-gender inequality is, of course, only one illustration of the advantages that the capability approach has. But it happens to be an illustration that is particularly important on its own as well, given the pervasive and stubborn nature of inequalities between women and men in different parts of the world.

Conclusion

The assessment of achievement and advantage of members of the society is a central part of development analysis. In this paper, I have tried to discuss how the capability approach may be used to substantiate the evaluative concerns of human development. The focus on human achievement and freedom, and on the need for reflective – rather than mechanical – evaluation, is an adaptation of an old tradition that can be fruitfully used in providing a conceptual basis for analysing the tasks of human development in the contemporary world. The foundational importance of human capabilities provides a firm basis for evaluating living standards and the quality of life, and also points to a general format in terms of which problems of efficiency and equality can both be discussed.

The concentration on distinct capabilities entails, by its very nature, pluralist approach. Indeed, it points to the necessity of seeing development as a combination of distinct processes, rather than as the expansion of some apparently homogeneous magnitude such as real income or utility. The things that people value doing or being can be quite diverse, and the valuable capabilities vary from such elementary freedoms as being free from hunger and undernourishment to such complex abilities as achieving self-respect and social participation. The challenge of human development demands attention being paid to a variety of sectoral concerns and a combination of social and economic processes.

In the collection of papers of which this one is a part, there are a number of specific studies dealing with such matters as education, health and nutrition, as well as the processes of agricultural expansion and industrial development. The problems of resource mobilization and participatory development are also addressed. Some of the subjects thus covered deal with variables that are direct determinants of human capability (e.g., education and health), while others relate to instrumental influences that operate through economic or social processes (e.g., the promotion of agricultural and industrial productivity). The uniting feature is the motivating concern with human development and its constitutive characteristics.

In the distinction between functionings and capabilities, emphasis was placed on the importance of having the freedom to choose one kind of life rather than another. This is an emphasis that distinguishes the capability approach from any accounting of only realized achievements. However, the ability to exercise freedom may, to a considerable extent, be directly dependent on the education we have received, and thus the development of the educational sector may have a foundational connection with the capability-based approach.

In fact, educational expansion has a variety of roles that have to be carefully distinguished. First, more education can help productivity. Secondly, wide sharing of educational advancement can contribute to a better distribution of the aggregate national income among different people. Thirdly, being better educated can help in the conversion of incomes and resources into various functionings and ways of living. Last (and by no means the least), education also helps in the intelligent choice between different types of lives that a person can lead. All these distinct influences can have important bearings on the development of valuable capabilities and thus on the process of human development.

There are also other interconnections between the different areas covered in the collection; for example, good health is an achievement in itself and also contributes both to higher productivity and to an enhanced ability to convert incomes and resources into good living. In focusing on human capabilities as the yardstick in terms of which successes and failures of human development are to be judged, attention is particularly invited to addressing these social interconnections. Given clarity regarding the ends (avoiding, in particular, the pitfall of treating human beings as means), the social and economic instrumentalities involved in the ends-means relations can be extensively explored.

One of the most important tasks of an evaluative system is to do justice to our deeply held human values. The challenge of "human development in the 1980s and beyond" cannot be fully grasped without consciously facing this issue and paying deliberate attention to the enhancement of those freedoms and capabilities that matter most in the lives that we can lead. To broaden the limited lives into which the majority of human beings are willy-nilly imprisoned by force of circumstances is the major challenge of human development in the contemporary world. Informed and intelligent evaluation both of the lives we are forced to lead and of the lives we would be able to choose to lead through bringing about social changes is the first step in confronting that challenge. It is a task that we must face.

Notes

1. *Grundlegung* (1785), sect. II; English translation *Fundamental Principles of the Metaphysics of Morals*, in *Kant's Critique of Practical Reason and Other Works on the Theory of Ethics*, 6th edition, T. K. Abbot, ed. (London, Longmans, 1909), p. 47.
2. Amartya Sen, "Equality of what?", in *Tanner Lectures on Human Values*. S. M. McMurring, ed., vol. I (Cambridge, Cambridge University Press, 1980 reprinted in *Choice, Welfare and Measurement* (Oxford, Blackwell; and Cambridge, Massachusetts, MIT Press, 1982)); *Resources, Values and Development* (Oxford, Blackwell; and Cambridge, Massachusetts Harvard University Press, 1984); *Commodities and Capabilities* (Amsterdam, North-Holland, 1985); "Well-being, agency and freedom: the Dewey lectures 1984", *Journal of Philosophy*, 82 (April 1985); and "Capability and well-being", WIDER conference paper, 1988.
3. Aristotle, *The Nicomachean Ethics*, book I, sect. 7; in the translation by David Ross, *World's Classics* (Oxford, Oxford University Press, 1980), pp. 12–14. Note that Aristotle's term "eudaimonia", which is often misleadingly translated simply as "happiness", stands for fulfilment of life in a way that goes well beyond the utilitarian perspective. Though pleasure may well result from fulfilment, that is seen as a consequence rather than the cause of valuing that fulfilment. For an examination of the Aristotelian approach and its relation to recent works on functionings and capabilities, see Martha Nussbaum "Nature, function and capability: Aristotle on political distribution", *Oxford Studies in Ancient Greek Philosophy*, supplementary volume 1988.
4. See Adam Smith *An Inquiry* into *the Nature and Causes of the Wealth of Nations* (1776), vol. I, book V sect. II; republished, R. H Campbell and A. S. Skinner, eds. (Oxford, Clarendon Press 1976), pp. 869–872; and Karl Marx, *Economic and Philosophic Manuscripts of 1844* (1844); English translation (Moscow Progressive Publishers, 1977).
5. See G. E. M. de Sainte Croix, *The Class Struggle in the Ancient Greek World* (London, Duckworth, 1981); and Martha Nussbaum, "Nature, function and capability . . ."
6. Karl Marx, *Economic and Philosophic Manuscripts of 1844* . . .
7. Aristotle, *op. cit.*, book I, sect. 5; in the translation by David Ross, p. 7.
8. Karl Marx, *Capital*, vol. I, English translation by S. Moore and E. Aveling (London, Sonnenschein 1887), chap. 1, sect. 4, pp. 41–55; see also Karl Marx, *Economic and Philosophic Manuscripts of 1844* . . .
9. See Adam Smith *op. cit.*, vol. II, book V, chap. II (section entitled "Taxes upon Consumable Commodities"); republished . . . , pp. 469–471.

10. There are several technical problems in the representation of functioning n-tuples and of capability as a set of alternative functioning n-tuples, any one n-tuple of which a person can choose. In this paper, I shall not be particularly concerned with these formal matters, for which see *Commodities and Capabilities* . . . , especially chaps. 2, 4 and 7.

11. Karl Marx and Friedrich Engels, *The German Ideology* (1846). The quoted passage is taken from the translation by David McLellan, *Karl Marx: Selected Writings* (Oxford, Oxford University Press, 1977), p. 190.

12. See Amartya Sen, "Well-being, agency and freedom . . ."; and *Commodities and Capabilities* . . .

13. It is sometimes presumed that to depart from a person's own actual desires or pleasures as the measuring rod of assessment would be to introduce paternalism into the evaluative exercise. This view overlooks the important fact that having pleasure and desiring are not themselves valuational activities, even though the latter (desire) can often result from valuing something, and the former (pleasure) can often result from getting what one values. A person's utility must not be confused with his or her own valuations, and thus tying the evaluative exercise to the person's own utility is quite different from judging a person's success in terms of the person's own valuation. The important distinction to note in this context is that a person may not have the courage to desire a big social change weighed down by the circumstances in which he or she lives, and yet given the opportunity to evaluate the situation, which is essentially a political exercise in this context, the person may well value a change. One advantage of valuing as opposed to feeling is that a proper evaluation has to be a reflective exercise – open to critical examination – in a way that feelings need not be (the requirement of critical examination does not apply in the same way to feelings as it does to reflective evaluations). These and related issues are discussed in "Well-being, agency and freedom . . ."

14. In many contexts, the formal representations will take the form of partial orderings, or of overdetermined rankings, or of "fuzzy" relations. This is, of course, not a special problem with the capability approach, and applies generally to conceptual frameworks in social theory; see Amartya Sen, *Collective Choice and Social Welfare* (San Francisco, Holden-Day, 1970 republished, Amsterdam, North-Holland, 1979); and *On Ethics and Economics* (Oxford, Blackwell, 1987); see also "Social choice theory", in *Handbook of Mathematical Economics*, K. J. Arrow and M. Intriligator, eds. (Amsterdam, North-Holland, 1985). The formal problems can be dealt with at different levels of precision (i.e., with varying extent of precise representation of ambiguities). The important general point to note here is that it may be, for substantive social theories, both terribly limiting and altogether unnecessary to shun ambiguities.

15. See Amartya Sen, *Choice, Welfare and Measurement* . . . , assays 17–20.

16. Bernard Williams raises this issue in his comments on the Tanner Lectures on the standard of living; see *The Standard of Living*, Tanner Lectures of Amartya Sen, with discussions by John Muellbauer, Ravi Kanbur, Keith Hart and Bemard Williams, edited by Geoffrey Hawthorn (Cambridge, Cambridge University Press, 1987), pp. 98–101 and 108–109.

17. I have tried to discuss some of the general methodological issues involved in description in "Description as choice", *Oxford Economic* Press, 32 (1980); reprinted in *Choice, Welfare and Measurement* . . .

18. See Amartya Sen, *Resources, Values and Development* . . . , chaps. 15, 19 and 20; and "The concept of development", in *Handbook of Development Economics*, H. Chenery and T. N. Srinivasan, eds. (Amsterdam, North-Holland, forthcoming).

19. The range of functionings and capabilities that may be of interest for the assessment of a person's well-being or agency can be very wide indeed; see Amartya Sen, "Well-being, agency and freedom . . .

20. See, among other contributions, Michael Lepton, *Assessing Economic Perfor*mance (London, Staples Press, 1968); Paul Streeten, *The Frontiers of Development Studies* (London, Macmillan, 1972); Irma Adelman and Cynthia Tuft Morris, *Economic Growth and Social Equity in Developing Countries* (Stanford, Stanford University Press, 1973); Amartya Sen, "On the development of Basic income indicators to supplement GNP measures", *Economic Bulletin for Asia and the Far East* (United Nations publication, Sates No. E.74.II.E4); H. Chenery and others, *Redistribution with Growth* (London, Oxford University Press, 1974); Irma Adelman, "Development economics: a reassessment of goals", *American Economic Review*, Papers and Proceedings, 66 (1975); James P. Grant, *Disparity Reduction Rates In Social Indicators* (Washington, D.C., Overseas Development Council, 1978); Keith Griffin and Azizur Rahman Khan, "Poverty in the third world; ugly facts and fancy models", *World Development*, 6 (1978); Paul Streeten and S. J. Burki, "Basic needs: some issues", *World Development*, 6 (1978); Morris D. Morris, *Measuring the Conditions of the World's Poor: The Physical Quality of Life* Index (Oxford, Pergamon, 1979); Paul Streeten, *Development Perspectives* (London, Macmillan, 1981); Paul Streeten and others, *First Things First: Meeting Basic Needs in Developing Countries* (New York, Oxford University Press, 1981); S. R. Osmani, *Economic Inequality and Group Welfare* (Oxford, Clarendon Press, 1982); and Frances Stewart, *Planning to Meet Basic Needs* (London, Macmillan, 1985).
21. This general question of foundations and informational bases is discussed in Amartya Sen, "Informational analysis of moral principles", in *Rational Action*, Ross Harrison, ed. (Cambridge, Cambridge University Press, 1979); and "Well-being, agency and freedom . . ." In the latter analysis, some distinctions are drawn (especially between agency and well-being and between achievement and freedom) that may be worth pursuing in a more elaborate treatment of this matter, but I shall resist the temptation to go into these issues here.
22. On this general question and on the relation between commodities, characteristics and functionings, see Amartya Sen, *Commodities and Capabilities* . . . , chap. 2.
23. On this question, see Amartya Sen, *Resources, Values and Development* . . . , essays 19 and 20; and Paul Streeten, "Basic needs: some unsettled questions", *World Development*, 12 (1984).
24. John Rawls, *.9 Theory of Justice* (Oxford, Clarendon Press; and Cambridge, Massachusetts, Harvard University Press, 1971), pp. 60–65.
25. See Amartya Sen, "Equality of what?" . . . , and *Resources, Values and Development* . . .
26. See Amartya Sen, "Freedom of choice: concept and content", Alfred Marshall Lecture at the European Economic Association, *European Economic Review*, 1988.
27. See Amartya Sen, *Commodities and Capabilities* . . . , pp. 60–67.
28. See Amartya Sen, "Well-being, agency and freedom . . ."; and "Freedom of choice: concept and content . . ."
29. On the question of the relation between achieved states and the extent of freedom and liberty, see Amartya Sen, "Liberty and social choice", *Journal of Philosophy*, 80 (1983).
30. Indeed, not to take note of the person's own evaluations of states of affairs in providing a measure of freedom can yield a very peculiar view of freedom, which would be seriously at odds with the tradition of seeing freedom as important. On this, see Amartya Sen, "Liberty as control: an appraisal", *Midwest Studies in Philosophy, 7* (1982); and "Liberty and social choice . . ."
31. See, for example, A. B, Atkinson, *Unequal Shares: Wealth in Britain* (London, Penguin, 1972); and *The Economics of Inequality* (Oxford, Clarendon Press, 1975).
32. I have discussed this question in *Commodities and Capabilities* . . . , appendix B, and also in *Resources, Values and Development* . . . , essays 15 and 16. The importance of perception biases in the continuation of inter-gender inequalities is discussed in

"Gender and cooperative conflicts", WIDER working paper, in *Persistent Inequalities*, Irene Tinker, ed. (forthcoming).

33. For an attempt to make such functioning-based comparisons between men and women, see Jocelyn Kynch and Amartya Sen, "Indian women: well-being and survival", *Cambridge Journal of Economics*, 7 (1983).

34. See Jocelyn Kynch, "How many women are enough: sex ratios and the right to life", *Third World Affairs* 1985 (London, Third World Foundation for Social and Economic Studies, 1985). The ratios of life expectancy seem to have turned in favour of women vis-à-vis men, according to reported statistics in most countries (see United Nations Children's Fund, *The State of the World's Children 1988* (New York, Oxford University Press, 1988), table 7), but the undoing of past biases against women in the sex composition of the population tends to be a stow process over the years.

35. United Nations Children's Fund, *The State of the World's Children 1988* . . . , table 4.

13

Capitals and Capabilities: Linking Structure and Agency to Reduce Health Inequalities

Thomas Abel and Katherine L. Frohlich

Introduction

Understanding and reducing social inequalities in health have been key issues and a central challenge in public health. Both structural conditions and individual agency have been identified for their roles in influencing these inequalities. Since the spawning of the Ottawa Charter (World Health Organization, 1986), promoting the public's health by enabling people to increase control over and improve their health is a laudable goal that has become the benchmark for health promotion. The Charter's proponents deliberately underscored the importance of structure and agency; social structural forces were believed to be creating and sustaining health inequalities and individuals were understood to be able to productively influence social structural conditions affecting their health through their actions (WHO-EURO, 1984). While some agreement has been reached since the writing of the Charter with regard to the importance of the structure–agency processes in the quest towards the reduction of inequalities in health, how exactly to enable people to act in favour of their health remains unresolved in health promotion.

We suggest that a better understanding of the basic dynamics behind the creation of health inequalities through social inequalities might help lead to an answer. More specifically, we examine the conditions and the

Source: *Social Science & Medicine*, 74(2) (2012): 236–244.

role of individual and collective agency in the social reproduction and modification of social inequalities in health. It is here that social theory, in particular medical sociological theory, can provide guiding insights. We draw and expand on two related literatures: the structure–agency debate within sociology over the past 30 years (Bourdieu, 1986; Frohlich, Corin, & Potvin, 2001; Giddens, 1984; Hays, 1994; Sewell, 1992) and research concerned with theoretical approaches to health inequalities (Abel, 2007; Cockerham, 2005; Frohlich, Potvin, Chabot, & Corin, 2002; Popay et al., 2003; G.H. Williams, 2003; S.J. Williams, 1995). We do not aim to engage in direct dialogue with the vast social scientific literature on the structure–agency debate, but instead, draw on a few fundamental issues from it in order to move towards an understanding of the mechanisms lying between social and health inequalities, including perspectives applicable for social change.

Structure and Agency in the Current Discourse on Health Inequalities

The discussion regarding the role of the social structure on shaping human activity has permitted for a strong understanding of its patterns and its potential relevance to health (Frohlich et al., 2002; Popay et al., 2003; Singh-Manoux & Marmot, 2005; Williams, 2003). First, structures are laden with differences in power and thus empower individuals and classes differentially. Second, the term social structure often implies stability (Sewell, 1992), which has led to sophisticated descriptions of how patterns of social life persist over time (Bourdieu's concept of habitus being a case in point), with less success noted in explanations of how these patterns change over time. Indeed, as Sewell (1992) remarks, many structural accounts of social transformation tend to introduce change only from outside the system. Third, the very essence of social structure has been questioned with disagreement arising with regard to its materiality. While more traditional notions of structure viewed the "material" as the "concrete" aspects of social life (e.g. financial resources, housing), contemporary discussions of the social structure have underlined both the material and nonmaterial (meaning, cultural schemes) aspect of it.

Medical sociology has also dealt in recent years with the structure–agency debate, most directly with regard to the issue of health inequalities (Cockerham, 2005; Frohlich et al., 2002; Popay et al., 2003; Williams, 2003). Today there is a near unanimous recognition that concern with the production and reproduction of health inequalities must take into account both the social structure and individual agency to be given credence. Yet we suggest that this literature, while helpful in moving forward the field by underlining and explicating the importance of both, has left some questions open with regard to how structure and agency are linked in the production, reproduction or reduction of health inequalities. For example, the work of Williams (2003) and Popay et al. (2003), through explorations of lay knowledge or knowledgeable narratives, demonstrates the power of causal narratives in their ability to contextualize risks. Their focus is on a better understanding of the social reproduction of health inequalities with less attention for the dynamics that might lead to social change. And in more common population level studies on health inequalities, researchers tend to rely on the power of statistically significant associations, without delineating, or even exploring, the conceptual and theoretical relationships between structure and agency.

In the present paper we examine the role of agency in the social production and reproduction of health inequalities and include potential contributions of agency for structural change. We focus most specifically on two recent theoretical concepts we think may offer some novelty to the current debate on the interplay of structure and agency in the production of health inequalities: capital interaction and capabilities. We discuss Pierre Bourdieu's capital inter-action theory in order to explain the importance of structurally-based resources while also considering the function of habitus as a structure-reproducing form of agency. We then extend the discussion to include the concept of structurally transformative agency as a critical component for the reduction of social inequalities. Moving from explanation to an applied perspective, we explore Amartya Sen's capability approach and draw attention to the range of options for health-relevant agency as a potentially helpful concept for understanding how social inequalities in health might be reduced through public health action. We begin with a discussion of the key concepts from both of these scholars and their contribution to the current discourse on social inequali-ties in health. Examples from a recent intervention project to reduce health inequalities are included to link our theoretical perspective to a practical expe-rience. First however we briefly re-visit Max Weber's theory on lifestyle which we use as the basis for our structure–agency perspective.

The Structuring of Choices

We begin our theoretical foray with some of the basic concepts Max Weber developed in his writing on lifestyles. Weber is particularly important here for two reasons. First, his discussion of *"Lebensführung"* (life conduct) helps us make the claim that agency and social structure are both critical for under-standing how health is unequally produced (Cockerham, Abel, & Lüschen, 1993). Second, and in a related manner, the *"Lebensführung"* concept helps us move beyond a notion of agency as being equivalent to 'risk behaviour', an approach that has been criticized by medical sociologists as being too lim-ited (Shim, 2002; Williams, 1995) and viewed by others to be thwarting the development of health promotion beyond its roots in health education (Abel, 2007; McQueen, 2007).

Max Weber (1978) focussed on life conduct to explain how individuals actively contribute to the social reproduction of status group distinctions through their dress codes, marriage patterns, eating habits, etc. A major contribution of Weber's work to a critical understanding of lifestyle is his acknowledgement that people's choices are constrained by the material resources and normative rules of the community or status group they belong to, thereby acknowledging both material and non-material aspects of structure. These resources and rules are all components of what Weber referred to as life chances, the structural part of lifestyle processes (Abel, 1991; Ruetten, 1995). Life chances thus refer to the structurally anchored probabilities of achieving one's goals (Cockerham et al., 1993; Dahrendorf, 1979).

The notion of *"Lebensführung"* therefore goes beyond behaviours to focus on people's active role in responding to demands and opportunities in everyday life. When it comes to behaviours, Weber was concerned with the social processes that link structural constraints and opportunities (life chances) on the one hand, and people's re-active or pro-active behaviours (life conduct), on the other. This issue has been at the root of the sociological discourse on lifestyles from either a Weberian or Bourdieusian tradition (Cockerham, Ruetten, & Abel, 1997). However, it is Weber's dualism of structure-based life chances and people's choice-based life conduct that provides *the* fundamental ground for thinking in terms of a duality of structure and agency with regard to social inequality.

Capitals and the (Re)Production of Social Inequalities

Weber's view on the dialectic interplay between life choices and life chances laid the ground for later analyses including some of Bourdieu's work (Bourdieu, 2007). However, Weber's analysis is insufficient in accounting for social differences in the contemporary patterning of health lifestyles (e.g. Cockerham et al., 1997). Moreover, Weber was concerned with status group formation and his definition of life chances (structurally anchored probabilities) does not clearly address the issue of how much individual freedom is left in the selection of lifestyles. This question is addressed by Bourdieu who argues for a strong link between the possession of different forms of capital, a class specific habitus and the choices individuals have. The significant advancement in Bourdieu's work for our argument is that it allows us to analyze key components of social inequality directly relevant for agency. We will argue that a closer consideration of the different forms of capital and their interactions opens a gateway into explorations of the role of the individual in the production and reproduction of health inequalities (Abel, 2007).

According to Bourdieu the unequal distribution of structurally-based resources (capitals) can be understood as part of the fundamental system of inequality in a given society; it is both the result and a key mechanism of the social reproduction of power and privilege. His concept of capital is based on the distinction of three forms: social, economic and cultural capital. These three forms of capital are interrelated and inextricably linked.

A major thrust of Bourdieu's theory is the elaborate account of the interaction between these three forms of capital in everyday life and the ways in which this interaction process contributes to the reproduction of social inequalities and power distribution in society (Bourdieu, 1984; Swartz, 1997). Since this interaction has not yet been given full attention in health research (Abel, 2008), we concentrate here on the interaction between the capitals in their different forms. Prior to this we briefly discuss Bourdieu's notion of economic, social and cultural capital (for more see Abel, 2007; Bourdieu, 1986; Williams, 1995).

According to Bourdieu (1986), "capital is accumulated labour (in its materialized or its 'incorporated', embodied form) [. . .]. It is a force inscribed in objective or subjective structures, but it is also the principle underlying the immanent regularities of the social world" (p. 241). A critical aspect to Bourdieu's capital theory is that no single one of the three forms of capital alone can fully explain the reproduction of social inequalities; it takes all three, and importantly, the interaction between the three to permit for social inequalities to endure over time.

Economic capital (or the lack thereof) in the form of money and material assets (income, property, financial stocks), is a decisive factor in social advantage and disadvantage. It is also "at the root of all the other types of capital" (p. 252). *Social capital*, from a Bourdieusian perspective, is located at the inter-individual level. As such, it refers to material and non-material resources which can be mobilized by virtue of many different kinds of social relationships. Social capital is thus understood as: ". . . [the] aggregate of the actual or potential resources which are linked to possession of a durable network of mutual acquaintance and recognition – or to membership in a group – which provides each of its members with the backing of the collectively-owned capital" (Bourdieu, 1986, p. 248f).

Lastly, *cultural capital* can be broadly defined as people's symbolic and informational resources for action (Bourdieu, 1986; Wacquant, 1992). Cultural capital exists in three different forms: incorporated (e.g. skills, knowledge); objectivized (e.g. books, tools, bicycles); and institutionalized (e.g. educational degrees, vocational certificates) (Bourdieu, 1986). It is acquired mostly through social learning, with learning conditions varying across social classes, status groups or milieus (Abel, 2007; Swartz, 1997; Veenstra, 2007; Williams, 1995). A person's educational level can be understood as an indicator representing cultural capital (for strengths and limitations of that measure see Bourdieu (1986)). Yet, cultural capital refers to more than a person's formal education to include different sets of cultural competencies. Acquisition and use of these is part of a broader comprehensive social learning (socialization) and thus depends heavily on "total, early, imperceptible learning, performed within the family from the earliest days of life" (Bourdieu, 1984, p. 66). In the form of knowledge and skills cultural capital is a precondition for most individual action and, as such, is a key component in people's capacity for agency, including that for health.

In the struggle over power and privileges the chances to acquire and apply different forms of capital are pre-determined and structured by people's habitus, an intangible concept observable only through individual practices, that is not itself an individual attribute. Moreover, habitus is itself dependent on the availability of different forms of capital for the individual and his or her family in the past and present. Tied – through socialization – to the different forms of capital allowing and supporting it, the habitus links structure and agency through collective strategies, a shared way in which groups of individuals from similar social classes tend to act.

Capital Interaction and Health Inequalities

There is a high degree of complexity among capitals in their different forms. Three of these relationships (conversion, accumulation and transmission) have been discussed by Bourdieu (1986) and here we add a fourth principle of interaction we call "conditionality". All four forms of interaction are important for the purposes of our argument.

First, different forms of capital, in their acquisition and use, are dependent and conditional on each other. For instance, cultural capital is essential in the acquisition of social capital; certain values, communication styles and behavioural skills are expected from all those who want to belong to, and participate in, powerful social networks. The (gainful) use of economic capital might depend on the authorizing properties of higher educational degrees and on the knowledge that comes with it.

Second, the different forms of capital can be converted one into another; economic capital, in the form of money, can be invested in order to improve one's education or cultural capital. *Third*, capital in these different forms can be accumulated; money can be invested in the stock market, for instance, in order to make more money. *Fourth*, the different forms of capital can be transmitted; children can inherit financial assets from parents and/or capital can be received through family socialization e.g. when knowledge and social skills are passed on from parents to their children.

Practical examples might serve to illustrate the importance of these four principles of interaction for the (re-)production of social inequalities in health: *Conditionality* occurs, for example, when knowledge with regard to health and health determinants (cultural capital) becomes a pre-requisite for spending one's income (economic capital) in a health promoting way. The conditionality relationship also accounts for the fact that some resources may lose their potential health benefits and turn into questionable assets from a health perspective, e.g. when excess income is spent for health compromising behaviours or when social capital is linked to participation in health damaging group behaviours. Spending money on health courses, books or other health education measures means *converting* parts of one's economic capital into health knowledge i.e. health-relevant individual or family cultural capital.

Health-relevant knowledge and skills can also be *accumulated* over time in individuals, families or peer groups through personal and collective investment such as social and cognitive learning, social exchange and support, all often part of individual's life long socialization. Finally, *transmission* of health-relevant cultural capital takes place in social networks (e.g. peer groups, work place or school settings) and, most important perhaps, in families, when they provide their children with environments conducive to the acquisition of certain health competencies and health promoting learning experiences (for a discussion and empirical findings of the particular role and function of the family for the transmission of cultural capital see Bourdieu (1984, 1986), DiMaggio (1982), and Georg (2004)).

The decisive meaning of the three forms of capital and their interaction leads us to acknowledge the active role of individuals who (beyond simply owning or consuming such resources) acquire and use, in some active way, health-relevant capital. The active acquisition and development of such capital is part of individual and collective agency, as is making health-relevant use of them. In other words, in order for cultural, social and economic capital to become health promoting, individuals have to actively use them. For instance, money is '*spent*' on health-relevant behaviours (such as physical activity classes), support in health matters is '*sought out*' (such as participating in self-help groups) and knowledge is '*applied*' by individuals in order for it to function actively to engender health (for instance decisions about what one eats).

In the shaping and reproduction of social inequalities these '*actions*' are related to each other through capital interaction that facilitates class or status-specific habitus and lifestyles. We therefore suggest that inequality goes beyond just the unequal distribution of capital. We argue that there is considerable social inequality also in the chances and ability for people to have the different forms of capital consistently support and complement each other with the end result of their interaction being a health advantage.

It should be emphasized here again that agency requires capital and that the unequal distribution of capital is first and foremost a matter of social structure: cultural capital is unequally distributed through stratified school systems as much as it is through milieus and families; access to social capital is regulated through class barriers as much as it is through language codes; and the unequal distribution of income is the primary marker of privilege or social disadvantage. Our point is that capital interaction needs the active individual and provides for a key role of agency in the conversion of social inequalities into health inequalities.

In Bourdieu's work the formation of the habitus depends on the availability of the different forms of capital, and in turn, the habitus affects the chances for the acquisition and use of capital. In the reproduction of unequal chances for good health, in particular, the habitus is a useful concept in so far as it explains how particular forms of agency contribute to the re-iteration of class differences in health disadvantages over time. Consequently, current work using Bourdieu's theory to explain social inequalities in health have stressed the "choice limiting" aspects in the links between structure (unequal distribution of capital) and agency (habitus and strategies) (Cockerham, 2005; Singh-Manoux & Marmot, 2005; Veenstra, 2007). With its focus on reproductive processes, the habitus concept however, does not attempt or mean to explain how structure-based patterns of behaviours are changed, nor does it point to ways in which conditions can be created to alter or increase individual agency for change. When theoretical guidance is sought in applied public health, and particularly in health promotion, for interventions to reduce social inequalities in health, the notion of habitus is not readily applicable; there is insufficient explanation of chances to expand opportunities or even on how agency might lead to structural change.

In fact, because its coherent frame and structure are clearly focussed on a comprehensive explanation of the reproduction of social inequality, it remains a major challenge to apply Bourdieu's capital and habitus theory when seeking guidance for action on reducing health inequalities. His capital theory has not yet been applied to explanations of how opportunity structures may change over time or how they may be improved by public health interventions to reduce social inequalities in health. However, although his theory stresses class specific habitus and habitus-conforming behaviours and actions, his explanations are not to be misread as "deterministic" as they refer to probabilities more than causal links (Bourdieu, 1977; Bourdieu & Wacquant, 1992; Cockerham, 2005). In fact as Wacquant (1992) has emphasized, there is considerable openness in Bourdieu's theory to accommodate a theory of change. As Swartz (1997) notes, Bourdieu showed that "the idea that structures reproduce and function as constraints is not incompatible with the idea that actors create structures" (Swartz, 1997, p. 290). Thus, the concept of habitus does clearly focus on the role of agency in the reproduction part of social action and has less to offer for our understanding of the dynamics of actual changes in the range of options to act for one's own benefit. However, we also find that Bourdieu's framework, while not fully accounting for explanations of change, provides options to add a theoretical component that might help identify and understand forms of agency for change including those that directly relate to health inequalities (for a more comprehensive discussion of the limits and potentials of Bourdieu's theory of change see Swartz, 1997).

From Understanding Social Reproduction of Health Inequalities to Reducing Them: The Capability Approach as a Nexus between Explanation and Public Health Action?

Drawing on Weber's dualism of life chances and choices, we view that despite material and normative constraints people do have some element of choice in their behaviours affecting their health. These choices arise out of the conditions they face and the opportunities they enjoy. In the production and maintenance of health, therefore, there is an important role for individual agency. The question can then be posed: "What theory of agency permits for individual and structural change"?

Following Hays (1994) we understand agency as "human social action involving choices among the alternatives made available by the enabling features of social structure, and made possible by solid grounding in structural constraints" (Hays, 1994, p. 64). For Hays, agency refers to social choices made and operating within the limits of social structures. As explained above, we suggest the dynamics of capital interaction as a key factor linking agency to structure. Applied to social inequalities in health we refer to agency as the socially structured development, acquisition and application of structural and personal resources by individuals in a given context. Agency is at work when individuals use the different forms of capital in the production of health and the improvement of their chances for being healthy (e.g. through certain lifestyle choices). Beyond individual health gains, in their roles as parents, neighbours, teachers etc., individuals (intentionally or not) are directly involved in the social patterning of health. As members of populations and subpopulations, unequally equipped with the different forms of capital, individuals – through their agency role – cannot but contribute to the reproduction, or to the modification, of both social and health inequalities.

The reproduction and modification of structures of inequality has been discussed by Hays (1994) who distinguishes between "structurally reproductive agency" and "structurally transformative agency". These two differentiate between agency resulting in empirically observable reproduction and that which modifies structural conditions. Bourdieu's agency concept is close to the first of these agency functions, explaining social reproduction through a link between structure, the unequal distribution and interaction of different forms of capital and class based habitus. As for health and social inequalities, certain health behaviours can acquire the form of "structurally reproductive agency", e.g. when sedentary lifestyles and unhealthy eating habits (rooted in unequal access to capital and empirically found to be more common in the lower social classes) contribute – through habitus – to a re-iteration of social patterns of health risks and to the reproduction of health disadvantages in lower social classes. In this respect health promotion has been rightly criticized for being vulnerable to practising inequality reenforcing measures, e.g. when not reaching deprived populations or when relying on interventions that are more attractive to, and effective in, certain population groups (Frohlich & Potvin, 2008). In such cases public health action might be contributing to structural health inequalities by promoting health advantages for those already privileged thereby reinforcing class differences in health through structurally reproductive agency.

Based on a solid understanding of the structure–agency processes at work alternative approaches might instead be focused on promoting "structurally transformative agency". Examples include new community structures that allow for citizen participation and increased autonomy in community health matters. Other examples might include the support of neighbourhood actions which lead to the construction of safer sidewalks or to better and more equal access to youth leisure time facilities, etc. When searching for theoretical guidance for interventions of this kind the habitus theory shows some considerable limitations as it is more concerned with the social reproduction of inequalities and structurally reproductive agency, and therefore imperfectly equipped to explain those forms of agency that can contribute to structural modifications and social change. Also, generally speaking, epidemiological and health promotion research on health inequalities tends to be lacking or devoid of discussions with regard to the notion of structurally transformative agency. In particular the theoretical links between meaningful explanation and practical public health action are lacking. We therefore propose the consideration of additional concepts that can be used to develop theoretically sound approaches to the reduction of social inequalities in health. We turn therefore, to Amartya Sen's capability approach as a potentially useful perspective when exploring issues of structurally transformative agency in health matters.

The Capability Approach

In the last 10 years or so, Sen's capability theory (CA) has been proposed as being potentially important for our understanding of inequalities in health and for public health action towards reducing social inequalities. This has recently been spearheaded by the WHO Commission on Social Determinants of Health (CSDH, 2008) and the writings of experts like Ruger (Marmot, 2010; Ruger, 2010). The CA puts the emphasis on the empowerment of individuals to be active agents of change in their own terms – both at the individual and collective levels (Ruger, 2004). Central to the CA is the idea of "the public as an active participant in change, rather than as a passive and docile recipient of instructions of dispensed assistance." (Sen, 1999, p. 281). This perspective on collective activity for change allows us to link the CA to the current sociological discourse on the role of agency and, in particular, to the concept of structurally transformative agency.

The core characteristic of the capability approach is its focus on what people are *effectively* able to do and be; that is, on their capabilities (Robeyns, 2005). For Sen, only the ends (called "doings" and "beings") have intrinsic importance. This distinction between means and ends provides the starting point of the capability approach (Robeyns, 2005; Sen, 1985, 1999). Individuals' effective opportunities to undertake the actions and activities that they want to engage in are what matter. These actions and activities ("doings") together with the "beings", or what Sen calls "functionings", constitute a valuable life. Functionings include, but are not limited to, being healthy, being active as a community member, working, resting, being literate, etc.

The distinction between realizable and realized functionings is crucial to the capabilities approach. "A functioning is an achievement, whereas a capability is the ability to achieve" (Sen, 1987, p. 36). Sen puts much emphasis on the distinction between functionings and capabilities because he believes that well-being should not only include realized functionings, but that the ability to choose from a set of alternative functionings is a freedom sui generis (Sen, 1999). Consequently, and in relation to inequalities in health, evaluations of social and health interventions based on the CA should include, on the structural side, not only the quality and quantity of available resources, or the realized doings and beings on the agency side, but also, the range of capabilities available to people. In other words we must take into account the "capability sets" from which individuals can draw (Sen, 1993). When it comes to evaluating the well-being of individuals or populations, therefore, the capability approach places people's chances to realize their life goals and plans as its focal point.

In contrast to utilitarian approaches, those that drive much of health promotion work specifically and public health more generally, resources according to Sen are not at the centre of the capability approach; instead, resources are understood as means to ends, namely to realize one's life goals. With the capability approach, then, the idea of fairness or justice does not apply to the availability of resources alone, but to the range of options for agency – the capabilities. This range of options, and one's ability to choose and actualize them, creates conditions for health-relevant agency from which, in turn, well-being, happiness and health may result. Moreover, beyond the freedom to be active for one's own health, capabilities can also be understood as a means for structurally transformative agency, namely when an improved range of capabilities allows for the actual realization of choices which yield consequences on the structural level.

The following example, taken from a recent health promotion project in Germany, BIG (Ruetten, Abu-Omar, Frahsa, & Morgan, 2009; Ruetten, Abu-Omar, Seidenstuecker, & Mayer, 2010), indicates similarly that agency in health promotion can be understood as going beyond social class differences in the uptake of healthier behaviours. In fact, from an empowerment oriented health promotion perspective, one's capabilities of living a healthy life should include options for structurally transformative agency, i.e. the chance to be active in relation to the structural conditions relevant for health. Conceptually based on the interactive processes between structure and agency, the BIG project demonstrates that these theoretical considerations can guide successful health promotion interventions.

Capitals and Capabilities in Health Promotion Practice: The BIG Project

BIG is short for "Bewegung als Investition in Gesundheit (BIG) – Movement as an Investment for Health", and is aimed at promoting health through the promotion of physical activity among women in difficult life situations (www.big-projekt.de). The city-based project addresses women of low socio-economic status known from national statistics to represent the population group most physically inactive with a high prevalence of sedentary lifestyles, and thus high levels of associated conditions such as obesity. Particular about this project is its cooperative planning approach in which the target group, the political level (i.e. mayor and city councillors) and operational level (public health practitioners, sport club representatives etc.) participate on an equal footing from the very beginning of the project. The researchers acted as perceived neutral facilitators and moderators of this academia-driven policy development process (Ruetten & Gelius, 2011). After funding for the research part of the project ended in 2008, the municipality took over responsibility for sustaining the local activities. BIG has been transferred to other municipalities in Germany. Currently, about 800 women weekly take part in BIG activities at ten locations across the country.

The BIG approach was designed to reduce social inequality in health in a low income community characterized by a high proportion of people of lower socio-economic position and ethnic minorities (Ruetten et al., 2010). Its participatory intervention approach focuses on women and their role as *social catalysts*, defined as empowered individuals acting in (in-) formal networks, mediating social institutions and organizational structures that might function as particularly important assets for the implementation of health promotion actions (Ruetten et al., 2008). Women from different ethnic backgrounds were provided the opportunity to get involved in, and were trained for, cooperative planning from the very early stages of the project. This yielded improvements in their abilities to engage in the development of new and more adequate health promotion programs (physical activity courses, health courses, women-only pool hours, access to school gyms and fitness studios, etc). Among the many positive results, the intervention led to the discovery of an unexpected demand for swimming courses by Muslim women. Not only were new swimming courses for women then offered, but the existing administrative rules of the local facilities were adjusted (extra opening hours were added, gender rules changed, culturally appropriate swimsuits/clothing in the pool was permitted). The women also successfully lobbied the city administration to change rules for school facilities and allow them access to a nearby school gym. Of course, such changes faced resistance on the side of the established stake holders, but significant structural changes were achieved through round tables and negotiations in which the women took an active part as social catalysts (Ruetten et al., 2008, 2009).

The interactive processes between structural opportunities and individual agency thus led to an increase in women's range of options for action that included not only their health behaviours (swimming courses becoming available to them according to their social cultural needs), but also their ability to actively socially engage towards improving health promoting conditions in their community. For instance, as a result of the high participation rates in the project, a new position was established within the local city administration in order to promote physical activity among the socially disadvantaged – an example of structurally transformative agency. There were also individual gains for the women by improving their chances for agency beyond health matters. One Muslim woman whose husband took care of her administrative tasks with local authorities in the past, said: "Now I can go myself when I have things to take care of at the city hall" (Röger, Ruetten, Frahsa, & Abu-Omar, 2011, p. 468). Women who were unemployed at the beginning of the project increased their chances of finding a job through the experience of exercising and improving their agency competencies. One was later hired by the city for a newly established position related to the project. The opportunity to establish new contacts through the project's community network was used by another woman to find a new job outside the health sector. A third woman was asked and agreed to run for city council. It should be noted, however, that the women who were most successful in achieving those *functionings* (e.g. better jobs) were those with the highest cultural capital (educational degrees) from the beginning. Some women, in particular Muslim women with low educational degrees, did however also become opinion leaders in their communities, contributing to and benefitting from the social capital in those groups (Ruetten et al., 2008).

The fact that many of the benefits became apparent only over the course of the intervention underlines the relevance of our theoretical approach. Linking capital to capability sets, rather than to pre-fixed and narrowly specified health targets, brings to bear a key aspect of agency that is often neglected in social inequality research and interventions, that is, having options to choose from. This range of options is a key feature of social inequality; a condition for being able to fulfil an active role towards individual or collective health gains and, capable of promoting structurally transformative agency. We argue, therefore, that policies should aim to increase resources that widen people's range of options in order to enable individual freedom as part of the empowerment process. The freedom to choose according to one's own needs is an important component of overcoming social inequalities, but may also increase the chances for sustainable success in health interventions.

The experiences and empirical findings from the BIG project support our main argument delineated above and link three particular facets of social inequality in a theoretically meaningful way: 1) membership in lower social classes is typically associated with low capital ownership, limited access to and control over structural resources for health and a lesser ability to convert capital, through their interaction, into health; 2) different forms of capital and their interaction provide for ranges of options (capabilities) from which individuals can choose in practising health-relevant agency; 3) depending on the capital available some forms of agency may yield improvements in the structural conditions of health enhancing behaviours and beyond.

Discussion and Conclusion

In this final section we first briefly summarize our line of argument and how it might lead to new questions in the current discourse in Medical Sociology. We then close our paper with a discussion of the relevance of our theoretical argument for public health and health promotion.

Previous theoretical explanations of social inequality and health have convincingly argued that the (re)-production of health inequalities cannot be explained on the basis of a theoretical or empirical divide between structural and behavioural effects, but instead, will always have to include complex and highly dynamic structure–agency processes. Drawing on theoretical contributions from Weber, Bourdieu and Sen, we suggest that major parts of the transformation of social inequality into health inequality can be understood in terms of capital interactions that shape people's range of options for health-relevant agency. While we draw mostly on Bourdieu's insights about the key role of the different forms of capital, we also observe that his habitus concept is primarily focused on the social reproductive processes between structure and agency. As such, it does not fully meet the need in public health and health promotion research for structurally transformative agency theories. We suggest, therefore, that the notion of *capabilities* should receive more attention in current debates on structure–agency explanations in health inequalities.

To those who look for theoretical guidance in reducing social inequalities in health, the capability approach can provide a missing link from explanation to action by accounting for individual agency that itself is dependent on structural factors, promoting or constraining agency in any given society and/or context. With regard to structural effects on health, it is the interplay of different forms of capital which needs to be better understood in order to explain the dynamics that lead to the unequal distribution of resources required for health-relevant agency. We further argue that our understanding of the choices in health-relevant agency should not be reduced to healthy behaviours or lifestyles among individuals. Instead, our focus should be on structurally-based choices which can and should include options for individual and collective action on the social conditions of health. These options would ideally allow for initiation and re-enforcement of processes that yield transformations or modifications of the structural conditions relevant for health and health behaviours.

We are aware that Bourdieu's theory, and in particular his notion of habitus, has been interpreted as a rather closed and deterministic explanation of social inequality (e.g. Archer, 2000) and that such a reading might preclude any attempts to link Bourdieu's work to perspectives of "change from inside" (Sewell, 1992) or "structurally transformative agency" (Hays, 1994). In response to these interpretations we refer to Bourdieu's original texts which are not totally deterministic and which do not exclude the option of change through agency (Bourdieu, 1977; Bourdieu & Wacquant, 1992).

We are also aware that Sen's theory has been viewed to be primarily individualistic (Stewart, 2005). Yet, while his capability approach is focused on individuals' freedom to achieve their personal goals, it does not exclude the idea that structural constraints and normative powers might structure or even determine the range of these choices. Sen's original texts do not, however, elaborate on the social structural factors that influence people's perception of options or their conscious and unconscious personal choices (Stuart, 2005). The link between capital and capabilities we have introduced here will need therefore, more theoretical exploration and empirical research. The aim for theoretical advances would be to probe into its explanatory potential. The guiding question for applications in health promotion would be to better understand if and how improving people's capabilities can lead to changes beyond individual health gains and thus contribute – through agency – to modifications in power relations (e.g. through the creation of active participation and negotiation for those in structurally disadvantaged positions such as the women described in the BIG project).

Capitals and Capabilities in Applied Public Health and Health Promotion

The capability approach, if solidly rooted in a structure–agency perspective as we have outlined above, could squarely address the basic principles of health promotion. The Ottawa Charter and later adherents to its fundamental ideas seek to ensure, through policy and other health promotion interventions, that the largest number of opportunities are made available to the largest number of people to enable them to make healthy choices for themselves, their families, communities etc. and eventually lead healthier lives. The focus here is on the individual's ability to choose and use resources which lead to health.

Indeed, unfortunately, the very notion of individual agency is too often dismissed from health promotion literature and research. Individuals targeted by health promotion are too often reduced to their behaviours directly relevant to health risks (e.g. smokers, obese, drinkers) and thus are reduced to *risk carriers*. These individuals are seen to have become so either because they do not know what is good for them or are in conditions in which they become exposed to risk. The health promotion consequences to date of this objectification of the individual is to either *teach* people to be healthier or to protect them; both approaches lack respect for individual autonomy and agency, and therefore disable people from realizing their true potential, and often, from becoming healthier. As we have indicated above there are alternatives to such *agency limiting* approaches.

A more solid grounding of health promotion in social theory might help to avoid such problems. Drawing on Weber we understand health-relevant behaviours as part of purposive human action that includes patterns of perception and decisions to (not) act, as well as the social patterning of preferences. Weber also helps recognize the interplay between individual's health-relevant choices and structurally anchored life chances (Abel, 1991; Cockerham et al., 1993). Both of these insights from Weber's work suggest that health promotion theory and practice would be better able to understand the relationship between social and health inequalities were we to accept and account for the significant role of both structural factors and agency in the production of social inequalities in health. Weber's early work has influenced Bourdieu's contemporary theory which has contributed significantly to better understanding how people's structural position is determined by the capital they have available and how this in turn limits and structures – through habitus – the lifestyle choices they make. Bourdieu, while defending the openness of his theory (Bourdieu, 1977; Bourdieu & Wacquant, 1992) however, did not elaborate on explanations of change nor did he provide explicit starting points for theories on how to reduce social inequalities through strengthening agency and individual empowerment.

Here, Sen's more recent insistence on a range of options (or "capability set") appears helpful and is critical for a modern health promotion that emphasizes equal opportunities for making healthy choices as well as people's control over the determinants of their health (Marmot, Allen, & Goldblatt, 2010; World Health Organization, 1986). However, while Sen stresses the interplay of resource availability and individual agency, he does not deal squarely with which resources, or what elements, should be distributed more equally in order to improve well-being or health. So, while Sen would suggest that the relationship between capabilities and realized functionings (e.g. living a health life) is one of constrained choice, he does not fully explain the nature of these constraints (Stewart, 2005). Recently, some authors have attempted to operationalize what these constraints might look like, suggesting conceptual models involving large numbers of concepts and variables (Ruger, 2010). To date, however, such attempts have been largely a-theoretical and somewhat difficult to imagine in practice.

Because Bourdieu's theory addresses which resources comprise social inequalities, it is a suitable accompaniment to Sen's capability theory. The fundamental tenet of Bourdieu's theory was to describe how the unequal distribution of economic, social and cultural capital functions as a root principle in the distribution of power and privileges. The interaction of the different forms of capital is a key factor that draws our attention not only to the absolute amount of health-relevant resources available to people, but to their chances and abilities to link material and non-material resources towards maximizing their individual or collective benefits. Indeed, we propose here that interactions of the different forms of health-relevant capital determine major parts of the dynamic processes in the (re-) production of health advantages and disadvantages: conversion of one form of health-relevant capital into another e.g. income into (health relevant) knowledge; accumulation and transmission of cultural and/or economic capital e.g. within families and conditional effects such as economic and cultural capital conditioning the acquisition of (health promoting) social capital; these processes to be successful in terms of health gains, require the active individual. They also render the different forms of capital key in the structure–agency interplay that leads to the social reproduction of health inequalities. It is in this way, indeed, that we think Bourdieu's capital interaction theory can be of utmost importance to our understanding of the potential and limits of health-relevant agency through capabilities.

Most individuals do have some choice with respect to the factors influencing their health and those choices are limited according to their socio-economic position, including those choices that relate to or include health behaviours. Next to the constraints of choice argument, normative pressures typical of modern societies are also important to consider. These include not only expectations towards certain healthy behaviours, but also the very basic question of choice-making in modernity. Increasingly individuals have, as Giddens puts it, "no choice but to choose" (Giddens, 1991, p. 81). So, choice has undoubtedly become an important feature in health inequalities (Cockerham, 2005). However, the concept of choice under these modern circumstances requires critical reflection (Donahue & McGuire, 1995; Leichter, 2003; Michailakis & Schirmer, 2010). In the context of capital and health-relevant capabilities we need to take into account that: 1) the range of options for any individual is limited by the amount of different forms of capital available to him or her; 2) the effectiveness of the application of the different forms of capital for health benefits depends on contexts and people's abilities to "play" their capital most effectively and; 3) the non-material aspects of the social structure shape individual preferences as well as what people find appropriate.

The capability approach might be vulnerable, however, to a somewhat uncritical understanding of choice. When Sen writes of people's abilities and options to choose "what they have reason to choose" (Sen, 1999, p. 148) he is proposing a somewhat simple assumption regarding the ability of individuals to simply resist normative pressures. His capability approach has been criticized accordingly by Stewart who argues that: "[. . .] society – and indeed particular groups within society – shapes *every* individual, influencing preferences and consequent choices" (Stewart, 2005, p. 189, italics in original). Indeed, by focussing largely on personal factors, rather than the structured and collective factors that lead to the application of capabilities for health, the social conditions and dynamics get insufficient attention in Sen's approach.

Here again, Weber and Bourdieu can be most helpful. As indicated above, Weber describes how behaviours are chosen according to status-specific norms and values as well as the resources available to people (Cockerham et al., 1993). And, referring to Bourdieu's theory on the reproduction of social inequality, the question arises as to what degree even the perception of potential choices is socially structured through class specific habitus and cultural capital. This leads us to conclude that while offering an important new focus on health-relevant capabilities Sen's approach cannot qualify as a "stand alone" application in a highly stratified world of social inequalities. Its application requires a social theory that can account for the social and cultural determinants of health and health-relevant behaviours in order to prevent approaches that might lead to more individual-level interventions in health promotion rather than more structural change. In this sense we propose that primary emphasis should be given to the promotion of the different forms of capital, rather than focus on personal factors, in order to facilitate structural change that improves the range of options for health promoting agency.

As an application oriented theory the capability approach points to ways to improve the probabilities for good health in individuals from lower classes by providing more (healthy) choices for those who did not have them before. For public health action aiming to reduce health inequalities at a structural level (i.e. the probabilities for being in good health associated with a certain social position in a given society) the capability approach needs certain conditions to be met: 1) the range of options needs to improve at a higher rate for the disadvantaged population groups (classes) compared to advantaged population groups; 2) the improved range of options must lead to the realization of more healthy choice options (functionings) and; 3) improved capabilities must yield effects beyond individual agency and personal health gains to improve – through agency – the structural conditions of health. While more theoretical and empirical work will have to follow, we belief that the capability approach can be helpful to move the structure – agency discourse on health inequalities from sound explanations to sound actions on their reduction . . . including people as active agents.

References

Abel, T. (1991). Measuring health lifestyles in a comparative analysis: theoretical issues and empirical findings. *Social Science & Medicine, 32*(8), 899–908.

Abel, T. (2007). Cultural capital in health promotion. In D. C. McQueen, & I. Kickbusch (Eds.), *Health and modernity. The role of theory in health promotion* (pp. 43–73). Berlin: Springer.

Abel, T. (2008). Cultural capital and social inequality in health. *Journal of Epidemiology and Community Health, 62*(7), e13.

Archer, M. (2000). *Being human. The problem of agency*. Cambridge: Cambridge University Press.

Bourdieu, P. (1977). *Outline of a theory of practice*. Cambridge: Cambridge University Press.

Bourdieu, P. (1984). *Distinction: A social critique of the judgement of taste*. Cambridge, MA: Harvard University Press.

Bourdieu, P. (1986). The forms of capital. In J. G. Richardson (Ed.), *Handbook of theory and research for the sociology of education*. Connecticut: Greenwood Press.

Bourdieu, P. (2007). *Sketch for a self-analysis*. Cambridge: Polity.

Bourdieu, P., & Wacquant, L. (1992). *An invitation to reflexive sociology*. Chicago: University of Chicago Press.

Cockerham, W. C. (2005). Health lifestyle theory and the convergence of agency and structure. *Journal of Health & Social Behavior, 46*(1), 51–67.

Cockerham, W. C., Abel, T., & Lüschen, G. (1993). Max Weber, formal rationality and health lifestyles. *Sociological Quarterly, 34*, 413–425.

Cockerham, W. C., Ruetten, A., & Abel, T. (1997). Conceptualizing contemporary health lifestyles: moving beyond Weber. *Sociological Quarterly, 38*, 321–342.

CSDH. (2008). *Closing the gap in a generation: Health equity through action on the social determinants of health*. Final report of the Commission on Social Determinants of Health. Geneva: WHO.

Dahrendorf, R. (1979). *Life chances: Approaches to social and political theory*. Chicago: University of Chicago Press.

DiMaggio, P. (1982). Cultural capital and school success: the impact of status culture participation on the grades of US high school students. *American Sociological Review, 47*(2), 189–201.

Donahue, J. M., & McGuire, M. B. (1995). The political economy of responsibility in health and illness. *Social Science & Medicine, 40*, 47–53.

Frohlich, K. L., Corin, E., & Potvin, L. (2001). A theoretical proposal for the relationship between context and disease. *Sociology of Health and Illness, 23*(6), 776–797.

Frohlich, K. L., & Potvin, L. (2008). Transcending the known in public health practice: the inequality paradox: the population approach and vulnerable populations. *American Journal of Public Health, 98*(2), 216–221.

Frohlich, K. L., Potvin, L., Chabot, P., & Corin, E. (2002). A theoretical and empirical analysis of context: neighbourhoods, smoking and youth. *Social Science & Medicine, 54*(9), 1401–1417.

Georg, W. (2004). Cultural capital and social inequality in the life course. *European Sociological Review, 20*(4), 333–344.

Giddens, A. (1984). *The constitution of society*. Cambridge: Polity Press.

Giddens, A. (1991). *Modernity and self-identity: Self and society in the late modern age*. Stanford: Stanford University Press.

Hays, S. (1994). Structure and agency and the sticky problem of culture. *Sociological Theory, 12*(1), 57–72.

Leichter, H. M. (2003). "Evil habits" and "personal choices": assigning responsibility for health in the 20th century. *Milbank Quarterly*, Wiley: Online Library.

Marmot, M. (2010). Fair society, healthy lives 2010. In M. Marmot (Ed.), *The Marmot review*.

Marmot, M., Allen, J., & Goldblatt, P. (2010). A social movement, based on evidence, to reduce inequalities in health. *Social Science & Medicine, 71*(7), 1254–1258.

McQueen, D. (2007). Critical issues in theory for health promotion. In D. V. McQueen, & I. Kickbusch (Eds.), *Health and modernity: The role of theory in health promotion* (pp. 21–42). New York: Springer.

Michailakis, D., & Schirmer, W. (2010). Agents of their health? How the Swedish welfare state introduces expectations of individual responsibility. *Sociology of Health & Illness, 32*(6), 930–947.

Popay, J., Thomas, C., Williams, G., Bennett, S., Gatrell, A., & Bostock, L. (2003). A proper place to live: health inequalities, agency and the normative dimensions of space. *Social Science & Medicine, 57*(1), 55–69.

Robeyns, I. (2005). The capability approach: a theoretical survey. *Journal of Human Development, 6*(1), 93–114.

Röger, U., Ruetten, A., Frahsa, A., & Abu-Omar, K. (2011). Differences in individual empowerment outcomes of socially disadvantaged women: effects of mode of participation and structural changes in a physical activity promotion program. *International Journal of Public Health, 56*, 465–473.

Ruetten, A. (1995). The implementation of health promotion: a new structural perspective. *Social Science & Medicine, 41*(12), 1627–1637.

Ruetten, A., Abu-Omar, K., Frahsa, A., & Morgan, A. (2009). Assets for policy making in health promotion: overcoming political barriers inhibiting women in difficult life situations to access sport facilities. *Social Science & Medicine, 69*, 1667–1673.

Ruetten, A., Abu-Omar, K., Levin, L., Morgan, A., Groce, N., & Stuart, J. (2008). Research note: social catalysts in health promotion implementation. *Journal of Epidemiology and Community Health, 62*, 560–565.

Ruetten, A., Abu-Omar, K., Seidenstuecker, S., & Mayer, S. (2010). Strengthening the assets of women living in disadvantaged situations – the German experience. In A. Morgan, E. Ziglio, & M. Davies (Eds.), *International health and development: Investing in assets of individuals, communities and organizations*. New York – Dordrecht – Heidelberg – London: Springer.

Ruetten, A., & Gelius, P. (2011). The interplay of structure and agency in health promotion: integrating a concept of structural change and the policy dimension into a multi-level model and applying it to health promotion principles and practice. *Social Science & Medicine, 73*(7), 953–959.

Ruger, J. P. (2004). Ethics of the social determinants of health. *The Lancet, 364*, 1092–1097.

Ruger, J. P. (2010). Health capability. Conceptualization and operationalization. *American Journal of Public Health, 100*(1), 41–49.

Sen, A. (1985). *Commodities and capabilities*. Amsterdam: Elsevier Science Publishers.

Sen, A. (1987). The standard of living. In G. Hawthorn (Ed.), *The standard of living*. Cambridge: Cambridge University Press.

Sen, A. (1993). Capability and well-being. In M. Nussbaum, & A. Sen (Eds.), *The quality of life* (pp. 30–53). Oxford: Clarendon Press.

Sen, A. (1999). *Development as freedom*. New York: Random Books.

Sewell, W. H. (1992). A theory of structure duality, agency, and transformation. *American Journal of Sociology, 98*, 1–29.

Shim, J. K. (2002). Understanding the routinised inclusion of race, socio-economic status and sex in epidemiology: the utility of concepts from technoscience studies. *Sociology of Health & Illness, 24*, 129–150.

Singh-Manoux, A., & Marmot, M. (2005). Role of socialization in explaining social inequalities in health. *Social Science & Medicine, 60*(9), 2129–2133.

Stewart, F. (2005). Groups and capabilities. *Journal of Human Development, 6*(2), 185–204.

Swartz, D. (1997). *Culture and power: The sociology of Pierre Bourdieu*. Chicago: University of Chicago Press.

Veenstra, G. (2007). Social space, social class and Bourdieu: health inequalities in British Columbia, Canada. *Health Place, 13*(1), 14–31.

Wacquant, L. J. D. (1992). Toward a social praxeology: the structure and logic of Bourdieu's sociology. In P. Bourdieu, & L. J. D. Wacquant (Eds.), *An invitation to reflexive sociology* (pp. 1–59). Chicago: University of Chicago Press.

Weber, M. (1978). *Economy and society: An outline of interpretive sociology*. Berkeley: University of California Press.

Williams, G. H. (2003). The determinants of health: structure, context and agency. *Sociology of Health & Illness, 25*(3), 131–154.

Williams, S. J. (1995). Theorising class, health and lifestyles: can Bourdieu help us? *Sociology of Health & Illness, 17*(5), 577–604.

World Health Organization. (1986). Ottawa charter for health promotion. *Health Promotion International, 1*(4), 405.

WHO-EURO. (1984). Health promotion: concepts and principles. In *A selection of papers presented at the working group on concept and principles*. Copenhagen, 9–13 July 1984. Copenhagen: WHO-EURO.

14

Social Determinants of Health Inequalities

Michael Marmot

There are gross inequalities in health between countries. Life expectancy at birth, to take one measure, ranges from 34 years in Sierra Leone to 81.9 years in Japan.[1] Within countries, too, there are large inequalities – a 20-year gap in life expectancy between the most and least advantaged populations in the USA, for example.[2] One welcome response to these health inequalities is to put more effort into the control of major diseases that kill and to improve health systems.[3,4]

A second belated response is to deal with poverty. This issue is the thrust of the Millennium Development Goals.[5,6] These goals challenge the world community to tackle poverty in the world's poorest countries. Included in these goals is reduction of child mortality, the health outcome most sensitive to the effects of absolute material deprivation.

To reduce inequalities in health across the world there is need for a third major thrust that is complementary to development of health systems and relief of poverty: to take action on the social determinants of health. Such action will include relief of poverty but it will have the broader aim of improving the circumstances in which people live and work. It will, therefore, address not only the major infectious diseases linked with poverty of material conditions but also non-communicable diseases – both physical and mental – and violent deaths that form the major burden of disease and death in every region of the world outside Africa and add substantially to the burden of communicable disease in sub-Saharan Africa.

Source: *The Lancet*, 365(9464) (2005): 1099–1104.

Panel 1: The Commission on Social Determinants of Health

The Commission will not only review existing knowledge but also raise societal debate and promote uptake of policies that will reduce inequalities in health within and between countries.

The Commission's aim is, within 3 years, to set solid foundations for its vision: the societal relations and factors that influence health and health systems will be visible, understood, and recognised as important. On this basis, the opportunities for policy and action and the costs of not acting on these social dimensions will be widely known and debated. Success will be achieved if institutions working in health at local, national, and global level will be using this knowledge to set and implement relevant public policy affecting health. The Commission will contribute to a long-term process of incorporating social determinants of health into planning, policy and technical work at WHO.

To understand the social determinants of health, how they operate, and how they can be changed to improve health and reduce health inequalities, WHO is setting up an independent Commission on Social Determinants of Health, with the mission to link knowledge with action (panel 1). Public policy – both national and global – should change to take into account the evidence on social determinants of health and interventions and policies that will address them.

This introduction to the Commission's task lays out the problems of inequalities in health that the Commission will address and the approach that it will take. This report will argue that health status should be of concern to all policy makers, not merely those within the health sector. If health of a population suffers it is an indicator that the set of social arrangements needs to change. Simply, the Commission will seek to have public policy based on a vision of the world where people matter and social justice is paramount.

Inequalities in Health Between and Within Countries: Poverty and Inequality

A catastrophe on the scale of the Indian Ocean tsunami rightly focuses attention on the susceptibility of poor and vulnerable populations to natural disasters. It is no less important to keep on the agenda the more enduring problem of inequalities in health among countries.

Children

Under-5 mortality varies from 316 per 1000 livebirths in Sierra Leone to 3 per 1000 livebirths in Iceland, 4 per 1000 livebirths in Finland, and 5 per 1000 livebirths in Japan.[1] In 16 countries (12 in Africa), child mortality rose in the 1990s,[7] by 43% in Zimbabwe, 52% in Botswana, and 75% in Iraq.[8]

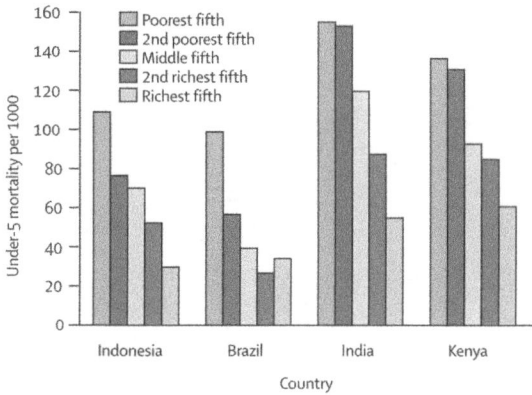

Figure 1: Under-5 mortality rates per 1000 livebirths by socioeconomic quintile of household

Figure 1 shows under-5 mortality rates for four countries with households classified according to socioeconomic quintile. Child mortality varies among countries.[9] Within countries, not only is child mortality highest among the poorest households but also there is a social gradient: the higher the socioeconomic level of the household the lower the mortality rate.

Adults

Differences in adult mortality among countries are large and growing. Figure 2 shows probability of death between age 15 and 60 years by region of the world between 1970 and 2002.[7] Mortality rose in Africa and in the countries of central and eastern Europe whereas it declined in the world as a whole. By 2002, for example, men in the high mortality countries of Europe had more than 40% probability of death between age 15 and 60 years compared to a 25% probability in southeast Asia. These data are for regions. Among countries, the differences are even more dramatic. The probability of a man dying between age 15 and 60 years is 8.3% in Sweden, 82.1% in Zimbabwe, and 90.2% in Lesotho.[7]

A particularly telling example of health inequalities within countries is the 20-year gap in life expectancy between Australian Aboriginal and Torres Strait Islander peoples – life expectancy is 56.3 years for men and 62.8 years for women – and the Australian average.[10] The men in this population would look unhealthy in India (male life expectancy 60.1 years) whereas Australian life expectancy is among the highest in the world, marginally behind Iceland, Sweden, and Japan. The poor health of Aboriginal and Torres Strait Islander peoples is not the result of a high rate of child deaths. Infant mortality is 12.7 per 1000 livebirths. This figure is high by Australian standards, but on a scale from Iceland to Sierra Leone, it is much closer to Iceland than to Sierra Leone. The shortened life expectancy of Aboriginal and Torres Strait Islander

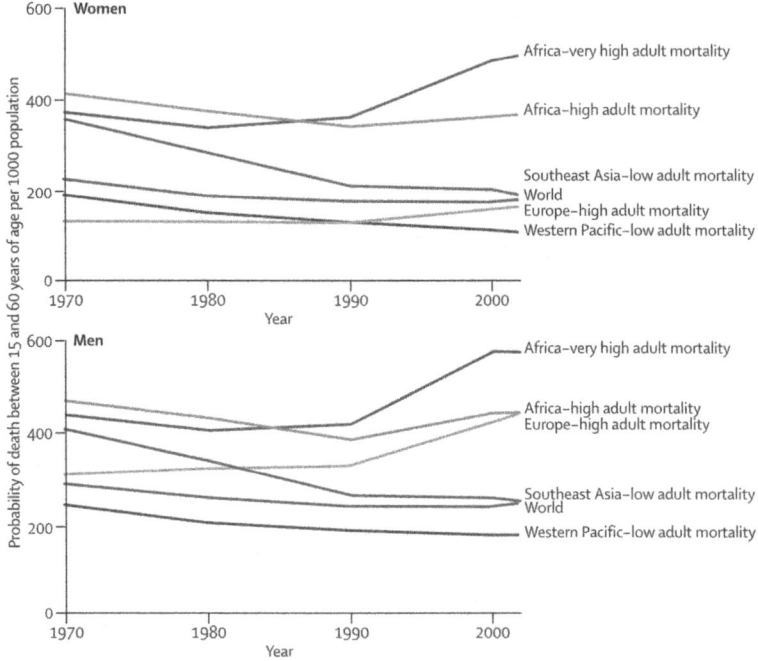

The graphs show the probability of death between 15 and 60 years of age per 1000 population. Reprinted from reference 7 with permission of the World Health Organization.
Reprinted with permission from World Health Organization (WHO).

Figure 2: Trends in adult mortality by sex in regions of the world, 1970–2002

peoples results from mortality in adults from non-communicable disease and injury. In this sense, the population is typical of the world health picture. Of the 45 million deaths among adults age 15 years and older in 2002, 32 million were due to non-communicable disease and a further 4.5 million to violent causes.[7]

Aboriginal and Torres Strait Islander peoples are a socially excluded minority within their country. But poor health is not confined to poor populations or those who are socially excluded. As with child mortality, there is a socioeconomic gradient in adult mortality rates within countries. Figure 3 shows that in Bangladesh, adult mortality rates vary inversely with level of education.[11] This gradient in mortality is quite remarkable. Within rich countries, with strikingly different material conditions from Bangladesh, there is a social gradient in mortality prompting consideration of the causal links between status and health.[12] Whether the social gradient in poor countries can be attributed to the same causal pathways is an urgent task for review. It is especially important because, in many countries, inequalities in health have been increasing.[13–15] In Russia for example, where life expectancy is low, social inequalities have grown (figure 4).[16]

Mortality statistics are readily available. They should not, however, lead to ignorance of the burden of non-fatal disease. In particular, mental illness causes much suffering but its effect is not clear by inspection of mortality

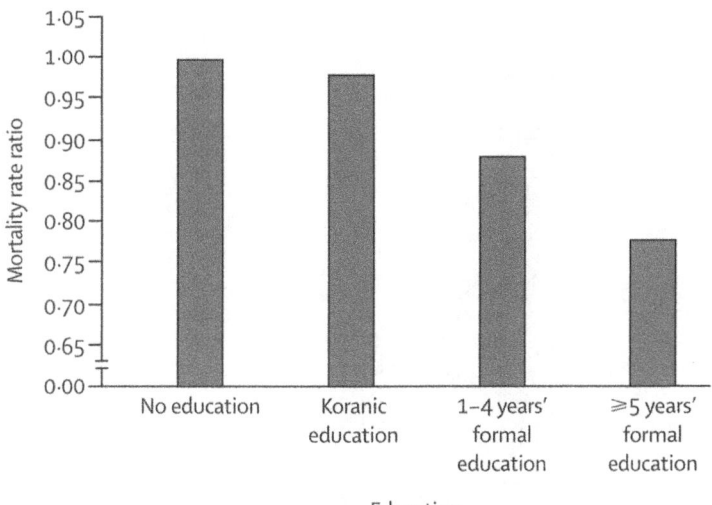

Figure 3: Mortality and education in men aged 45–90 years in Matlab, Bangladesh, 1982–98[11]

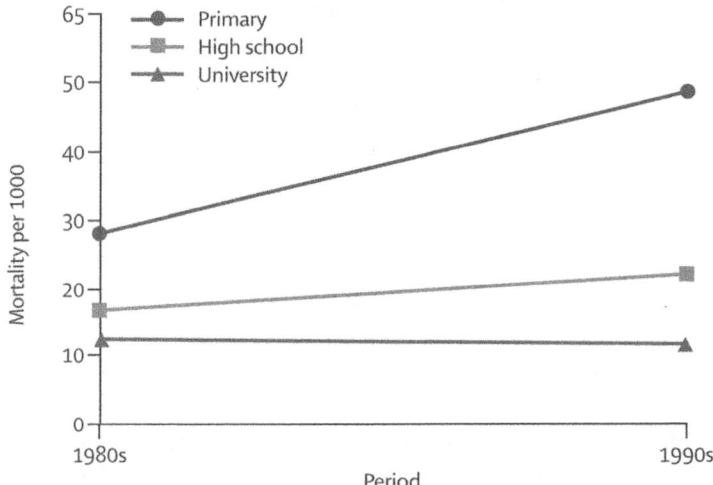

Figure 4: Increase in educational differentials in mortality between the 1980s and 1990s in St Petersburg men[16]

data. Worldwide, the second highest cause of disease burden among adults age 15–59 years is unipolar depressive disorder.[7]

The Ageing of the World's Population

It is convenient, but quite wrong, to think that the greying of the world's population is an issue only for the rich countries. Figure 5 shows the projected increase between 2000 and 2030 in the population older than 65 years in

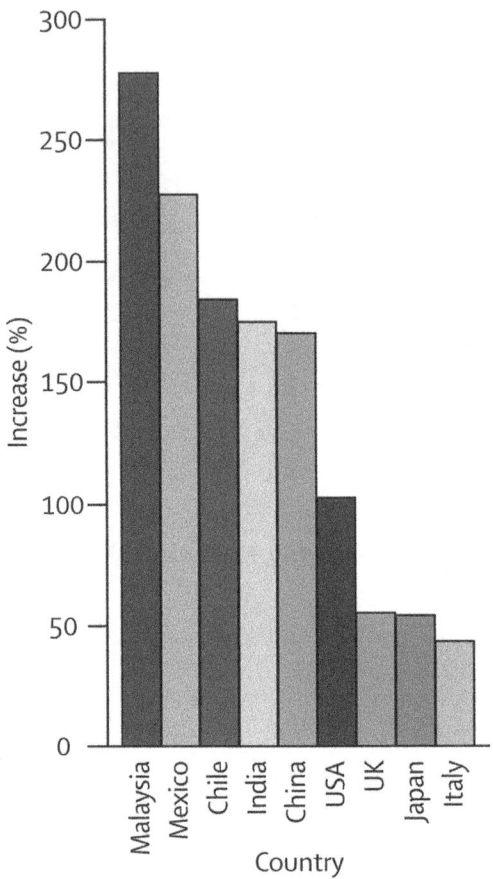

Adapted from reference 17 with permission of the US Census Bureau.

Figure 5: Projected percentage increase in the elderly population (older than 65 years) from 2000 to 2030 in selected countries

selected countries.[17] The fastest rates of increase are in countries at an intermediate level of human development, starting from a low base. The social determinants of the health of older people claim attention alongside those of health at younger ages.

Social Determinants: Poverty, Inequality, and the Causes of the Causes

In consulting widely in developing the plan for the Commission on Social Determinants of Health, a common question was: "What's new? We know that poverty is bad for health. Does that need a Commission?"

It is not difficult to understand how poverty in the form of material deprivation – dirty water, poor nutrition – allied to lack of quality medical care can account for the tragically foreshortened lives of people in Sierra Leone. Such understanding is insufficient in two important ways. First, it fails properly to take into account that relief of such material deprivation is not simply a technical matter of providing clean water or better medical care. Who gets these resources is socially determined.[18] Second, and related, international policies have not been pursued as if they had people's basic needs in mind. The critics of the policies pursued by the International Monetary Fund in the global South have argued eloquently that the economic policies pursued under structural adjustment have not benefited disadvantaged people in poor countries.[19] Recognising the health effects of poverty is one thing. Taking action to relieve its effects entails a richer understanding of the health effects of social and economic policies.

Dirty water, lack of calories, and poor antenatal care cannot account for the 20-year deficit in life expectancy of Australian Aboriginal and Torres Strait Islander peoples. On a world scale, their infant mortality rate, at 12.7 per 1000 livebirths, is low. Their high rate of adult mortality is from cardiovascular diseases, cancers, endocrine nutritional and metabolic diseases (including diabetes), external causes (violence), respiratory disorders, and digestive diseases.[10] This fact is not to deny that poverty is important. But the form that poverty takes and its health consequences are quite different when considering chronic disease and violent deaths in adults, compared to deaths from infectious disease in children. It entails a richer understanding of the social determinants of health.

The health experience of Aboriginal and Torres Strait Islander peoples has relevance for the health of disadvantaged people worldwide. While in Africa the major contributor to premature mortality is communicable disease, in every other region of the world it is non-communicable disease.[1] Careful analysis of the global burden of disease has pointed to the importance of risk factors, such as being overweight, smoking, alcohol, and poor diet.[20] These are indeed potent causes. But would it be helpful to go into a deprived Australian Aboriginal population and point out that they should really take better care of themselves – that their smoking and obesity were killing them; and if they must drink, please do so in moderation? Unlikely. To borrow Geoffrey Rose's term, we need to examine the causes of the causes:[21] the social conditions that give rise to high risk of non-communicable disease whether acting through unhealthy behaviours or through the effects of impossibly stressful lives[12] (panel 2).

A further answer to the what's new question: although it might be obvious that poverty is at the root of much of the problem of infectious disease, and needs to be solved, it is less obvious how to break the link between poverty and disease. Income poverty provides, at best, an incomplete explanation of differences in mortality among countries or among subgroups within countries. It is well known that among rich countries, there is little correlation between gross national product (GNP) per person and life expectancy.

Panel 2: The Solid Facts

Because the causes of the causes are not obvious, the WHO Regional Office for Europe asked a group at University College London to summarise the evidence on the social determinants of health, published as *The Solid Facts*.[22] It had ten messages on the social determinants of health based on:

- the social gradient
- stress
- early life
- social exclusion
- work
- unemployment
- social support
- addiction
- food
- transport

As an indication that there was a ready audience for these messages, in the first 12 months after publication of the second edition it was downloaded from the internet 218 000 times.

The Solid Facts reviewed evidence from Europe, aimed mainly at reducing inequalities in health within countries. The task of the Commission will be to review evidence on the social determinants of health that are relevant to global health: inequalities among countries and within.

Greece for example, with a GNP at purchasing power parities of just more than US$17 000, has a life expectancy of 78.1 years; the USA, with a GNP of more than $34 000, has a life expectancy of 76.9 years. Costa Rica and Cuba stand out as countries with GNPs less than $10 000 and yet life expectancies of 77.9 years and 76.5 years.[23]

There are many examples of relatively poor populations with similar incomes but strikingly different health records.[8] Kerala and China, famously, have good health, despite low incomes.[24] The social processes that lead to this beneficial state of health need not wait for the world order to be changed to relieve poverty in the worst-off countries. A social determinants perspective is crucial. It is also important to enquire whether the action that is taking place to relieve poverty is having the desired effect not only on average incomes but also on income distribution and hence on the poorest people.

The social gradient in health is a particular challenge. Where material deprivation is severe, a social gradient in mortality could arise from degrees of absolute deprivation. In rich countries with low levels of material deprivation the gradient changes the focus from absolute to relative deprivation.[25] Relative deprivation relates to a broader approach to social functioning and meeting of human needs[12] – capabilities in the words of Amartya Sen,[26] spiritual resources to use Robert Fogel's term.[27] It is likely that both material or physical needs and capability, spiritual, or psychosocial needs are important to the gradient in health, which will, therefore, be an important focus.

A focus on material conditions and control of infectious disease must not be to the exclusion of social determinants. The circumstances in which people live and work are as important for communicable as they are for non-communicable disease. Social conditions powerfully influence both the onset and response to treatment of the major infectious diseases that kill.[28,29]

The Commission on Social Determinants of Health will need to have in its sights poverty of the sub-Saharan African sort and the social determinants that account for Bolivia having 14 fewer years of life expectancy than Costa Rica or Aboriginal and Torres Strait Islander peoples having 20 years fewer than other Australians. As these examples illustrate, it will examine inequalities in health between countries and inequalities within.

Action Is Possible and Necessary

A review of policies in European countries identified several that took action on the social determinants of health.[30] Although the reason for the policies was not necessarily to improve health they were nevertheless relevant to health: taxation and tax credits, old-age pensions, sickness or rehabilitation benefits, maternity or child benefits, unemployment benefits, housing policies, labour markets, communities, and care facilities.

In Sweden, the new strategy for public health is "to create social conditions that will ensure good health for the entire population".[31] Of 11 policy domains, five relate to social determinants: participation in society, economic and social security, conditions in childhood and adolescence, healthier working life, and environment and products. These are in addition to health promoting medical care and the usual health behaviours. The UK set reduction of health inequalities as a key aim of health policy. It assembled evidence and expert judgments on areas suitable for policy development.[32] These then formed the basis of a plan of action to reduce health inequalities.[33]

These are examples from rich countries. There are further encouraging examples. Familias en Accion in Colombia transfers cash to poor families. To qualify, families must ensure their children receive preventive health care, enrol in school, and attend classes. The results are encouraging: favourable growth of children and fewer episodes of diarrhoea.[34] The Oportunidades programme in Mexico had somewhat similar aims with similarly encouraging results.[35]

Meeting Human Needs

Two linked themes provide the rationale for the Commission on Social Determinants of Health. First, there is no choice. If the major determinants of health are social, so must be the remedies. Treating existing disease is urgent and will always receive high priority but should not be to the exclusion of taking action on the underlying social determinants of health. Disease control, properly planned and directed, has a good history, but so too does social and economic development in combating major disease and improving population health. Wider social policy will be crucial to reduction of inequalities in health.

There is a second theme that relates to the question of how one can tell if a population is thriving. One standard answer is to measure economic wellbeing with measures such as GNP, average income, or consumption patterns. A better answer is to measure health status.[36] There is no difficulty in convincing medical and health personnel that health is important – that is what we do. It is more challenging, but necessary, to convince policy makers and others that the health of the population is important precisely because it is a measure of whether, in the end, a population is benefiting as a result of a set of social arrangements.

In other words, action on the social determinants of health is necessary not only to improve health but also because such improvement will indicate that society has moved in a direction of meeting human needs.[37] There is a great deal of dogmatic dispute about the rights and wrongs of economic and social policies. People use labels – globalisation, neoliberal economic policies – as badges of allegiance and terms of abuse. The Commission will have one basic dogma: policies that harm human health need to be identified and, where possible, changed. From this perspective, globalisation and markets are good or bad in so far as the way they are operated affects health.

Inequalities in health between and within countries are avoidable.[38] There is no necessary biological reason why life expectancy should be 48 years longer in Japan than in Sierra Leone or 20 years shorter in Australian Aboriginal and Torres Strait Islander peoples than in other Australians. Reducing these social inequalities in health, and thus meeting human needs, is an issue of social justice.

References

1. WHO. The World Health Report 2004: changing history. Geneva: World Health Organization, 2004.
2. Murray CJL, Michaud CM, McKenna MT, Marks JS. US patterns of mortality by county and race: 1965–94. Cambridge: Harvard Center for Population and Development Studies, 1998.
3. WHO. Treating 3 million by 2005: making it happen: the WHO strategy – the WHO and UNAIDS global initiative to provide retroviral therapy to 3 million people with HIV/AIDS in developing countries by the end of 2005. Geneva: World Health Organization, 2003.
4. The Global Fund to fight AIDS, tuberculosis, and malaria. http://www.theglobalfund.org.
5. United Nations Development Group. Millennium Development Goals, http://www.developmentgoals.org.
6. Sachs JD, McArthur JW. The Millennium Project: a plan for meeting the Millennium Development Goals. *Lancet* 2005; 365: 347–53.

7. WHO. World Health Report 2003: shaping the future. Geneva: World Health Organization, 2003.
8. United Nations. Human Development Report 2004. New York: United Nations Development Programme, 2004.
9. Victora CG, Wagstaff A, Schellenberg JA, Gwatkin D, Claeson M, Habicht JP. Applying an equity lens to child health and mortality: more of the same is not enough. *Lancet* 2003; 362: 233–41.
10. Aboriginal and Torres Strait Commissioner, Statistics Human Rights and Equal Opportunity Commission. A statistical overview of Aboriginal and Torres Strait Islander peoples in Australia. http://www.humanrights.gov.au/social_justice/statistics/index.html (accessed Oct 28, 2004).
11. Hurt LS, Ronsmans C, Saha S. Effects of education and other socioeconomic factors on middle age mortality in rural Bangladesh. *J Epidemiol Community Health* 2004; 58: 315–20.
12. Marmot M. Status syndrome. London: Bloomsbury, 2004.
13. Donkin A, Goldblatt P, Lynch K. Inequalities in life expectancy by social class 1972–1999. *Health Stat Q* 2002; 15: 5–15.
14. Mackenbach JP, Bos V, Andersen O, et al. Widening socioeconomic inequalities in mortality in six Western European countries. *Int J Epidemiol* 2003; 32: 830–37.
15. Crimmins EM, Saito Y. Trends in healthy life expectancy in the United States, 1970–1990: gender, racial, and educational differences. *Soc Sci Med* 2001; 52: 1629–41.
16. Plavinski SL, Plavinskaya SI, Klimov AN. Social factors and increase in mortality in Russia in the 1990s: prospective cohort study. *BMJ* 2003; 326: 1240–42.
17. Kinsella K, Velkoff VA, US Census Bureau. An aging world: 2001 – series P95/01-1. Washington: US Government Printing Office, 2001.
18. Kim JY, Millen JV, Irwin A, Gershman J. Dying for growth: global inequality and the health of the poor. Monroe: Common Courage Press, 2000.
19. Stiglitz JE. Globalization and its discontents. London: Allen Lane, 2002.
20. WHO. Reducing risks, promoting healthy life: World Health Report 2002. Geneva: World Health Organization, 2002.
21. Rose G. Strategy of preventive medicine. Oxford: Oxford University Press, 1992.
22. Wilkinson R, Marmot M. The Solid Facts. Copenhagen: World Health Organization, 2003.
23. United Nations Development Programme. Human development report. New York: Oxford University Press, 2003.
24. Sen A. Development as freedom. New York: Alfred A Knopf, 1999.
25. Wilkinson RG. The impact of inequality: how to make sick societies healthier. London: Routledge, 2005.
26. Sen A. Inequality reexamined. Oxford: Oxford University Press, 1992.
27. Fogel RW. The fourth great awakening and the future of egalitarianism. Chicago: University of Chicago Press, 2000.
28. Farmer P. Infections and inequalities. Berkeley: University of California Press, 1999.
29. Farmer P. Pathologies of power: health, human rights, and the new war on the poor. Berkeley: University of California Press, 2003.
30. Crombie IK, Irvine L, Elliott L, Wallace H. Closing the health inequalities gap: an international perspective. Dundee: NHS Health Scotland and University of Dundee, 2004.
31. Hogstedt H, Lundgren B, Moberg H, Pettersson B, Agren G. The Swedish public health policy and the National Institute of Public Health. *Scan J Public Health* 2004; 32 (suppl 64): 1–64.
32. Acheson D. Inequalities in health: report of an independent inquiry. London; HMSO, 1998.

33. Department of Health. Tackling health inequalities: a programme for action. London, Department of Health, 2003.
34. Attanasio O, Vera-Hernandez M. Medium and long run effects of nutrition and child care: evaluation of a community nursery programme in rural Colombia – IFS working papers EWP04/06. London: Institute for Fiscal Studies, 2004.
35. World Bank. Mexico's Oportunidades program. http://www.worldbank.org/wbi/reducingpoverty/case-Mexico-OPORTUNIDADES.html (accessed Feb 9, 2005).
36. Sen A. Mortality as an indicator of success and failure: Innocenti Inaugural Lecture 1995. Instituto degli Innocenti, Florence, Italy; March 3, 1995.
37. Doyal L, Gough I. A theory of human need. London: Macmillan, 1991.
38. Whitehead M. The concepts and principles of equity and health. Copenhagen: World Health Organization, 1990.

15

Income Inequality and Population Health: A Review and Explanation of the Evidence

Richard G. Wilkinson and Kate E. Pickett

Introduction

Whether or not the extent of income inequality in a society is a determinant of population health remains a controversial issue despite a large body of research. Although the findings of a substantial majority of studies suggest that more egalitarian societies do have better health and longevity (Lynch, Smith, & Harper, 2004a; Subramanian & Kawachi, 2004), a minority conclude otherwise and several authorities remain skeptical as to whether inequality has any implications for population health (Deaton, 2003; Lynch et al., 2004a). To gain a clearer understanding of the evidence and the nature of the disagreement, we decided to review all the research reports published in peer reviewed journals, and then to see if we could arrive at an interpretation of them which made sense of both the supportive and unsupportive findings.

The Review

We compiled a list of 155 published peer reviewed reports of research on the relation between income distribution and measures of population health. This is much the most comprehensive list of studies yet compiled: as well as containing all the eligible studies listed in three previous reviews of parts

Source: *Social Science & Medicine*, 62(7) (2006): 1768–1784.

of the literature (Hsieh & Pugh, 1993; Lynch et al., 2004a; Subramanian & Kawachi, 2004), we also found 37 additional papers either by using electronic searches or through informal contacts. Several papers contained analyses at more than one level of aggregation (for example analyses of international data and data from states or regions within one country). The 155 papers contained 168 separate analyses. To facilitate comparison, we decided to classify findings using the same criteria as Lynch et al. (2004a). Thus, we classified analyses into three categories according to their findings after adjustment for whatever control variables authors thought appropriate. We classified them as "wholly supportive" if they reported *only* statistically significant associations between greater income inequality and poorer population health; as "unsupportive" if they found *no* statistically significant positive associations; and as "partially supportive" or "mixed" if some, but not all, of the associations they reported showed significant positive associations. [These categories correspond to those labeled "positive", "negative" and "mixed" in Lynch et al. (2004a)]

Findings

Table 1 provides a summary of the 168 analyses according to classification and the type of area over which inequality was measured. Table 2 lists all the analyses included according to their classification. A tally of numbers showed 87 wholly supportive analyses, 44 partially supportive, and 37 unsupportive. Almost three-quarters of all analyses were classified as either wholly or partially supportive. Of all analyses classified as wholly supportive or unsupportive, 70 per cent were wholly supportive. Given that almost every paper reported many different measures of association, for different health variables in different age, sex or ethnic subgroups, it is notable that only nine (classified as partially or unsupportive) contained any measures of an association suggesting a health variable was significantly better where inequality was greater. In no analysis were such associations the predominant finding.

Table 1 Summary of results of 169 analyses of the relation between income distribution and population health contained in 155 papers (in parentheses: homicide studies)

	Wholly supportive	Partially supportive	Unsupportive	Total	Wholly supportive as per cent of all analyses
	Only sig. positive findings	Some sig. positive and some null	No sig. positive findings	All studies	excluding partially supportive (%)
Nations	30 (11)	9	6	45 (11)	83
States, regions, cities	45 (13)	21	17	83 (13)	73
Counties, tracts, parishes	12 (2)	14	14 (1)	40 (3)	45
Total	87 (26)	44	37 (1)	168 (27)	70

Table 2: Studies of income inequality and health

	Supportive studies	Mixed studies	Unsupportive studies
International studies	Avison and Loring (1986[HT])	Ellison (2002)	Beckfield (2004[N,U])
	Davey Smith and Egger (1996)	Judge et al. (1998)	Bobak et al. (2000)[M,U]
	De Vogli et al. (2005[N])	Lester (1987[N])	Gravelle et al. (2002)
	Drain, Smith, Hughes, Halperin, and Holmes (2004[N])	Lobmayer and Wilkinson (2000)	Judge (1995)
	Duleep (1995)	Lynch et al. (2001[RC])	Mellor and Milyo (2001)
	Fajnzylber et al. (2002[N,H])	Pampel (2002[M])	Wildman et al. (2003)
	Flegg (1979, 1982)	Pampel and Pillai (1986)	
	Groves et al. (1985[H])	Ross et al. (2005[N])	**6 studies**
	Hales, Howden-Chapman, Salmond, Woodward, and Mackenbach (1999)	Weatherby et al. (1983)	
	Hansmann and Quigley (1982[H])	**9 studies**	
	Kick and LaFree (1985[H]) Krahn et al. (1986[H])		
	Krohn (1976[H])		
	Lee and Bankston (1999[H])		
	Legrand (1987)		
	Macinko et al. (2004)		
	Marmot and Bobak (2000)		
	McIsaac and Wilkinson (1997)		
	Messner (1980[H], 1982[H], 1989[H])		
	Pampel and Zimmer (1989)		
	Pickett, Kelly, Lobstein, Brunner, and Wilkinson (2005[N])		
	Pickett, Mookherjee, and Wilkinson (2005[N])		
	Rodgers (2002)		
	Steckel (1983)		
	Waldmann (1992)		
	Wennemo (1993)		
	Wilkinson (1992)		
	30 studies, including 11 of homicide only		
States, regions, metropolitan areas			
1. US states	Baron and Straus (1988[H])	Blakely et al. (2002)	Deaton and Lubotsky (2003[U])
	Blakely, Kennedy, Glass, and Kawachi (2000[M])	Daly, Duncan et al. (1998[M])	Henderson, Liu, Diez Roux, Link, and Hasin (2004[M,U])
	Blakely et al. (2001[M])	Diez-Roux et al. (2000[M])	McLeod et al. (2004[N,U])

(Continued)

Table 2: (*Continued*)

	Supportive studies	Mixed studies	Unsupportive studies
	Daly et al. (2001)[H] Gold, Kennedy, Connell, and Kawachi (2002)[N] Holtgrave and Crosby (2004)[N] Huff-Corzine et al. (1986)[H] Kahn, Wise, Kennedy, and Kawachi (2000)[M] Kaplan, Pamuk, Lynch, Cohen, and Balfour (1996) Kawachi and Kennedy (1997) Kawachi, Kennedy, Lochner, and Prothrow-Stith (1997) Kennedy, Kawachi, Glass, and Prothrow-Stith (1998)[H] Kennedy et al. (1996) Kennedy et al. (1998)[M] Loftin and Hill (1974)[H] Pickett, Mookherjee, and Wilkinson (2005)[N] Ross et al. (2000)[RC] Shi (1999) Smith and Parker (1980)[H] Subramanian, Blackely, and Kawachi (2003), Subramanian and Kawachi (2003)[M] Subramanian and Kawachi (2003[M], 2004[N]) Wilkinson et al. (1998) Wolfson, Kaplan, Lynch, Ross, and Backlund (1999)[N] **25 studies, including 6 of homicide only**	Holtgrave and Crosby (2003) Kahn, Tatham et al. (1998) Laporte (2002)[RC] Lochner, Pamuk, Makuc, Kennedy, and Kawachi (2001)[M] Lynch, Smith, Harper et al. (2004N) Mayer and Sarin (2005[N,M]) Mellor and Milyo (2001) Reagan and Salsberry (2005[N,M]) Shi et al. (2004N) Subramanian et al. (2001)[M] **13 studies**	Mellor and Milyo (2002[M,U], 2003[M,U]) Muller (2002[U]) Shi et al. (2003[U]) **7 studies**
2. Canadian provinces	Daly et al. (2001)[H]		Laporte and Ferguson (2003)
3. UK regions			Ross, Wolfson, Dunn et al. (2000)
4. Spanish regions		Weich, Lewis, and Jenkins (2001[M], 2002[M])	Regidor, Calle, Navarro, and Dominguez (2003) Regidor, Navarro, Dominguez, and Rodriguez (1997)
5. Brazilian states	De Vogli et al. (2005)[N]		Messias (2003[U])
6. Italian regions	Walberg et al. (1998)		
7. Russian regions			

Area			
8. Chilean regions	Subramanian et al. (2003[M])		
9. Israeli regions		Shmueli (2004[N])	
10. Finnish regions			Blomgren, Martikainen, Makela, and Valkonen (2004 [NM])
11. Ecuadorean regions Metropolitan areas and cities – studies are of USA unless otherwise noted	Larrea and Kawachi (2005[NM]) Bailey (1984[H]) Balkwell (1990[H]) Blau and Blau (1982[H]) Chiang (1999) – *Taiwan* Cooper et al. (2001) Kahn et al (1999[N]) Kennedy et al. (1991[H]) – *Canada* Loftin and Parker (1985[H]) Lopez (2004[N]) Lynch et al. (1998) Messner (1983[H]) Ronzio (2003[N]) Ronzio et al. (2004[N]) Sanmartin et al. (2003[N]) Shi and Starfield (2000[M], 2001[RC]) **16 studies, including 6 of homicide only**	Blakely et al. (2002[MRC]) Larrea and Kawachi (2005[NM]) – *Ecuador* Lobmayer and Wilkinson (2002) Sanmartin et al. (2003[N]) – *Canada* Shi, Starfield, Politzer, and Regan (2002[M]) **5 studies**	Deaton and Lubotsky (2003[U]) McLeod, Lavis, Mustard, and Stoddart (2003[M]) – *Canada* Mellor and Milyo (2002[MU]) Sturm and Gresenz (2002[MU]) **4 studies**
Small areas: Counties, Neighborhoods, Zip codes, Census Tracts, Parishes, etc. – studies are of USA unless otherwise noted	Baldani et al (2004[N]) Galea et al. (2003[M]) Gold, Kawachi, Kennedy, Lynch, and Connell (2001[N]) Messner and Tardiff (1986[H]) Massing et al. (2004[N])	Brodish et al. (2000) Fiscella and Franks (2000[M]) Franzini et al. (2001)	

(Continued)

Table 2: (*Continued*)

Supportive studies	Mixed studies	Unsupportive studies
Muramatsu (2003[N])	Gold et al. (2004[N,M])	Blakely, Atkinson, and O'Dea (2003[M,U]) – *New Zealand*
Pattussi, Marcenes, Croucher, and Sheiham (2001[N]) – *Brazil*	Hou and Chen (2003)	Blakely et al. (2002[U])
Shi et al. (2005[N])	LeClere and Soobader (2000[M])	Drukker et al. (2004[N]) – *Netherlands*
Soobader and LeClere (1999[M])	McLaughlin and Stokes (2002)	Fiscella and Franks (1997[M,U])
Stanistreet et al. (1999) – *UK*	McLaughlin et al. (2001)	Franzini and Spears (2003[M])
Subramanian, Delgado, Jadue et al. (2003) – *Chile*	Osler et al. (2003[M]) – *Denmark*	Hou and Myles (2005[N,M,U]) – *Canada*
Szwarcwald, Bastos, Viacava et al. (1999[H]) – *Brazil*	Osler et al. (2002[M]) – *Denmark*	Larrea and Kawachi (2005[N,M,U]) – *Ecuador*
	Robert and Reither (2004[N,M])	Lorant, Thomas, Deliege, and Tonglet (2001) – *Belgium*
	Sohler et al. (2003)	Muntaner et al. (2004[N,M,U])
	Szwarcwald, Andrade, and Bastos (2002[RC]) – *Brazil*	Osler et al. (2002[M,U]) – *Denmark*
	Veenstra (2002a) – *Canada*	Shibuya et al. (2002[M]) – *Japan*
		Szwarcwald, Bastos, Viacava et al., (1999) – *Brazil*
		Veenstra (2002b[U]) – *Canada*
		Wen et al. (2003[M])
12 studies, including 2 of homicide only	**14 studies**	**14 studies, including 1 of homicide only**

This table contains the 98 studies reviewed by Lynch et al. (2004a, b) (including all peer-reviewed studies in Subramanian and Kawachi, 2004) and the 20 studies of income inequality and homicide reviewed by Hsieh and Pugh (1993), as well as 37 additional studies which were omitted from earlier reviews or have been published more recently. These 155 papers contain the 168 separate analyses listed here. To aid comparisons with other reviews the following notation is used in the table.
[N]Denotes new and previously omitted studies (n = 37).
[H]Denotes studies where homicide was the only outcome (n = 25).
[RC]Denotes studies that we re-classified from the Lynch et al. (2004) review, as results did not match the classification given previously (n = 6).
[M]Denotes multi-level, rather than ecological studies, although not all of these use appropriate multi-level statistical techniques.
[U]Denotes "unsupportive" studies that reported an unadjusted association between income inequality and health that was removed when inappropriate control variables were added to models.

However, our aim in this paper is to go beyond these crude categories, which take no account of methodological quality or statistical power, and to try to gain a theoretically coherent overview of the literature as a whole. We shall proceed by drawing attention to various patterns in the findings and the likely implications of each. In doing so we will outline a consistent interpretation of most of the evidence, both supportive and unsupportive. We shall end by discussing possible criticisms and alternative interpretations.

Explanations of Findings

The Size of Area

Table 1 shows the percentage of analyses classified as either wholly supportive or unsupportive according to whether they were international analyses using data for whole countries, whether their data were for large subnational areas such as states, regions and metropolitan areas, or whether they were for smaller units such as counties, census tracts or parishes. The proportion of analyses classified as wholly supportive falls from 83 per cent (of all wholly supportive or unsupportive) in the international studies, to 73 per cent in the large subnational areas, to 45 per cent among the smallest units.

The tendency towards more positive findings in the largest areas compared to the smallest is important and has already received some attention in the literature (Subramanian & Kawachi, 2004). The same pattern was observed in a review of studies of the relation between homicide and inequality (Hsieh & Pugh, 1993). It was also shown in a study by Franzini, Ribble, and Spears (2001) comparing the strength of association among the counties of Texas according to population size. Wilkinson (1997) has argued that income inequality in small areas is affected by the degree of residential segregation of rich and poor and that the health of people in deprived neighbourhoods is poorer not because of the inequality *within* their neighbourhoods, but because they are deprived in relation to the wider society. If that is what matters, then it is to be expected that inequality will only be sensitive to this broader pattern of deprivation if inequality is measured across the wider framework in which the relevant social comparisons are made. The fact that measures of inequality made across larger areas are more closely related to health bears out this point.

This takes us to a familiar and difficult question: if inequality is important, what are the relevant social comparisons? Rather than suggesting any new causal processes or framework of comparisons which affect health, it is more parsimonious to suggest that inequality is related to health insofar as it serves as a measure of the extent of the same processes of class differentiation and social distances in a society which are responsible for class differences in health. The processes which lead to class differences in health are likely to be closely related to those which explain why greater inequality is related to worse health. If that is right, then the question becomes one of the scale of the social units in relation to which one's class position is defined. The broad impression is that social class stratification establishes itself primarily as a national social structure, though there are perhaps also some more local civic hierarchies – for instance within cities and US states. But it should go without saying that classes are defined in relation to each other: one is higher because the other is lower, and vice versa. The lower class identity of people in a poor neighbourhood is inevitably defined in relation to a hierarchy which includes a knowledge of the existence of superior classes who may live in other areas some distance away.

Control Variables

If, in the association between income inequality and health, we are seeing the effects of the scale of social class stratification, of bigger or smaller class differences, then it is hard to decide what are legitimate variables to use as controls when analysing that association. What is part of class and what is not? If we had classified analyses by their findings *before* the use of control variables, 21 of the 37 studies we have listed as unsupportive of an association between income distribution and health actually started off with supportive findings but then lost them as a result of the various control variables they used.

A wide variety of control variables have been used, including the per cent without a high school education (Muller, 2002), individual income (reviewed in Subramanian & Kawachi, 2004), perceived control (Bobak, Pikhart, & Rose, 2000), ethnicity (Blakely, Atkinson, & O'Dea, 2003; Deaton & Lubotsky, 2003), social capital (Veenstra, 2002a), and unemployment (Shi & Starfield, 2000). Subramanian and Kawachi (2004) have a useful discussion of possible confounding by education, individual income, race, and regional effects. To know which are genuine confounders and which are pathway, or mediating, variables means – for us – knowing what is part of social class and what is not. If ethnicity is related to health because it is a proxy for a classification by class, then perhaps we should not control for ethnicity.

Similarly, if Sahlins (1974) was right to say "Poverty is not a certain small amount of goods (but) . . . a relation between people . . . a social status . . . an invidious distinction between classes . . ." (p. 37), then it may be misconceived even to control for individual income. It has been suggested a number of times that the social gradient in health within countries is primarily a gradient in relative income, or social status, rather than a reflection of absolute material living standards. Marmot (2004) and others (Charlesworth, Gilfillan, & Wilkinson, 2004; Singh-Manoux, Adler, & Marmot, 2003; Wilkinson, 2005) have argued that the relation between health and social status may be primarily a reflection of the effects of social position itself. This view is strongly supported by the fact that the international relation between

Table 3: Partial correlation coefficients showing the independent relations of income inequality and gross domestic product per capita to life expectancy (M&F) among 21 rich countries

	Partial corr. with life expectancy	Significance p
GDP per capita	−0.034	0.887
Income inequality	−0.512	0.021

Note: Gross domestic product was measured at purchasing power parities. Income inequality was measured as the ratio of the top 20% to the bottom 20% of incomes. The 21 countries included are all those with populations over 3 million (to exclude tax havens) among the richest 50 countries for which GDP and income data were available from the World Bank World Development Indicators Database 2004. Life expectancy data came from UN World Population Prospects 2003.

Gross National Income per capita and life expectancy not only grows progressively weaker as countries get richer, but disappears altogether among the richest (Marmot & Wilkinson, 2001; Wilkinson, 1997; see also Discussion section below and Table 3). Although there are clearly aspects of rising material living standards which contribute directly to better health even in the richest countries, we believe their effects are not apparent in international comparisons either because they are relatively small, or because they are offset by other factors. It is surely unwise to ignore this evidence and conduct analyses as if any given level of individual income had the same effect on health regardless of the social status it buys. If a person's income is a marker of their social position, then adjusting inequality effects for individual income may be like controlling measures of class stratification for individual social status differentiation.

However, even if this objection to controlling for individual income is ignored, it appears that despite often using small areas, analyses of inequality which use multilevel methods have usually been able to identify inequality effects even after controlling out the effects of individual income (Subramanian & Kawachi, 2004).

That there are so many correlates of income distribution is consistent with our view that income inequality is an indicator of the extent of social stratification and points to the need to think carefully about which factors are confounders and which are mediators in this relation.

Discussion

Taking account of the size of the area and the use of control variables reveals a high degree of consistency in the research findings. Thus, if we confine our attention to the 128 analyses which use data for areas the size of metropolitan areas or larger, only 23 fail to find some support for the hypothesis. If we were to reclassify analyses on the basis of results before the use of potentially problematic control variables (including individual income in multilevel models), then only eight (6 per cent) of the 128 analyses would remain classified as unsupportive.

The extent of social class divisions may vary substantially from country to country: we know that human beings have lived in every kind of society from the most egalitarian (Erdal & Whiten, 1996) to the most tyrannical. Given the importance of the social class gradient in health, the societal differences in the extent of social class inequality are not something we can ignore. Several variables may provide rough measures of the extent of social class differentiation. These might include educational differences, inequalities in the distribution of power or wealth, and perhaps scores on the Social Dominance Orientation Scale (Sidanius & Pratto, 1999). However, income inequality is likely to be one of the most widely applicable. Although it may not be the best measure of social hierarchy in all cultures, the fact that dominance hierarchies (in human societies as among animals) are fundamentally about privileged access to scarce resources, may mean that differences in income and/or wealth are particularly apposite indicators of rank difference across cultures. But even if there are better measures – perhaps ones which include measures of ownership of assets, income inequality has the substantial advantage that it is collected for numerous other purposes and so can be used in secondary data analyses. We hope that this interpretation may bring us closer to the thinking of others working in this field and narrow the area of controversy (Deaton, 2003; Lynch et al., 2004a, b).

The interpretation of the evidence which we have put forward has the advantage of simplicity. Instead of suggesting that inequality is a new risk factor for health, it may be telling us more about the already widely recognised health effects of socioeconomic status and class. It may simply be that larger class differences lead to a steeper social gradient in health, but it could also be that a more unequal society becomes more dominated by status competition and class differentiation and suffers a more widespread health disadvantage as a result. It is already clear from studies designed to illuminate this issue, that the health disadvantages of inequality are not confined to the poorest (Kennedy, Kawachi, & Glass, 1998; Lochner, Pamuk, & Makuc, 2001).

Counter Arguments

What are the objections and counter arguments to these interpretations? The most important is undoubtedly that income is related to health because it is a determinant, not of class differences or social position, but of material living standards which it is claimed continue to exert a major direct influence on health. However, although raising absolute material living standards continues to be important in developing countries, among the 25 or 30 richest countries there is no relation between Gross National Income per capita and health (Marmot & Wilkinson, 2001; Wilkinson, 1997) – even though curves are sometimes still fitted to the data to suggest otherwise (Lynch et al., 2004a). If absolute living standards were overwhelmingly important, it would be difficult to understand why, despite having a median income four times as high, life expectancy among black men in the USA was 9 years shorter than for men in Costa Rica (Marmot & Wilkinson, 2001). Similarly Greece, with half the average real income of the US has, like many other developed countries, better life expectancy. Indeed, looking at the relations between life expectancy, income distribution, and Gross National Product per capita among 21 rich countries, we found only income distribution had significant independent effects on life expectancy. Gross National Product per capita showed no sign at all – regardless of statistical significance – of an independent association. The unweighted partial correlations are shown in Table 3.

We emphasise that rather than meaning material factors can be ignored, psychosocial pathways provide a major new route through which they affect health. But the psychosocial link changes the nature of the relationship, particularly making us more sensitive to relativities and to the social connotations of material differences. And insofar as psychosocial risk factors have drawn attention to the importance of the social environment to health, we believe that the social structure is built substantially on material foundations – hence the importance of inequality.

Alongside reasons such as these for discounting the continued primacy of material influences on health in the rich countries, the relation between income inequality and homicide (Hsieh & Pugh, 1993) shows that inequality has powerful psychosocial and behavioural effects (Wilkinson, 2004). Indeed, the relation with homicide appears to be part of a more general effect which inequality has on social capital and the quality of social relations, both of which might be expected to influence health. As well as having higher levels of violence, people in more unequal societies also seem less likely to trust others and less likely to be involved in community life (Wilkinson, 2005). Other examples of behavioural effects of inequality include higher teenage pregnancy rates (Gold, Connell, & Heagerty, 2004; Pickett, Mookherjee, & Wilkinson, 2005a) and more obesity (Pickett, Kelly, & Lobstein, 2005b). While it is easy to understand why inequality and increased status differentiation should affect both health and behaviour through psychosocial stress (Wilkinson, 2005), it would be harder to argue that material factors could affect behaviour directly.

A second potential criticism of our interpretation of the studies of income inequality and health is the view, derived from relative deprivation theory, that people compare themselves with near equals (Runciman, 1966). No doubt this explains why some have chosen to measure inequality in small areas, intending to capture the effects of these social comparisons rather than focusing on the wider structure of inequality. However, people's judgment of who their near equals are is dependent on a prior recognition of their class identity and where they fit into the wider class structure. The *logic* of what is happening in a dominance hierarchy when comparisons appear to be between near equals was spelt out by Sapolsky when describing conflict over rank among baboons in the Serengeti. They also seem to compare themselves with near equals:

> A pattern emerged that has grown familiar to me over the years. When you look at the frequencies of dominance interactions, the typical pattern you see is that, for example, number 4 is having his most interactions with 3 and 5, losing to the former, defeating the latter. Number 17 mostly interacts with 16 and 18 (Sapolsky, 2001, p. 95).

As Sapolsky points out, there is no point in animals fighting those which are clearly much higher or lower in the dominance hierarchy: because the outcome is predictable the subordinate recognises its inferiority and avoids making a challenge. The point concerns the logic of ranking systems, not whether humans are like baboons. So, for instance, when the results of a race are announced, a competitor placed second might claim he was really first, or the eighth might claim to have come seventh, but the eighth is unlikely to claim he really came first. However, that rank is only contestable among near equals does not mean that the rest of the social hierarchy is irrelevant: ignoring it may result in ridicule or injury. Who counts as a near equal is merely the converse of recognising who is not a near equal. Similarly, our recognition of our class status is constituted primarily by our recognition of uncontestable status differences.

The logic of ranking systems which leads baboons to conflict mainly with near neighbours, is the same logic which leads us to exclaim of other people "Who do they think they are?" – not so much at those we accept as higher status, but at our equals who pretend to superiority. To maintain rank we have to pay attention to the fine grain of social status: that means keeping up with the Joneses. And even if we live in a neighbourhood in which everyone is poor, that does not mean that we are unaware of those in richer neighbourhoods whose existence defines our lower status and relative poverty. Sociologists agree that class identity is defined by position in the wider society (Bourdieu, 1984; Canadine, 1998). Because classes are mutually defining, researchers cannot identify the effects of social status differentiation, and of our class identity within it, in a statistical context (such as a small, residentially segregated, neighbourhood) which excludes much richer or poorer neighbourhoods.

Sometimes discussion of these issues is further confused by arbitrarily labelling an income variable as absolute rather than relative income. This is important because it is often assumed that a relation between health and absolute income reflects the direct effects of material living standard on health – regardless of the rest of society, whereas any effects of relative income are assumed to reflect psychosocial processes contingent on social status or social comparisons. But in different analyses income differences which are called absolute in one context may be called relative in another (Drukker, Feron, & van Os, 2004; Hsieh & Pugh, 1993). Income differences which make up income inequality within large areas can of course be broken down into inequalities within and between smaller constituent areas (Franzini et al., 2001; Lobmayer & Wilkinson, 2000; Soobader & LeClere, 1999). The smaller and more numerous the constituent areas used, the more of the income inequality in the larger areas gets converted into income differences *between* the small areas and the less that remains as inequality *within* them. That conversion can be done almost *ad infinitum* until, at the limit, all inequality is reduced to differences in income between the smallest (single household) areas. How much of the income differences in a society are analysed as differences in relative income within, and how as differences in absolute income between, small areas is inevitably a by-product of the choice of units of analysis.

In this situation what tends to happen is that people look for an effect of income inequality in small areas, find it weak or non-existent, and report an association between health and the average income of the small areas. Instead of interpreting that as an effect of low income *relative* to the wider society, it is interpreted as evidence of a direct effect of material living standards.

If health among the developed countries is unrelated to the big differences in material living standards between countries, then why should the same differences in living standards have an effect on health when they occur within the same society? The truth is surely that income is related to health where – as within countries – it serves as a marker for position in the national structure of class inequality.

For those who still prefer to believe in the primacy of the direct – over the psychosocially mediated – effects of material living standards on health, and interpret the association between income inequality and health as a reflection of a curvilinear relation between individual income and health (Gravelle, 1998), there is one more major obstacle. Although it once seemed plausible that the curvilinear relation between individual income and health reflected a tendency towards diminishing health returns to increased income, so that any given sum of money made more difference to the health of the poor than the rich, such a pattern cannot explain the findings of this review. If the relation between individual income and health resulted simply from the healthfulness of whatever material standard of living a given income can buy, then these effects would be just as apparent if inequality was measured in small areas as in large areas. According to that interpretation it is the incomes themselves which count – regardless of their social meaning. We would then be left with no explanation of why the overwhelming majority of studies which measure inequality in large areas do report associations whereas only a minority of those using data for small areas do. This confirms our view that it is mistaken to control for individual income in multilevel models because it amounts to controlling the effects of income (and class) inequality for the effects of individual social status. It also seems likely that if we were dealing with the material effects of individual income they would be harder to control away than the effects of class inequality and social differentiation for which we believe income inequality is merely a proxy.

International Analyses

A group of results which our interpretation does not explain, is a small group of international studies using data from between the later 1980s and the mid 1990s. Although 30 of the 45 international studies are classified as wholly supportive and a further nine as partially supportive, there are nevertheless four classified as unsupportive which cannot be attributed to the use of inappropriate control variables, and so run counter to the interpretation we have advanced in this paper.

During the 1980s and early 1990s, when income differences were widening particularly rapidly in many countries, much of the relation between inequality and mortality among rich countries temporarily disappeared (Gravelle, Wildman, & Sutton, 2002; Judge, 1995; Mellor & Milyo, 2001; Wildman, Gravelle, & Sutton, 2003). What happened varied by age group (Judge, Mulligan, & Benzeval, 1998; Lobmayer & Wilkinson, 2000; Lynch, Smith, & Hillemeier, 2001). While infant mortality rates remained consistently related to inequality, the relation was entirely lost with death rates among the middle aged and elderly. The relation was clear earlier (Wilkinson, 1992) and has now reappeared (De Vogli, Mistry, & Gnesotto, 2005). It is noticeable that the publication dates of the positive international studies tends to be earlier than the mixed and negative ones.

The three most likely explanations of why the international relationship temporarily disappeared – except with health in the youngest age groups – are: first, that it was affected by the downward shift in the age distribution of relative poverty which took place in many countries. Although relative poverty had been more common among the elderly, it became more common among young families with children (Kangas & Palme, 2000). Second; in one country after another, death rates among older people began an unprecedentedly rapid decline, particularly from cardiovascular mortality. Perhaps as a result of differences in the uptake of both primary and secondary forms of prevention, the timing of the onset of the decline was earlier in some countries than others (Menotti et al., 2003). It may therefore have affected international comparisons of mortality before the decline became general. Interestingly, three of the very few statistically significant *negative* associations (greater inequality related to better health) reported in any of the studies, were international studies among rich countries of death rates among the elderly. The third possible explanation is that the changes in income distribution may have had lagged effects on mortality, particularly on mortality at later ages (Subramanian & Kawachi, 2004). Although income differences widened particularly rapidly in many countries during the 1980s and 1990s, the relationship between income distribution and infant mortality remained throughout, perhaps because lag times are shortest at youngest ages. Mayer and Sarin (2005) found that neonatal mortality rates are significantly more closely related to current inequality than to inequality 5 years earlier, whereas Subramanian and Kawachi (2004) found the strongest associations with adult health after a lag of 10 or 15 years. Health among adults may then reflect the inequalities of the past.

Mechanism

Low social status and the quality of the social environment are both known to affect health (Berkman & Kawachi, 2000; Marmot & Wilkinson, 1999). Not only are more unequal societies likely to have a bigger problem of low social status, but there is now substantial evidence to suggest that inequality is socially corrosive, leading to more violence, lower levels of trust, and lower social capital (Wilkinson, 2005). Psychosocial factors, many of which are associated with low social status, are known to affect health partly through direct physiological effects of chronic stress (Brunner & Marmot, 1999), and partly through their influence on health related behaviour. Marmot (2004) has argued that low social status is stressful because it reduces people's control over their lives and work. Others have argued that low social status is stressful because people are made to feel looked down on, devalued and inferior (Charlesworth et al., 2004; Wilkinson, 2005). Both suggestions are borne out by a recent review of the most salient stressors affecting cortisol responses (Dickerson & Kemeny, 2005). What matters most are uncontrollable threats to ones social esteem, value and status. As well as explaining the relationship between health and inequality, this approach is concordant with the suggestion that inequality is related to violence because the increased burden of low social status makes more people feel disrespected. Feeling disrespected, put down and humiliated is much the most frequent trigger to violence (Gilligan, 1996; Wilkinson, 2004).

Conclusions

Our interpretation of 168 analyses of the relationship between income inequality and health is that income distribution is related to health where it serves as a measure of the scale of social class differences in a society. In small areas, where income inequality is unlikely to reflect the degree of social stratification in the wider society, it is – as Table 1 shows – less likely to be related to health. The overwhelmingly positive evidence from studies of larger areas suggests that this interpretation is correct. The fact that social stratification is such a fundamental feature of social organisation explains why there are so many socioeconomic factors correlated with inequality.

Many will function, like income inequality itself, as other proxies for the extent of social stratification or socioeconomic inequality. Others still may be mediating or pathway variables.

The methods researchers have used to test the hypothesis that greater inequality is associated with poorer population health have reflected many different assumptions about the mechanisms involved. In effect, a whole family of quite different hypotheses about income distribution and health have been tested. The two most important kinds of differences between the tests are those we have discussed: first, the different sizes of areas in which people have thought inequality most likely to be salient, and second, in what are regarded as legitimate control variables. Studies which have analysed data for areas as small as parishes, and controlled for things as closely related to class as education differentials, have helped clarify how income inequality does not work. Similarly, the suggestion that the per cent of the population who are black explains away the income inequality relation at state (if not county) level in the USA, has been regarded by some as a falsification of the inequality hypothesis (Deaton, 2003); but we think it comes closer to being a confirmation of the underlying view that what matters is the extent of social class differentiation. No one suggests that it is blackness itself which matters. Rather it is the social meaning attached to it – the fact that it serves as a marker for class and attracts class prejudice – which leads both to worse health and to wider income differences. Future tests of the theory that the extent of class inequality is a determinant of population health must test its most plausible form.

One of the most important points to come out of this analysis is that it looks as if there are fundamentally important and measurable differences in the extent of socioeconomic stratification in different societies. While income distribution is a convenient and widely applicable measure, we hope that better ones may be found.

We recognise that some have been reluctant to accept the involvement of psychosocial pathways to ill health and argue that differences in health – such as the social gradients in health or the relation between population health and inequality – are the direct and unmediated effects of exposure to different material circumstances. However, if the argument were to move on to explanations of the range of *behavioural* outcomes which also show social gradients and relations with inequality, the psychosocial mediation of behaviour is undeniable. So, for instance, in the very well-established relation between violence and inequality, the causal chain must run all the way from the material facts of inequality to the psychosocial effects which lead to violence. The same is presumably true of other behavioural outcomes – such as teenage pregnancy, obesity and trust – which seem to be related to inequality as well as showing social gradients. If psychosocial processes consequent on low social status are recognised as having health effects, then it seems likely that they will also have behavioural consequences. Perhaps there are some pathways common to health and many of the social problems known to be more common in poorer areas. The possibility that we shall arrive at a general theory of social gradients capable of explaining (and policy prescriptions capable of reducing) a wide range of social ills, is obviously a worthwhile objective.

References

Avison, W. R., & Loring, P. L. (1986). Population diversity and cross-national homicide. *Criminology, 24*, 733–749.

Bailey, W. C. (1984). Poverty, inequality and city homicide rates. *Criminology, 22*, 531–550.

Baldani, M. H., Vasconcelos, A. G., & Antunes, J. L. (2004). Association of the DMFT index with socioeconomic and dental services indicators in the state of Parana, Brazil. *Cad Saude Publica, 20*(1), 143–152.

Balkwell, J. W. (1990). Ethnic inequality and the rate of homicide. *Social Forces, 69*(1), 53–70.

Baron, L., & Straus, M. A. (1988). Cultural and economic sources of homicide in the United States. *The Sociological Quarterly, 29*(3), 371–390.

Beckfield, J. (2004). Does income inequality harm health? *Journal of Health and Social Behavior, 45*(3), 231–248.

Berkman, L., & Kawachi, I. (Eds.). (2000). *Social epidemiology*. New York: Oxford University Press.

Blakely, T., Atkinson, J., & O'Dea, D. (2003). No association of income inequality with adult mortality within New Zealand. *Journal of Epidemiology and Community Health, 57*(4), 279–284.

Blakely, T. A., Kennedy, B. P., Glass, R., Kawachi, et al. (2000). What is the lag time between income inequality and health status? *Journal of Epidemiology and Community Health, 54*(4), 318–319.

Blakely, T. A., Kennedy, B. P., & Kawachi, I. (2001). Socioeconomic inequality in voting participation and self-rated health. *American Journal of Public Health, 91*(1), 99–104.

Blakely, T. A., Lochner, K., & Kawachi, I. (2002). Metropolitan area income inequality and self-rated health. *Social Science & Medicine, 54*(1), 65–77.

Blau, J. R., & Blau, P. M. (1982). The cost of inequality: Metropolitan structure and violent crime. *American Sociological Review, 47*, 114–129.

Blomgren, J., Martikainen, P., Makela, P., Valkonen, et al. (2004). The effects of regional characteristics on alcohol-related mortality. *Social Science & Medicine, 58*(12), 2523–2535.

Bobak, M., Pikhart, H., Rose, R., et al. (2000). Socioeconomic factors, material inequalities, and perceived control in self-rated health. *Social Science & Medicine, 51*(9), 1343–1350.

Bourdieu, P. (1984). *Distinction: A social critique of the judgement of taste*. London: Routledge.

Brodish, P. H., Massing, M., & Tyroler, H. A. (2000). Income inequality and all-cause mortality in the 100 counties of North Carolina. *Southern Medical Journal, 93*(4), 386–391.

Brunner, E., & Marmot, M. (1999). Social organization, stress & health. In M. Marmot, & R. G. Wilkinson (Eds.), *Social determinants of health*. Oxford: Oxford University Press.

Canadine, D. (1998). *Class in Britain*. Yale University Press.

Charlesworth, S. J., Gilfillan, P., & Wilkinson, R. (2004). Living inferiority. *British Medical Bulletin, 69*, 49–60.

Chiang, T. (1999). Economic transition and changing relation between income inequality and mortality in Taiwan. *British Medical Journal, 319*(7218), 1162–1165.

Cooper, R. S., Kennelly, J. F., Durazo-Arvizu, R., Oh, H. J., Kaplan, G., Lynch, J., et al. (2001). Relationship between premature mortality and socioeconomic factors in black and white populations of US metropolitan areas. *Public Health Reports, 116*(5), 464–473.

Daly, M., Wilson, M., & Vasdev, S. (2001). Income inequality and homicide rates in Canada and the United States. *Canadian Journal of Public Health, 43*(2), 219–236.

Daly, M. C., Duncan, G. J., Kaplan, G. A., Lynch, J., et al. (1998). Macro-to-micro links in the relation between income inequality and mortality. *Milbank Quarterly, 76*(3), 315–339 303–314.

Davey Smith, G., & Egger, M. (1996). Commentary: Understanding it all – health, meta-theories, and mortality trends. *British Medical Journal, 313*(7072), 1584–1585.

De Vogli, R., Mistry, R., Gnesotto, R., et al. (2005). Income inequality and life expectancy: Evidence from Italy. *Journal of Epidemiology and Community Health, 59*, 158–162.

Deaton, A. (2003). Health, inequality, and economic development. *Journal of Economic Literature, 41*, 113–158.

Deaton, A., & Lubotsky, D. (2003). Mortality, inequality and race in American cities and states. *Social Science & Medicine, 56*(6), 1139–1153.

Dickerson, S. S., & Kemeny, M. E. (2005). Acute stressors and cortisol responses. *Psychological Bulletin, 130*, 355–391.

Diez-Roux, A. V., Link, B. G., & Northridge, M. E. (2000). A multilevel analysis of income inequality and cardiovascular disease risk factors. *Social Science & Medicine, 50*(5), 673–687.

Drain, P. K., Smith, J. S., Hughes, J. P., Halperin, Holmes, et al. (2004). Correlates of National HIV Seroprevalence: An ecologic analysis of 122 developing countries. *Journal of the Acquired Immune Deficiency Syndrome, 35*(4), 407–420.

Drukker, M., Feron, F. J., & van Os, J. (2004). Income inequality at neighbourhood level and quality of life. *Social Psychiatry and Psychiatric Epidemiology, 39*(6), 457–463.

Duleep, H. O. (1995). Mortality and income inequality among economically developed countries. *Social Security Bulletin, 58*(2), 34–50.

Ellison, G. T. (2002). Letting the Gini out of the bottle? Challenges facing the relative income hypothesis. *Social Science & Medicine, 54*(4), 561–576.

Erdal, D., & Whiten, A. (1996). Egalitarianism and Machiavellian intelligence in human evolution. In P. Mellars, & K. Gibson (Eds.), *Modelling the early human mind.* Cambridge: McDonald Institute Mongraphs.

Fajnzylber, P., Lederman, D., & Loayza, N. (2002). Inequality and violent crime. *Journal of Law and Economics, 45*, 1–40.

Fiscella, K., & Franks, P. (1997). Poverty or income inequality as predictor of mortality. *British Medical Journal, 314*(7096), 1724–1727.

Fiscella, K., & Franks, P. (2000). Individual income, income inequality, health, and mortality: What are the relationships? *Health Services Research, 35*(1 Pt 2), 307–318.

Flegg, A. T. (1979). Role of inequality of income in the determination of birth-rates. *Population Studies – A Journal of Demography, 33*(3), 457–477.

Flegg, A. T. (1982). Inequality of income, illiteracy and medical-care as determinants of infant-mortality in underdeveloped-countries. *Population Studies–A Journal of Demography, 36*(3), 441–458.

Franzini, L., Ribble, J., & Spears, W. (2001). The effects of income inequality and income level on mortality vary by population size in Texas counties. *Journal of Health and Social Behavior, 42*(4), 373–387.

Franzini, L., & Spears, W. (2003). Contributions of social context to inequalities in years of life lost to heart disease in Texas, USA. *Social Science & Medicine, 57*(10), 1847–1861.

Galea, S., Ahern, J., Vlahov, D., Coffin, P. O., Fuller, C., Leon, A. C., et al. (2003). Income distribution and risk of fatal drug overdose in New York City neighborhoods. *Drug and Alcohol Dependence, 70*(2), 139–148.

Gilligan, J. (1996). *Violence: Our deadly epidemic and its causes.* New York: Putnam.

Gold, R., Connell, F. A., Heagerty, P., Bezruchka, Davis, Cawthon, et al. (2004). Income inequality and pregnancy spacing. *Social Science & Medicine, 59*, 1117–1126.

Gold, R., Kawachi, I., Kennedy, B. P., Lynch, J., Connell, F. A., et al. (2001). Ecological analysis of teen birth rates: Association with community income and income inequality. *Maternal and Child Health Journal, 5*(3), 161–167.

Gold, R., Kennedy, B., Connell, F., Kawachi, I., et al. (2002). Teen births, income inequality, and social capital. *Health and Place, 8*(2), 77–83.

Gravelle, H. (1998). How much of the relationship between population mortality and inequality is a statistical artefact? *British Medical Journal, 316*, 382–385.

Gravelle, H., Wildman, J., & Sutton, M. (2002). Income, income inequality and health: What can we learn from aggregate data? *Social Science & Medicine, 54*(4), 577–589.

Groves, W. B., McCleary, R., & Newman, G. R. (1985). Religion, modernization, and world crime. *Comparative Social Research, 8*, 59–78.

Hales, S., Howden-Chapman, P., Salmond, C., Woodward, A., Mackenbach, J., et al. (1999). National infant mortality rates in relation to gross national product and distribution of income. *The Lancet, 354*(9195), 2047.

Hansmann, H. B., & Quigley, J. M. (1982). Population heterogeneity and the sociogenesis of homicide. *Social Forces, 61*, 206–244.

Henderson, C., Liu, X., Diez Roux, A. V., Link, B. G., Hasin, D., et al. (2004). The effects of US state income inequality and alcohol policies on symptoms of depression and alcohol dependence. *Social Science & Medicine, 58*(3), 565–575.

Holtgrave, D. R., & Crosby, R. A. (2003). Social capital, poverty, and income inequality as predictors of gonorrhoea, syphilis, chlamydia and AIDS case rates in the United States. *Sexually Transmitted Infections, 79*(1), 62–64.

Holtgrave, D. R., & Crosby, R. A. (2004). Social determinants of tuberculosis case rates in the United States. *American Journal of Preventive Medicine, 26*(2), 159–162.

Hou, F., & Chen, J. (2003). Neighbourhood low income, income inequality and health in Toronto. *Health Reports, 14*(2), 21–34.

Hou, F., & Myles, J. (2005). Neighborhood inequality, neighborhood affluence and population health. *Social Science & Medicine, 60*, 1557–1569.

Hsieh, C. C., & Pugh, M. D. (1993). Poverty, income inequality, and violent crime: A meta-analysis of recent aggregate data studies. *Criminal Justice Review, 18*, 182–202.

Huff-Corzine, L., Corzine, J., & Moore, D. C. (1986). Southern exposure: Deciphering the South's influence on homicide rates. *Social Forces, 64*, 906–924.

Judge, K. (1995). Income distribution and life expectancy: A critical appraisal. *British Medical Journal, 311*(7015), 1282–1285 (discussion 1285–1287).

Judge, K., Mulligan, J. A., & Benzeval, M. (1998). Income inequality and population health. *Social Science & Medicine, 46*(4–5), 567–579.

Kahn, H. S., Patel, A. V., Jacobs, E. J., Calle, M. E., Kennedy, P., Kawachi, I., et al. (1999). Pathways between area-level income inequality and increased mortality in US men. *Annals of the New York Academy of Sciences, 896*, 332–334.

Kahn, H. S., Tatham, L. M., Pamuk, E. R., Heath, J. R., et al. (1998). Are geographic regions with high income inequality associated with risk of abdominal weight gain? *Social Science & Medicine, 47*(1), 1–6.

Kahn, R. S., Wise, P. H., Kennedy, B. P., Kawachi, I., et al. (2000). State income inequality, household income, and maternal mental and physical health. *British Medical Journal, 321*(7272), 1311–1315.

Kangas, O., & Palme, J. (2000). Does social policy matter? Poverty cycles in OECD countries. *International Journal of Health Services, 30*(2), 335–352.

Kaplan, G. A., Pamuk, E. R., Lynch, J. W., Cohen, Balfour, et al. (1996). Inequality in income and mortality in the United States. *British Medical Journal, 312*(7037), 999–1003.

Kawachi, I., & Kennedy, B. P. (1997). The relationship of income inequality to mortality: Does the choice of indicator matter? *Social Science & Medicine, 45*(7), 1121–1127.

Kawachi, I., Kennedy, B. P., Lochner, K., Prothrow-Stith, D., et al. (1997). Social capital, income inequality, and mortality. *American Journal of Public Health, 87*(9), 1491–1498.

Kennedy, B. P., Kawachi, I., Glass, R., Prothrow-Stith, D., et al. (1998). Income distribution, socioeconomic status, and self rated health in the United States. *British Medical Journal, 317*(7163), 917–921.

Kennedy, B. P., Kawachi, I., & Prothrow-Stith, D. (1996). Income distribution and mortality: Cross sectional ecological study of the Robin Hood index in the United States. *British Medical Journal, 312*(7037), 1004–1007.

Kennedy, B. P., Kawachi, I., Prothrow-Stith, D., Lochner, K., Gupta, V., et al. (1998). Social capital, income inequality, and firearm violent crime. *Social Science & Medicine, 47*(1), 7–17.

Kennedy, L. W., Silverman, R. A., & Forde, D. R. (1991). Homicide in urban Canada. *Canadian Journal of Sociology, 16*(4), 397–410.

Kick, E. L., & LaFree, G. D. (1985). Development and the social context of murder and theft. *Comparative Social Research, 8*, 37–58.

Krahn, H., Hartnagel, T. F., & Gartrell, J. W. (1986). Income inequality and homicide rates. *The Sociological Quarterly, 17*, 303–313.

Krohn, M. D. (1976). Inequality, unemployment and crime. *The Sociological Quarterly, 17*, 303–313.

Laporte, A. (2002). A note on the use of a single inequality index in testing the effect of income distribution on mortality. *Social Science & Medicine, 55*(9), 1561–1570.

Laporte, A., & Ferguson, B. S. (2003). Income inequality and mortality: Time series evidence from Canada. *Health Policy, 66*(1), 107–117.

Larrea, C., & Kawachi, I. (2005). Does economic inequality affect child malnutrition? The case of Ecuador. *Social Science & Medicine, 60*(1), 165–178.

LeClere, F. B., & Soobader, M. J. (2000). The effect of income inequality on the health of selected US demographic groups. *American Journal of Public Health, 90*(12), 1892–1897.

Lee, M. R., & Bankston, W. B. (1999). Political structure, economic inequality, and homicide. *Deviant Behavior, 20*(1), 27–55.

Legrand, J. (1987). Inequalities in health – some international comparisons. *European Economic Review, 31*(1–2), 182–191.

Lester, D. (1987). Relation of income inequality to suicide and homicide rates. *Journal of Social Psychology, 127*(1), 101–102.

Lobmayer, P., & Wilkinson, R. G. (2000). Income, inequality and mortality in 14 developed countries. *Sociology of Health and Illness, 22*(4), 401–414.

Lobmayer, P., & Wilkinson, R. G. (2002). Inequality, residential segregation by income, and mortality in US cities. *Journal of Epidemiology and Community Health, 56*(3), 183–187.

Lochner, K., Pamuk, E., Makuc, D., Kennedy, B. P., Kawachi, I., et al. (2001). State-level income inequality and individual mortality risk. *American Journal of Public Health, 91*(3), 385–391.

Loftin, C., & Hill, R. H. (1974). Regional subculture and homicide. *American Sociological Review, 39*, 714–724.

Loftin, C., & Parker, R. N. (1985). An errors-in-variable model of the effect of poverty on urban homicide rates. *Criminology, 23*, 269–287.

Lopez, R. (2004). Income inequality and self-rated health in US metropolitan areas: A multi-level analysis. *Social Science & Medicine, 59*(12), 2409–2419.

Lorant, V., Thomas, I., Deliege, D., Tonglet, et al. (2001). Deprivation and mortality: The implications of spatial autocorrelation for health resources allocation. *Social Science & Medicine, 53*(12), 1711–1719.

Lynch, J., Smith, G. D., Harper, S., Hillemeier, M., et al. (2004a). Is income inequality a determinant of population health? Part 1. A Systematic Review. *Milbank Quarterly, 82*(1), 5–99.

Lynch, J., Smith, G. D., Harper, S., Hillemeier, M., et al. (2004b). Is income inequality a determinant of population health? Part 2. US National and regional trends in income inequality and age- and cause-specific mortality. *Milbank Quarterly, 82*(2), 355–400.

Lynch, J., Smith, G. D., Hillemeier, M., Shaw, M., Raghunathan, T., Kaplam, G., et al. (2001). Income inequality, the psychosocial environment, and health: Comparisons of wealthy nations. *The Lancet, 358*(9277), 194–200.

Lynch, J. W., Kaplan, G. A., Pamuk, E. R., Cohen, R. D., Heck, K. E., Balfour, J. L., et al. (1998). Income inequality and mortality in metropolitan areas of the United States. *American Journal of Public Health, 88*(7), 1074–1080.

Macinko, J. A., Shi, L., & Starfield, B. (2004). Wage inequality, the health system, and infant mortality in wealthy industrialized countries, 1970–1996. *Social Science & Medicine, 58*(2), 279–292.

Marmot, M. (2004). *Status syndrome: How your social standing directly affects your health and life expectancy*. London: Bloomsbury.

Marmot, M., & Bobak, M. (2000). International comparators and poverty and health in Europe. *British Medical Journal, 321*(7269), 1124–1128.

Marmot, M., & Wilkinson, R. G. (Eds.). (1999). *Social determinants of health*. Oxford: Oxford University Press.

Marmot, M., & Wilkinson, R. G. (2001). Psychosocial and material pathways in the relation between income and health. *British Medical Journal, 322*, 1233–1236.

Massing, M. W., Rosamund, W. D., Wing, S. B., Suchindran, C. M., Kaplan, B. H., Tyroler, H. A., et al. (2004). Income, income inequality, and cardiovascular disease mortality: Relations among county populations of the United States, 1985–1994. *Southern Medical Journal, 97*(5), 475–484.

Mayer, S. E., & Sarin, A. (2005). Some mechanisms linking economic inequality and infant mortality. *Social Science & Medicine, 60*, 439–455.

McIsaac, S., & Wilkinson, R. G. (1997). Income distribution and cause-specific mortality. *European Journal of Public Health, 7*, 45–53.

McLaughlin, D. K., & Stokes, C. S. (2002). Income inequality and mortality in US counties: Does minority racial concentration matter? *American Journal of Public Health, 92*(1), 99–104.

McLaughlin, D. K., Stokes, C. S., & Nonyama, A. (2001). Residence and income inequality: Effects on mortality among US Countries. *Rural Sociology, 66*(4), 579–598.

McLeod, C. B., Lavis, J. N., Mustard, C. A., & Stoddart, G. L. (2003). Income inequality, household income, and health status in Canada. *American Journal of Public Health, 93*(8), 1287–1293.

McLeod, J. D., Nonnemaker, J. M., & Call, K. T. (2004). Income inequality, race, and child well-being. *Journal of Health and Social Behavior, 45*(3), 249–264.

Mellor, J. M., & Milyo, J. (2001). Reexamining the evidence of an ecological association between income inequality and health. *Journal of Health, Politics, Policy and Law, 26*(3), 487–522.

Mellor, J. M., & Milyo, J. (2002). Income inequality and individual health: Evidence from the Current Population Survey. *Journal of Human Resources, 37*(3), 510–539.

Mellor, J. M., & Milyo, J. (2003). Is exposure to income inequality a public health concern? Lagged effects of income inequality on individual and population health. *Health Services Research, 38*(1 Pt. 1), 137–151.

Menotti, A., Puddu, P. E., Lanti, M., Kromhout, D., Blackburn, H., & Nissinen, A. (2003). Twenty-five-year coronary mortality trends in the seven countries study using the accelerated failure time model. *European Journal of Epidemiology, 18*(2), 113–122.

Messias, E. (2003). Income inequality, illiteracy rate, and life expectancy in Brazil. *American Journal of Public Health, 93*(8), 1294–1296.

Messner, S. F. (1980). Income inequality and murder rates: Some cross-national findings. *Comparative Social Research, 3*, 185–198.

Messner, S. F. (1982). Societal development, social equality, and homicide: A cross-national test of a Durkheimian model. *Social Forces, 61,* 225–240.

Messner, S. F. (1983). Regional and racial effects on the urban homicide rate: The subculture of violence revisited. *American Journal of Sociology, 88,* 997–1007.

Messner, S. F. (1989). Economic discrimination and societal homicide rates. *American Sociological Review, 54,* 597–611.

Messner, S. F., & Tardiff, K. (1986). Economic inequality and levels of homicide: An analysis of urban neighborhoods. *Criminology, 24,* 297–317.

Muller, A. (2002). Education, income inequality, and mortality: A multiple regression analysis. *British Medical Journal, 324*(7328), 23–25.

Muntaner, C., Li, Y., Xue, X., O'Campo, Chung, Eaton, et al. (2004). Work organization, area labor-market characteristics, and depression among US nursing home workers. *International Journal of Occupational and Environmental Health, 10*(4), 392–400.

Muramatsu, N. (2003). County-level income inequality and depression among older Americans. *Health Services Research, 38*(6 Pt. 2), 1863–1883.

Osler, M., Christensen, U., Due, P., Lund, R., Andersen, I., Diderichsen, F., et al. (2003). Income inequality and ischaemic heart disease in Danish men and women. *International Journal of Epidemiology, 32*(3), 375–380.

Osler, M., Prescott, E., Gronbaek, M., Christensen, U., Due, P., & Engholm, G. (2002). Income inequality, individual income, and mortality in Danish adults. *British Medical Journal, 324*(7328), 13–16.

Pampel, F. C., Jr. (2002). Inequality, diffusion, and the status gradient of smoking. *Social Problems, 49*(1), 35–57.

Pampel, F. C., Jr., & Pillai, V. K. (1986). Patterns and determinants of infant mortality in developed nations, 1950–1975. *Demography, 23*(4), 525–542.

Pampel, F. C., & Zimmer, C. (1989). Female labour force activity and the sex differential in mortality: Comparisons across developed nations, 1950–1980. *European Journal of Population, 5*(3), 281–304.

Pattussi, M. P., Marcenes, W., Croucher, R., Sheiham, A., et al. (2001). Social deprivation, income inequality, social cohesion and dental caries in Brazilian school children. *Social Science & Medicine, 53*(7), 915–925.

Pickett, K. E., Kelly, S., Lobstein, T., Brunner, E., & Wilkinson, R. G. (2005). Wider income gaps, wider waist-bands? An ecological study of income inequality and obesity. *Journal of Epidemiology and Community Health.*

Pickett, K. E., Mookherjee, J., & Wilkinson, R. G. (2005). Teenage births and violence are related to income inequality among rich countries. *American Journal of Public Health.*

Reagan, P. B., & Salsberry, P. J. (2005). Race and ethnic differences in determinants of preterm birth in the USA: Broadening the social context. *Social Science & Medicine, 60*(10), 2217–2228.

Regidor, E., Calle, M. E., Navarro, P., & Dominguez, V. (2003). Trends in the association between average income, poverty and income inequality and life expectancy in Spain. *Social Science & Medicine, 56*(5), 961–971.

Regidor, E., Navarro, P., Dominguez, V., & Rodriguez, C. (1997). Inequalities in income and long-term disability in Spain: Analysis of recent hypotheses using cross sectional study based on individual data. *British Medical Journal, 315*(7116), 1130–1135.

Robert, S. A., & Reither, E. N. (2004). A multilevel analysis of race, community disadvantage, and body mass index among adults in the US. *Social Science & Medicine, 59*(12), 2421–2434.

Rodgers, G. B. (2002). Income and inequality as determinants of mortality: An international cross-section analysis, 1979. *International Journal of Epidemiology, 31*(3), 533–538.

Ronzio, C. R. (2003). Urban premature mortality in the US between 1980 and 1990: Changing roles of income inequality and social spending. *Journal of Public Health Policy, 24*(3–4), 386–400.

Ronzio, C. R., Pamuk, E., & Squires, G. D. (2004). The politics of preventable deaths: Local spending, income inequality, and premature mortality in US cities. *Journal of Epidemiology Community Health, 58*(3), 175–179.

Ross, N. A., Dorling, D., Dunn, J. R., Hendricksson, G., Glover, J., Lynch, J., et al. (2005). Metropolitan income inequality and working-age mortality: A cross-sectional analysis using comparable data from five countries. *Journal of Urban Health, 82*(1), 101–110.

Ross, N. A., Wolfson, M. C., Dunn, J. R., Berthelot, J.-M., Kaplan, G. A., Lynch, J. W., et al. (2000). Relation between income inequality and mortality in Canada and in the United States. *British Medical Journal, 320*(7239), 898–902.

Runciman, W. G. (1966). *Relative deprivation and social justice: A study of attitudes to social inequality in twentieth-century England.* London: Routledge.

Sahlins, M. (1974). *Stone age economics.* London: Tavistock.

Sanmartin, C., Ross, N. A., Tremblay, S., Wolfson, J. R., & Lynch, J. (2003). Labour market income inequality and mortality in North American metropolitan areas. *Journal of Epidemiology and Community Health, 57*(10), 792–797.

Sapolsky, R. (2001). *A primate's memoir.* London: Jonathan Cape.

Shi, L., Macinko, J., Starfield, B., Xu, J., & Politzer, R. (2003). Primary care, income inequality, and stroke mortality in the United States: A longitudinal analysis, 1985–1995. *Stroke, 34*(8), 1958–1964.

Shi, L., Macinko, J., Starfield, B., Politzer, R., Wulu, J., Xu, J., et al. (2004). Primary care, infant mortality, and low birth weight in the states of the USA. *Journal of Epidemiology and Community Health, 58*(5), 374–380.

Shi, L., Macinko, J., Starfield, B., Xu, J., Regan, J., Politzer, R., et al. (2005). Primary care, social inequalities, and all-cause, heart disease, and cancer mortality in US countries, 1990. *American Journal of Public Health, 95*(4), 674–680.

Shi, L., & Starfield, B. (2000). Primary care, income inequality, and self-rated health in the United States: A mixed-level analysis. *International Journal of Health Services, 30*(3), 541–555.

Shi, L., & Starfield, B. (2001). The effect of primary care physician supply and income inequality on mortality among blacks and whites in US metropolitan areas. *American Journal of Public Health, 91*(8), 1246–1250.

Shi, L., Starfield, B., Politzer, R., Reagan, J., et al. (2002). Primary care, self-rated health, and reductions in social disparities in health. *Health Services Research, 37*(3), 529–550.

Shi, L. Y. (1999). Income inequality, primary care, and health indicators. *Journal of Family Practice, 48*(4), 275–284.

Shibuya, K., Hashimoto, H., & Yano, E. (2002). Individual income, income distribution, and self rated health in Japan. *British Medical Journal, 324*(7328), 16–19.

Shmueli, A. (2004). Population health and income inequality: New evidence from Israeli time-series analysis. *International Journal of Epidemiology, 33*(2), 311–317.

Sidanius, J., & Pratto, F. (1999). *Social dominance: An intergroup theory of social hierarchy and oppression.* Cambridge: Cambridge University Press.

Singh-Manoux, A., Adler, N. E., & Marmot, M. G. (2003). Subjective social status: Its determinants and its association with ill-health in the Whitehall II study. *Social Science & Medicine, 56,* 1321–1333.

Smith, M. D., & Parker, R. N. (1980). Type of homicide and variation in regional rates. *Social Forces, 59,* 136–147.

Sohler, N. L., Arno, P. S., Chang, C. J., Fang, J., & Schechter, C. (2003). Income inequality and infant mortality in New York City. *Journal of Urban Health, 80*(4), 650–657.

Soobader, M. J., & LeClere, F. B. (1999). Aggregation and the measurement of income inequality: Effects on morbidity. *Social Science & Medicine, 48*(6), 733–744.

Stanistreet, D., Scott-Samuel, A., & Bellis, M. A. (1999). Income inequality and mortality in England. *Journal of Public Health Medicine, 21*(2), 205–207.

Steckel, R. H. (1983). Height and per-capita income. *Historical Methods, 16*(1), 1–7.

Sturm, R., & Gresenz, C. R. (2002). Relations of income inequality and family income to chronic medical conditions and mental health disorders. *British Medical Journal, 324*(7328), 20–23.

Subramanian, S. V., Blakely, T., & Kawachi, I. (2003). Income inequality as a public health concern: Where do we stand? Commentary on "Is exposure to income inequality a public health concern?". *Health Services Research, 38*(1 Pt 1), 153–167.

Subramanian, S. V., Delgado, I., Jadue, L., Vega, J., & Kawachi, I. (2003). Income inequality and health: Multilevel analysis of Chilean communities. *Journal of Epidemiology and Community Health, 57*(11), 844–848.

Subramanian, S. V., & Kawachi, I. (2003). The association between state income inequality and worse health is not confounded by race. *International Journal of Epidemiology, 32*(6), 1022–1028.

Subramanian, S. V., & Kawachi, I. (2004). Income inequality and health: What have we learned so far? *Epidemiologic Reviews, 26*, 78–91.

Subramanian, S. V., Kawachi, I., & Kennedy, B. P. (2001). Does the state you live in make a difference? Multilevel analysis of self-rated health in the US. *Social Science & Medicine, 53*(1), 9–19.

Szwarcwald, C. L., Andrade, C. L., & Bastos, F. I. (2002). Income inequality, residential poverty clustering and infant mortality: A study in Rio de Janeiro, Brazil. *Social Science & Medicine, 55*(12), 2083–2092.

Szwarcwald, C. L., Bastos, F. I., Viacava, F., De Andrade, C. L. T., et al. (1999). Income inequality and homicide rates in Rio de Janeiro, Brazil. *American Journal of Public Health, 89*(6), 845–850.

Veenstra, G. (2002a). Social capital and health (plus wealth, income inequality and regional health governance). *Social Science & Medicine, 54*(6), 849–868.

Veenstra, G. (2002b). Income inequality and health. Coastal communities in British Columbia, Canada. *Canadian Journal of Public Health, 93*(5), 374–379.

Walberg, P., McKee, M., Shkolnikov, V., Chenet, L., & Leon, D. A. (1998). Economic change, crime, and mortality crisis in Russia. *British Medical Journal, 317*(7154), 312–318.

Waldmann, R. J. (1992). Income-distribution and infant-mortality. *Quarterly Journal of Economics, 107*(4), 1283–1302.

Weatherby, N. L., Nam, C. B., & Isaac, L. W. (1983). Development, inequality, health care, and mortality at the older ages: A cross-national analysis. *Demography, 20*(1), 27–43.

Weich, S., Lewis, G., & Jenkins, S. P. (2001). Income inequality and the prevalence of common mental disorders in Britain. *British Journal of Psychiatry, 178*, 222–227.

Weich, S., Lewis, G., & Jenkins, S. P. (2002). Income inequality and self rated health in Britain. *Journal of Epidemiology and Community Health, 56*(6), 436–441.

Wen, M., Browning, C. R., & Cagney, K. A. (2003). Poverty, affluence, and income inequality: Neighborhood economic structure and its implications for health. *Social Science & Medicine, 57*(5), 843–860.

Wennemo, I. (1993). Infant-mortality, public-policy and inequality – A comparison of 18 industrialized countries 1950–85. *Sociology of Health and Illness, 15*(4), 429–446.

Wildman, J., Gravelle, H., & Sutton, M. (2003). Health and income inequality: Attempting to avoid the aggregation problem. *Applied Economics, 35*(9), 999–1004.

Wilkinson, R. G. (1992). Income distribution and life expectancy. *British Medical Journal, 304*(6820), 165–168.

Wilkinson, R. G. (1997). Income, inequality and social cohesion. *American Journal of Public Health, 87*, 104–106.

Wilkinson, R. G. (2005). *The impact of inequality: How to make sick societies healthier*. NY: New Press.

Wilkinson, R. G. (2004). Why is violence more common where inequality is greater? *Annals of the New York Academy of Sciences, 1036*, 1–12.

Wilkinson, R. G., Kawachi, I., & Kennedy, B. (1998). Mortality, the social environment, crime and violence. *Social Health and Illness, 20*(5), 578–597.

Wolfson, M., Kaplan, G., Lynch, J., Ross, Backlund, et al. (1999). Relation between income inequality and mortality: Empirical demonstration. *British Medical Journal, 319*, 953–957.

16

Inequalities in Health: Some International Comparisons

Julian Le Grand

1. Introduction

'The desire for the prolongation of life . . . we may take to be one of the most universal of all human motives.' *Kenneth Arrow*[1]

There have been many attempts to compare various dimensions of inequality between different countries, including, particularly, those relating to income and wealth. However, there have been relatively few international studies of inequalities in health.[2] Given that, as has been noted by economists as eminent as Kenneth Arrow and Amartya Sen,[3] individuals' states of health are contributors to welfare comparable in impact to income or wealth holdings, this relative poverty of health studies is surprising. It cannot be explained by lack of data. Such studies of health inequality as do exist generally concern mortality and the data situation here is significantly better than that for other welfare indicators, such as income. There are extensive data, available for a large number of countries; although the quality varies [see United Nations (1982)], it is likely to be generally higher than that for income data. Moreover, there are fewer methodological problems associated with the data; for obvious reasons, mortality is easier to define than income, and there is no problem equivalent to that of defining the income unit.

A more plausible explanation for the relative shortage of international comparative studies of health and concerns differences in the standard

Source: *European Economic Review*, 31(1–2) (1987): 182–191.

methods for measuring inequality. In the case of health, the conventional procedure has been to compare the health (usually, mortality) experience of different social or occupational classes [see Black (1980) for extensive discussion and application]. Any attempt to apply this procedure to international comparisons, however, encounters the obvious difficulty that definitions of social class, and of the occupational classifications that underlie them, vary widely from country to country. The fact that the only 'successful' comparisons have been undertaken between countries with great similarities in culture (and therefore in occupational classifications) reinforces the point [Le Grand (1986)].

In the case of income and wealth, the generally accepted procedure has been to calculate individually based summary statistic measures of inequality, such as the Gini coefficient, or the Atkinson index. These are calculated using individual differences in the indicator concerned (income or wealth), hence they do not depend on the vagaries of classification schemes. International comparisons are thus more easily made and more readily interpretable.

The summary statistic methodology has been applied to the problem of measuring changes in inequalities in health within the same country over time [Illsley and Le Grand (1986), Le Grand and Rabin (1986)]. However, as yet, the techniques have not been systematically applied to international comparisons; the research summarised in this paper illustrates how this might be done.

2. Methodology

The first task is to select an indicator of health for which a distribution can be derived. If the focus is on mortality, then an appropriate indicator would seem to be length of life, or age-at-death. To illustrate the kind of distribution involved, fig. 1 shows the (age-standardised) distributions of age-at-death in 1982 for England and Wales, and for Poland. It will be observed that they are rather different in shape from the kind that economists normally explore. In particular, they are bimodal, with a low peak at a very early age (representing infant mortality) and a much higher peak in later years. Moreover, each distribution is bounded by biological limits, in a way that has no parallel in, for example, income distributions. This has implications for the choice of inequality measure, as will be seen shortly.

The second task is to choose an inequality measure. Two procedures for doing this have been suggested in the inequality measurement literature [Cowell (1977) Chs. 2 and 3]. One is to list the set of properties that it would be desirable for a measure to possess for the task in hand, and then to choose the measure that has as many of these properties as possible. The other stems from Atkinson's path-breaking article (1970) and bases the choice of measure directly on the welfare function of the researcher.

More detailed discussion of the properties that it would be desirable for an inequality measure defined over age-at-death to possess can be found in

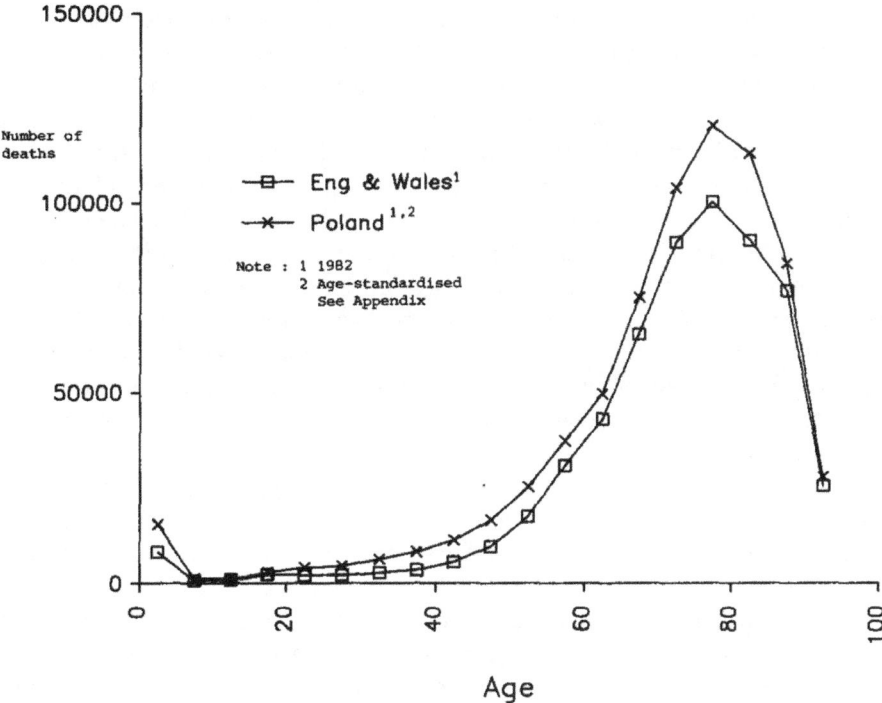

Figure 1: Age-at-death distribution

Le Grand (1986). Here there is space to draw attention only to two issues. The first concerns the property of *scale independence*. In the income distribution literature, it is argued that if everyone's income has changed by the same proportion, there has been no essential change in the distribution and hence there should be no change in the inequality measure. The equivalent in our context is the requirement that the distribution of age-at-death in one country differs from that in another simply by a scale factor there should be no difference in measured inequality.

The desirability of scale independence has been challenged, even within the income distribution literature [Kolm (1976)]; for the rationale for concentrating on proportional differences in income, rather than on, say, absolute differences, is not immediately apparent. Moreover, as fig. 1 illustrated, the types of distribution encountered in the health area are rather different from standard income distributions. One country's distribution is likely to differ from another's, not by a scale factor, but through different means (and modes), all within the biological boundaries that determine human life-span.

This suggests that an alternative to scale independence that might be more useful in the health context is *translation* independence. This is the requirement that the inequality measure should be invariant to the translation of a distribution by a constant. Thus, for example, measured inequality would not differ between two countries with different mean ages-at-death, but with the same absolute differences between individuals' ages-at-death.

Among the well-known inequality measures, the coefficient of variation, the Gini coefficient and the Atkinson index are scale, but not translation, independent. However, the first two have a translation-independent analogue: the variance and the Absolute Mean Difference (AMD). The distance concept employed in the class of measures that include the variance and the coefficient of variation (differences in age-at-death from mean age-at-death) seems rather arbitrary, and less satisfactory than the distance concept incorporated in the class of measures including the AMD and the Gini coefficient (the differences between *every pair* of ages-at-death). Hence the measures used were the AMD, the Gini coefficient and, to represent the welfare measures, the Atkinson index.

The second issue concerns the question of age-standardisation. One country's distribution of age-at-death may differ from another's, not because of differences in mortality rates at different ages, but because of differences in the distribution of the population between ages. A country with a relatively young age distribution, for example, will have more deaths at lower ages and less at higher ages, and hence a 'flatter' distribution, ceteris paribus, than one with a relatively elderly age distribution. Demographers and others working in the area routinely age-standardise their data to overcome this problem; here we have followed their example.

3. Results

Data provided by the World Health Organisation on age-at-death for 32 developed countries were used to calculate three sets of inequality measures. For the reasons given above, the measures chosen were the AMD, the Gini coefficient, and the Atkinson index. For the Atkinson index, estimates were calculated for two values of the inequality aversion parameter, 0.75 and 1.25. The results presented here are for age-standardised aggregate (males plus females) deaths. Further details of the methods of calculation can be found in the appendix, as can the country-by-country results (table A.1). Estimates for males and females separately, and for the actual (as distinct from the age-standardised) distributions can be found in Le Grand (1986).

Table 1 summarises the results for aggregate deaths. It was constructed by ranking the countries according to their AMD value and grouping them into three broad categories; where relevant, footnotes indicate how their categorisation would change if they were ranked according to either of the other measures. It is notable that the rankings, admittedly in terms of these very broad categorisations, are fairly insensitive to the choice of inequality measure. The rankings are the same for over two-thirds of the countries (22); the Gini coefficient gives the same ranking as the AMD for all but two, while

Table 1: International health inequality: A summary[a]

Ranking by AMD of age-at-death		
Lowest third (10)	Czechoslovakia[b]	Luxembourg
	Eire	Netherlands
	England & Wales	N. Ireland[c]
	Finland	Scotland
	GDR	Sweden
Middle third (11)	Australia	Iceland[d]
	Belgium	Italy[e]
	Bulgaria	Japan[f]
	Canada	Norway
	Denmark[d]	Switzerland
	FRG	
Highest third (11)	Austria[c]	Portugal
	France[c]	Romania
	Greece	Spain
	Hungary	USA
	New Zealand	Yugoslavia
	Poland	

[a]AMD = Absolute Mean Difference, FRG = Federal Republic of Germany, GDR = German Democratic Republic.
[b]Middle third by Gini Coefficient and Atkinson Index (ε = 0.75, 1.25).
[c]Middle third by Atkinson Index (ε = 0.75, 1.25).
[d]Lowest third by Atkinson Index (ε = 0.75, 1.25).
[e]Highest third by Atkinson Index (ε = 0.75, 1.25).
[f]Lowest third by Gini Coefficient and Atkinson Index (ε = 0.75, 1.25).

the Atkinson index, for both parameter values, gives the same ranking as the AMD for 22 countries.

Looking at the results for specific countries, we can make the following observations. Of the Scandinavian countries Sweden and Finland are in the lowest third, according to all three measures. However, the estimates for Norway for all three measures and, according to the AMD and the Gini, for Denmark and Iceland puts these countries down a category. All the countries in the British Isles – England and Wales, Scotland, Northern Ireland and Eire – appear to have relatively low inequality, except for the Atkinson index for Northern Ireland. Belgium is the only one of the Benelux countries not to appear in the lowest category. Of the two Germanies, the GDR appears to be more equal than the FRG; of the other socialist countries, Czechoslovakia is relatively equal, Bulgaria is in the middle category and Romania, Poland and Hungary are relatively unequal.

Looking nearer the bottom of the table, the southern European countries of Greece, Portugal, Spain and Yugoslavia all show relatively high inequality, regardless of inequality measure. Austria and France are in the highest cate gory, except according to the Atkinson index; Italy, on the other hand, is in the middle category, except according to the Atkinson index; Switzerland is also in the middle category. Of the Australasian countries, New Zealand is consistently unequal, but Australia is in the middle category. Japan is in the lowest category, according to the Gini coefficient and the Atkinson index, but in the middle, according to the AMD. Finally, of the North American countries, the United States is relatively unequal, but Canada is in the middle category.

It is of interest to explore the relationship between inequality in health, as measured in this way, and other features of the countries concerned, such as their medical systems, their levels of GNP and of overall economic inequality. To this end a simple regression analysis was undertaken, correlating mean mortality and AMD with health care expenditures (public and private), per capita GDP and a measure of income inequality.

The analysis was necessarily limited by the availability of data. The dependent variables were age-standardised mean age-at-death and AMD. For the independent variables, where possible, 1982 was selected as the data year, because the mortality data for the majority of countries was for that year. Data were available from the OECD (1985) on per capita expenditures on medical care (public and private), measured at current GDP purchasing power parity rates (PCE); on public expenditure on medical care as a proportion of total medical care expenditures (PPHE); and on per capita GDP, at purchasing power parity rates (PCGDP). This gave data for 23 countries.[4]

It proved impossible to obtain data of comparable quality and coverage on economic inequality. In the end, data on the share of the bottom 20% in National Income for 17 of the 23 countries were taken from World Bank tables (INEQ).[5] Two sets of regressions were run, therefore, one using the set of 23 countries (the Large Set) and omitting INEQ, and the other the set of 17 countries (the Small Set) and including INEQ. Also, to test for the possibility that the dispersion and mean of age-at-death were not independent, the AMD regressions were run with and without the mean as an independent variable.

The results are given in table 2. They contain a number of surprises. The equations with the mean as dependent variable were quite unsatisfactory

Table 2: Regression results[a]

| Independent variable | Mean age-at-death | | AMD | | | |
	Large set	Small set	Large set		Small set	
C	72.708	69.164	14.615	10.0495	22.544	11.211
	(73.02)	(62.34)	(1.71)	(20.15)	(2.97)	(25.33)
Mean	–	–	–0.0628	–	–0.1639	–
			(–0.54)		(–1.49)	
PCE	–1.588	–0.987	1.2065	1.3062	1.1798	1.3416
	(–1.14)	(–0.88)	(1.64)	(1.87)	(2.67)	(2.98)
PCGDP	0.0922	0.0488	–0.1649	–0.1707	–0.2082	–0.2162
	(0.56)	(0.35)	(–1.95)	(–2.08)	(–3.90)	(–3.88)
PPHE	0.1304	1.0781	–0.5667	–0.5749	0.1143	–0.0624
	(0.17)	(1.74)	(–1.43)	(–1.48)	(0.43)	(–0.25)
INEQ	–	0.4521	–	–	–0.0958	–0.1698
		(4.13)			(–1.48)	(–3.89)
S	0.7974	0.5552	0.4071	0.3994	0.2109	0.2215
R^2	0.0	0.589	0.13	0.16	0.67	0.64
n	23	17	23	23	17	17
Column	(1)	(2)	(3)	(4)	(5)	(6)

[a]t-statistics in brackets under estimated coefficients,
C – constant,
S – standard error of the regression,
R^2 – adjusted for degrees of freedom,
n – number of observations.

(columns [1] and [2]). None of the independent variables were significant for the Large Set; only *INEQ* was significant for the Small Set. The AMD equations, however, were more promising. Inclusion of the mean revealed the latter to be insignificant (Columns [3] and [5]). If it were omitted, then, using the Large Set (Column [4]), *PCE* was positive and significant at the 90% level and *PCGDP* was negative and significant at the 95% level. This suggests that increasing health inequality was associated with decreasing per capita GDP, which does not seem unreasonable, but that it was associated with *increasing* per capita medical care, which seems rather less intuitive. Of course, these estimates give no indication of the direction of causality; rather than indicating a failure of health spending to reduce inequality, the correlation could be the result of countries with high health inequality spending more on health care partially to compensate. The coefficient for the proportion of medical care expenditures that are publicly financed (*PPHE*) was negative, as proponents of public health care might hope; unfortunately, it was not significant. These results were replicated in the Small Set regressions, (Column [6]) with the addition that the *INEQ* variable was negative and highly significant. Thus the results suggest that the higher the share of the poor in the National Income, the less will be health inequality.

Given the weaknesses of some of the data (particularly, those for income inequality), the small sample sizes and, perhaps most crucially, the absence of an underlying theoretical structure within which to interpret them, too much should not be made of these results. Nonetheless they seem to be of some interest, if only because they raise some intriguing possibilities for future research.

4. Concluding Comments

It is hoped that this paper has demonstrated that it is of interest to apply some of the techniques of inequality measurement developed in the income distribution literature to the problem of international comparisons of health inequality. Moreover, it is hoped also to have shown that, because the techniques do not rely on the vagaries of classification systems, they are, in this respect at least, easier to apply to this task than measures of health inequality relying on, for instance, comparisons between social classes. However, it must be emphasised that they do not answer the same kinds of questions as those comparisons, and therefore should be viewed as complement to them, not as a substitute.

Appendix

Country-by-country results are provided in table A.1. The data for each country were for the most recent year available at the time of the calculalations. The data were grouped in age ranges; it was assumed that the average age-at-death was in the middle of the age-range. The last range (85+) was open-ended; it was assumed that it 'closed' at 100. Except for the Atkinson index, two grouping assumptions

Table A1: Inequality in age-at-death by country: Age-standardised

Country	Year	Mean	AMD	Gini	Atkinson 0.75	1.25
Australia	1981	72.09	9.02	0.125	0.047	0.146
Austria	1981	72.63	9.12	0.126	0.046	0.144
Belgium	1978	72.59	8.91	0.123	0.045	0.141
Bulgaria	1981	72.05	9.04	0.125	0.055	0.179
Canada	1982	71.99	9.01	0.125	0.045	0.139
Czechoslovakia	1981	72.39	8.68	0.120	0.046	0.150
Denmark	1982	72.27	8.75	0.121	0.040	0.121
Eire	1979	72.18	8.38	0.115	0.039	0.126
England and Wales	1982	72.82	8.54	0.117	0.042	0.138
Finland	1981	72.46	8.58	0.118	0.034	0.100
FRG	1982	72.72	8.91	0.123	0.045	0.142
France	1981	71.80	9.53	0.133	0.055	0.168
GDR	1983	73.28	8.56	0.117	0.040	0.126
Greece	1981	73.19	9.20	0.126	0.058	0.194
Hungary	1982	71.12	9.19	0.129	0.052	0.168
Iceland	1979	72.56	8.94	0.123	0.041	0.126
Italy	1979	72.63	9.06	0.125	0.052	0.170
Japan	1982	73.92	8.71	0.118	0.040	0.120
Luxembourg	1982	71.56	7.66	0.107	0.038	0.122
Netherlands	1982	73.18	8.57	0.117	0.041	0.129
New Zealand	1982	71.63	9.19	0.128	0.050	0.157
N. Ireland	1982	72.31	8.59	0.119	0.045	0.145
Norway	1982	72.97	8.77	0.120	0.042	0.131
Poland	1982	70.92	9.51	0.134	0.060	0.194
Portugal	1979	73.13	9.56	0.130	0.063	0.209
Romania	1982	71.53	10.08	0.141	0.072	0.234
Scotland	1982	72.13	8.51	0.118	0.040	0.128
Spain	1979	72.88	9.24	0.127	0.055	0.179
Sweden	1981	73.63	8.53	0.116	0.037	0.112
Switzerland	1981	73.32	9.03	0.123	0.043	0.130
United States	1982	70.22	9.67	0.138	0.056	0.172
Yugoslavia	1980	71.68	9.69	0.135	0.072	0.242

were used. One, the upper bound, assumed that deaths were divided equally between the upper and lower ends of each age range; the other, the lower bound, assumed that all deaths took place at the middle of the range. The mean of the two was taken as the final figure. The upper bound calculation is not appropriate for the Atkinson index, because the procedure would generate observations of value zero for the lowest age-range. Hence only the lower bound calculation was used.

The age-standardisation procedure was to calculate the deaths that would have occurred at each age for a particular country if its own age-specific mortality rates were applied to the population distribution of England and Wales. The resulting distribution gives the number of deaths that would have occurred at each age in the country concerned, if it had had the same population distribution as England and Wales.

Notes

1. (1963, p.75).
2. For references to studies of both kinds, see Le Grand (1986).
3. For Arrow's views, see the quotation at the beginning of the paper. Sen is concerned with 'capabilities' or 'functionings' rather than conventional measures of utility, but re-

fers extensively to several health and health-related factors as indicators of capabilities. See Sen (1985), especially Appendix A.

4. All the countries in table 1 except for the socialist countries, and including the United Kingdom but excluding England and Wales, Scotland and Northern Ireland.

5. The countries in note 4, except Australia, Belgium, Eire, Iceland, Luxembourg and Switzerland. *Source*: World Bank World Tables, vol. 1, 1983.

References

Arrow, K.J., 1963, Social choice and individual values, second ed. (Wiley, New York).

Atkinson, A.B., 1970, On the measurement of inequality, Journal of Economic Theory 2, 244–263.

Black, D., 1980, Inequalities in health, Report of a research working group chaired by Sir Douglas Black (Department of Health and Social Security, London).

Cowell, F.A., 1977, Measuring inequality (Allan, Oxford).

Illsley, R. and J. Le Grand, 1986, The measurement of health inequality, Paper presented to the British Association for the Advancement of Science (Bristol), Sept.

Kolm, S.-C., 1976, Unequal inequalities, Journal of Economic Theory 12, 416–442; 13, 82–11.

Le Grand, J., 1986, International comparisons of inequality in health, Welfare State Programme discussion paper (London School of Economics, London).

Le Grand, J. and M. Rabin, 1986, Trends in British health inequality, 1931–1983, in: A.J. Culyer and B. Jönsson, eds., Public and private health services (Blackwell, Oxford).

Organisation for Economic Co-operation and Development, 1985, Measuring health care 1960–1983 (OECD, Paris).

Sen. A.K., 1985, Commodities and capabilities (North-Holland, Amsterdam).

United Nations, 1982, Levels and trends in mortality since 1950 (United Nations, New York).

17

Health Inequalities and Welfare State Regimes: Theoretical Insights on a Public Health 'Puzzle'

Clare Bambra

Background

It is now widely acknowledged that welfare states are important determinants of health as they mediate the social determinants of health.[1] Welfare state provision varies extensively, but typologies have been put forward to categorise them into distinctive types – *welfare state regimes*.[1-3] Welfare state regimes have increasingly been used within social epidemiology to analyse cross-national differences in population health.[4-8] These studies have almost invariably concluded that population health is enhanced by the relatively generous and universal welfare provision of the Social Democratic *Scandinavian* countries, especially when contrasted to the *Anglo-Saxon* welfare states.[4-8] The different types of welfare state and their constituent countries are described in box 1. However, in contrast to their comparatively strong performance in terms of overall health, data from most, but not all, of the recent comparative studies of health inequalities in the general population suggest that the Scandinavian welfare states do not have the smallest health inequalities.[9-12] For example, Mackenbach *et al*[9]'s Europe wide study of inequalities in mortality found 'no evidence for systematically smaller inequalities in health in countries in northern Europe (Scandinavia)'. Indeed, relative inequalities in mortality were smaller in the

Source: *Journal of Epidemiology & Community Health*, 65(9) (2011): 740–745.

Box 1: Welfare State Regimes[1]

Liberal/Residual

In the welfare states of the liberal regime (UK, USA, Ireland, Canada, Australia), state provision of welfare is minimal; social transfers are modest and often attract strict entitlement criteria; and recipients are usually means-tested and stigmatised. In this model, the dominance of the market is encouraged both passively, by guaranteeing only a minimum, and actively, by subsidising private welfare schemes. The liberal welfare state regime thereby minimises the decommodification effects of the welfare state, and a stark division exists between those, largely the poor, who rely on state aid and those who are able to afford private provision.

Conservative/Corporatist/Bismarckian

The conservative welfare state regime (Germany, France, Austria, Belgium, Italy and, to a lesser extent, the Netherlands) is distinguished by its 'status differentiating' welfare programs in which benefits are often earnings related, administered through the employer and geared towards maintaining existing social patterns. The role of the family is also emphasised and the redistributive impact is minimal. However, the role of the market is marginalised.

Social Democratic/Scandinavian

The Social Democratic regime type (Nordic countries) is characterised by universalism, comparatively generous social transfers, a commitment to full employment and income protection and a strongly interventionist state. The state is used to promote social equality through a redistributive social security system. Unlike the other welfare state regimes, the Social Democratic regime type promotes an equality of the highest standards, not an equality of minimal needs and it provides highly decommodifying programs.

Southern/Latin

It has been proposed that the Southern European welfare states (Italy, Greece, Portugal and Spain) comprise a distinctive southern welfare state regime. The southern welfare states are described as 'rudimentary' because they are characterised by their fragmented system of welfare provision, which consists of diverse income maintenance schemes that range from the meagre to the generous and welfare services, particularly, the healthcare system, that provide only limited and partial coverage. Reliance on the family and voluntary sector is also a prominent feature.

Southern (Italy, Spain, Portugal) and *Bismarckian* (Netherlands, Belgium, Germany, France) countries.[9] Data are provided from three other example studies of health inequalities in Europe in table 1.[10][12] Given the higher levels of social expenditure in the Scandinavian welfare states, the smaller income inequalities, and the commitment to equality underpinning the Social

Democratic welfare model in Scandinavia, it has long been something of a 'puzzle' in public health as to why the Scandinavian countries do not have the smallest health inequalities.[13–15] This essay draws upon the theories of health inequalities to scrutinise this puzzle.[16–20]

Theoretical Insights on Comparative Health Inequalities

Box 2 outlines the main theories of health inequalities.[16–20] These are commonly used to explain socioeconomic health inequalities within countries. In this paper, they are applied to cross-national differences in the magnitude of socioeconomic health inequalities and used to offer insights into the puzzle as to why health inequalities are not the smallest in the Scandinavian welfare states.

Artefact

The artefact explanation questions the existence of health inequalities, considering them to be a mere artefact of data collection and measurement (box 2). Applying it to the issue of comparative health inequalities leads to the conclusion that the 'public health puzzle' – of why health inequalities are not the smallest in the Scandinavian countries – is not in fact a real puzzle, but simply the result of the data and methods used. Certainly, the application of different indicators of social inequality (eg, income, occupation and education) and the use of different data sets has produced divergent results (see table 1). Different cross-national patterns also emerge in terms of the different ways in which specific indicators of inequality are calculated. For example, studies of educational inequalities can compare those with average years of education to those with 1SD below the national average[12] or the difference between those with no education or only primary education compared with those with tertiary education (box 2).[21] There are also more general issues in terms of making cross-national comparisons of health inequalities as it is not clear whether the bottom groups are the same in each country and whether their composition changes over time.[12 22] The use of relative or absolute measures of health inequalities is also an important issue (see 'Discussion' section). There is of course another clear measurement problem, which is the use of 'welfare state regimes', a concept that assumes a homogeneous approach to welfare provision within and between the countries of any particular regime type.[1 23]

Health Selection

The health selection approach asserts that health determines socioeconomic class status rather than socioeconomic class determining health (box 2). This would imply that the social consequences of ill health would need to be

Box 2: Theories of Health Inequalities[16–20]

Artefact

The artefact approach suggests that socioeconomic inequalities do not really exist but are a result of the data used and methods of measurement: that difference in health by socioeconomic class can be explained by differences in measurement and that the size of the inequalities observed is due to differences in data measurement tools.

Health Selection

The health selection approach asserts that health determines socioeconomic class status rather than socioeconomic class determining health. Individuals who are 'fitter' are more likely to move up the social hierarchy. In contrast, people with ill health are downwardly mobile (or less upwardly mobile) and are therefore concentrated within the lower socioeconomic classes.

Cultural–Behavioural

The cultural–behavioural approach asserts that the link between socioeconomic class and health is a result of differences between socioeconomic class in terms of their health-related behaviour: smoking rates, alcohol and drug consumption, dietary intake, physical activity levels, risky sexual behaviour and health service usage. Such differences in health behaviour, it is argued, are themselves a consequence of disadvantage, and unhealthy behaviours may be more culturally acceptable among lower socioeconomic class.

Materialist

The (neo)materialist explanation focuses on income and what income enables such as access to goods and services and the limitation of exposures to physical, and psychosocial, risk factors. Materialist approaches give primacy to structure in their explanation of health and health inequalities, looking beyond individual-level factors (agency), in favour of the role of public policy and services such as schools, transport and welfare in the social patterning of inequality.

Psychosocial

Psychosocial explanations focus on how social inequality makes people feel and their biological and health consequences. Social inequality leads to long-term feelings of subordination or inferiority, which in turn stimulate chronic stress responses that have profound consequences for physical and mental health. The socioeconomic class gradient is therefore explained by the unequal social and economic distribution of psychosocial risk factors.

Life Course

The life course approach combines aspects of the other explanations, thereby allowing different causal mechanisms and processes to explain the social gradient in different diseases. Health inequality between socioeconomic classes is the result of inequalities in the accumulation of social, psychological and biological advantages and disadvantages over time.

greater in the Scandinavian countries and that people who have ill health are more likely to be concentrated in the lower socioeconomic groups. Instinctively, such direct selection seems unlikely given the extensive employment protection for people with ill health within the Nordic countries and their comparatively high replacement rates for people out of the labour market due to sickness or disability.[18] Selection is also considered to be more influential in respect to income-related inequalities than educational ones and so it is unlikely to explain the results of the comparative studies of educational inequalities in health.[23]

Culture and Behaviour

The cultural–behavioural approach asserts that the link between socioeconomic class and health is a result of differences between socioeconomic classes in terms of their health-related behaviour (box 2). In terms of physical activity and diet, there is no evidence of larger inequalities in the Scandinavian countries, at least as measured by educational inequalities in obesity.[9] However, socioeconomic inequalities in smoking are much higher in the Nordic countries than in the other welfare state regimes.[9 14] Similarly, inequalities in deaths from cardiovascular disease are higher in the Scandinavian countries (except Denmark) as compared with other European countries.[9] This, it is argued, is because the Scandinavian countries are at a very mature stage of the smoking epidemic with the majority of smoking behaviour concentrated in the least educated groups.[24] This suggests that one consequence of the Social Democratic welfare states is that the universal health messages and health promotion interventions are taken up primarily by the middle classes.[14] This results in what has been referred to as 'intervention generated inequalities', as while the health of everyone improves, that of the middle classes does so at a faster rate.[25]

Materialist Explanations

The (neo)materialist explanation focuses on income and what income enables such as access to goods and services and the limitation of exposures to physical, and psychosocial, risk factors (box 2). Applying a materialist perspective

Table 1: Summary findings from three example comparative studies of socioeconomic inequalities in self-reported health (bad/poor vs fair/good/very good) by welfare state regime

Study	Measure of inequality		Summary of results*			
			Men		Women	
			Absolute prevalence rate difference	Relative prevalence, OR (95% CI)	Absolute prevalence rate difference	Relative prevalence, OR (95% CI)
Eikemo et al[2]	Education – average education versus 1SD below average	Bismarckian	6.4	1.19 (1.14 to 1.24)	5.7	1.25 (1.20 to 1.30)
		Anglo-Saxon	9.6	1.35 (1.23 to 1.48)	8.2	1.29 (1.18 to 1.41)
		Scandinavian	10.5	1.44 (1.35 to 1.53)	12.1	1.54 (1.44 to 1.64)
		Southern	14.8	1.57 (1.47 to 1.69)	17.3	1.69 (1.58 to 1.81)
Eikemo et al[10]	Income – top versus bottom income tertiles	Bismarckian	9.8	1.68 (1.50 to 1.89)	11.6	1.81 (1.62 to 2.03)
		Southern	10.9	1.79 (1.46 to 2.19)	14.8	2.14 (1.77 to 2.57)
		Scandinavian	13.0	1.97 (1.70 to 2.27)	15.8	2.14 (1.84 to 2.49)
		Anglo-Saxon	17.4	2.86 (2.12 to 3.70)	17.4	2.73 (2.17 to 3.44)
Espelt et al[11]	Social class (education aspects = secondary or more vs less than secondary)	Christian Democratic	11.2	1.24 (1.12 to 1.37)	12.7	1.31 (1.19 to 1.45)
		Social Democratic	13.3	1.43 (1.26 to 1.63)	13.7	1.36 (1.21 to 1.52)
		Late Democracies	18.9	1.87 (1.45 to 2.42)	24.2	1.75 (1.39 to 2.21)

*Age-standardised differences between the top and bottom socioeconomic groups in each analysis.

may initially seem somewhat limited as the Scandinavian countries have the smallest income inequalities and offer largely universal welfare services.[26] However, as Diderichsen[27] has commented, lower levels of income inequality do not negate inequalities in exposure to the other material determinants of health. Furthermore, as has consistently been shown, social inequalities in access to services remain even within universal systems, for example, the inverse care law in relation to nationalised health services.[28 29] There is certainly tentative evidence to suggest that inequalities in total avoidable mortality (as a result of diseases amenable to medical intervention) are higher in the Scandinavian countries than elsewhere.[30] From a slightly different angle, there have been longstanding criticisms that the Social Democratic welfare states operate on an insider/outsider basis with vulnerable 'outsider' groups, such as immigrants, often marginalised and without entitlement to the full benefits of the universalist system.[31]

Psychosocial

Psychosocial explanations focus on the biological and health consequences of how social inequality makes people feel (box 2). From a psychosocial perspective then, it has been speculated that 'relative deprivation' may be a factor behind the larger than expected relative health inequalities in the Scandinavian welfare states.[14] Relative deprivation will occur in all unequal societies, including the Nordic welfare states. Following Dahl and colleagues,[14] it is possible to speculate that the effects of relative deprivation may be more extensive in the Nordic welfare states because of the high levels of expectation of upward social mobility and prosperity that they generate among the less privileged expectations that are seldom met.[15 32] This may increase health inequalities especially in stress-related conditions, such as heart disease, or indeed self-assessed health.[32]

Life Course

Life course epidemiology has highlighted how different causal mechanisms and processes may lie behind the social gradient in different diseases (box 2).[16] This may also be the case in terms of the inequalities in different welfare state regimes. For example, a study found that in both Britain and Sweden, lone mothers were more likely to report poor health than couple mothers.[33] However, the pathways leading to the health disadvantage of lone mothers were very different in the two countries: poverty and worklessness were the primary issues in Britain but not in Sweden.[33] Extrapolating from this example, it is possible to suggest that the same outcomes – socioeconomic health inequalities – may be present in all welfare state regimes to a greater or lesser extent, but as a result of different causal mechanisms. This suggests that the welfare state regimes approach is perhaps too generalised and only able to offer a rough guide to inequalities.[34]

Discussion

These theoretical insights are rather limited and somewhat speculative: none of the theories alone can provide a wholly convincing explanation. While there appears to be some power to the cultural – behavioural perspective, really, beyond issues of artefact, it is very difficult to explain why health inequalities are not smaller in the Scandinavian countries through reference to existing theories of health inequalities. This is perhaps because all the other theories (selection, psychosocial, materialist, life course) to a greater or lesser extent expect health inequalities to be smaller in the Scandinavian countries. This may indicate that the existing theoretical explanations are lacking and need to be combined and developed. Certainly, no single theory is able to empirically explain within-country inequalities, never mind between country ones.

Alternatively, of course, it may be that the contrasting performances of the Scandinavian welfare states in regards to overall health versus health inequalities cannot really be considered to be a puzzle at all. First, there have only been a small number of cross-national comparative studies conducted to date and these have focused on the health gap rather than the social gradient.[9–12] Second, the use of welfare state typologies has been extensively critiqued not least on the grounds that it obscures important policy differences between welfare states (eg, the flexicurity of Denmark compared with the protectionism of Sweden or Norway).[34][35] Furthermore, some have argued that there is a need to move beyond Scandinavian welfare state exceptionalism and to acknowledge the commonalities that there are between, say, the Bismarckian and Scandinavian models, particularly in terms of the status of the lowest socioeconomic groups, as well as the progress of other welfare states, such as Japan, in terms of creating healthy environments.[36] This suggests that comparative social epidemiology should shift focus and conduct comparisons of more precise policy areas and specific social determinants (such as the work environment) instead.[37][38] This could enable a deeper and more nuanced understanding of how particular national policies, or the shared policies of specific welfare state regimes, impact on health inequalities.[34]

Another factor that needs to be taken into consideration is that the puzzle has emerged partly as a result of the focus of comparative epidemiological research on relative, as opposed to absolute, measures of health and inequality. This has meant that the Scandinavian countries are effectively victims of their own success, as while they have substantially improved the health of all, the high level of health of the middle classes has meant that relative social inequalities remain.[34] This, it could be argued, is the real issue in terms of why the Scandinavian countries perform comparatively poorly in terms of relative health inequalities, and, as Lundberg[34] has pointed out, this is an achievement, not something to be criticised. The lowest socioeconomic groups in the Scandinavian countries are objectively better off in absolute terms than the lowest socioeconomic groups in the other welfare state regimes. For example, the absolute mortality risk difference between manual and non-manual is lowest in Sweden and Norway.[39] There is also emerging evidence to suggest that among the most vulnerable social groups – the old, the sick and children – there are smaller socioeconomic inequalities in the Social Democratic welfare states.[40–42] Indeed, there is by no means an accepted research consensus that relative health inequalities among the general population are not the smallest in the Nordic countries as, for example, Borrell and colleagues'[43] analysis of data from individual country health interview surveys suggested that the Social Democratic countries did exhibit the smallest adult health inequalities. Furthermore, it has been shown that relative measures of inequalities are negatively associated with total population health: countries with lower overall mortality tend to experience larger inequalities in mortality.[44] This is perhaps because the social determinants of population health differ from the determinants of health inequalities.[45]

The use of absolute or relative measures of health inequality also raises important normative and political issues about whether the role of the welfare state is to improve the status of those at the very bottom of society or whether it is about promoting general equality. Implicitly, cross-national research to date has tended to favour the latter view; however, it is possible to suggest that it should move beyond relative comparisons and focus instead on absolute ones. This would perhaps also enhance the policy relevance of such research,[46] after all, as Rose[46] famously commented, 'relative risk is not what decision-taking requires . . . relative risk is only for researchers; decisions call for absolute measures'. Future comparative research could therefore benefit from examining the absolute health of the most marginalised, poorest and vulnerable within different types of welfare state.

The limits of the study of the formal welfare state are also perhaps exposed by the puzzle. Comparative social epidemiology has to date largely focused on analysing the influence on health and health inequalities of the formal and the public – the state, the economy, politics, public policies, welfare services and social benefits. In contrast, there has been relatively little attention paid to the potential influence on differences in cross-national health and health inequalities of the informal and the private side of welfare capitalism – unpaid care, the family, community and social support and different constructions of gender roles.[47–49] For example, some studies have suggested that those countries with a higher proportion of unpaid family care and domestic labour by women have smaller health inequalities.[49] Such social differences in the informal welfare sector could therefore be a factor behind the smaller than expected health inequalities found by some studies in the Southern and Bismarckian welfare states[49] (C Alvarez-Dardet, personal communication, 2011).

The impact of the social – the private and the informal welfare sphere – on comparative health inequalities is underexplored in public health and might provide important insights. However, as Raphael and Bryant's[50] research has noted, women's health is more sensitive to public welfare and is improved by high levels of state social welfare, so Bartley's[49] assertion that analysing the social sphere is challenging and complex is therefore well made. The intersectional nature of inequality – gender, social class and ethnic stratifications – is therefore also something that needs to be considered in future cross-national research on health.[51]

Conclusions

The existence, extent, interpretation and causes of the Scandinavian public health puzzle remain controversial. On the one hand, the puzzle highlights the limitations of existing theories of health inequalities and thereby challenges conventional public health thinking. On the other hand, it has been seen to act as a distraction away from the real potential of comparative

---

What Is Already Known on This Subject

- Population health is enhanced by the relatively generous and universal welfare provision of the Scandinavian countries.
- However, some international studies of socioeconomic inequalities in health have thrown up a public health 'puzzle' as the Scandinavian welfare states do not, as would generally be expected, have the smallest health inequalities.

What This Study Adds

- This paper outlines and interrogates this 'puzzle' by drawing upon existing theories of health inequalities – artefact, selection, cultural – behavioural, materialist, psychosocial, and life course.
- It finds that these theories provide little insight into the issue and that while this may be a result of poor theory development in public health, it may also demonstrate the limitations – both methodological and conceptual – of contemporary comparative social epidemiology.

Policy Implications

- The paper raises normative issues about whether the role of the welfare state and public health policy is about improving the status of those at the very bottom of society (absolute measures of health) or about promoting general equality (relative measures of health).
- A focus on the absolute health of the most vulnerable as well as an awareness of the social sphere and intersectionality could enhance the policy relevance of comparative health research.

social epidemiology in providing detailed assessments of the public policies of different welfare states and how the social determinants vary. However, the issue of the puzzle highlights the strong, and often unacknowledged, normative tensions within comparative social epidemiology in terms of whether the welfare state is about creating overall equality or improving the situation of the poorest and most vulnerable or both. The future of comparative social epidemiology research will be largely determined by the shifting balance of power in this debate both in terms of the empirical research agenda and the extent of theoretical evolution. The latter may well benefit from an increased interaction with social policy, social theory and political economy perspectives.[52]

References

1. Bambra C. Going beyond the three worlds of welfare capitalism: regime theory and public health research. *J Epidemiol Community Health* 2007;**61**:1098–102.
2. Ferrera M. The southern model of welfare in social Europe. *J Eur Soc Policy* 1996;**6**:17–37.
3. Esping-Andersen G. *The Three Worlds of Welfare Capitalism*. London: Polity, 1990.
4. Navarro V, Muntaner C, Borrell C, *et al*. Politics and health outcomes. *Lancet* 2006;**368**:1033–7.
5. Coburn D. Beyond the income inequality hypothesis: class, neo-liberalism, and health inequalities. *Soc Sci Med* 2004;**58**:41–56.
6. Bambra C. Health status and the worlds of welfare. *Soc Policy Society* 2006;**5**:53–62.
7. Chung H, Muntaner C. Welfare state matters: a typological multilevel analysis of wealthy countries. *Health Policy* 2007;**80**:328–39.
8. Eikemo TA, Bambra C, Judge K, *et al*. welfare state regimes and differences in selfperceived health in Europe: a multi-level analysis. *Soc Sci Med* 2008;**66**:2281–95.
9. Mackenbach J, Stirbu I, Roskam A, *et al*. Socioeconomic inequalities in health in 22 European countries. *N Engl J Med* 2008;**358**:2468–81.
10. Eikemo T, Bambra C, Joyce K, *et al*. Welfare state regimes and income related health inequalities: a comparison of 23 European countries. *Eur J Public Health* 2008;**18**:593–9.
11. Espelt A, Borrell C, Rodríguez-Sanz M, *et al*. Inequalities in health by social class dimensions in European countries of different political traditions. *Int J Epidemiol* 2008;**37**:1095–105.
12. Eikemo TA, Huisman M, Bambra C, *et al*. Health inequalities according to educational level under different welfare regimes: a comparison of 23 European countries. *Sociol Health Illn* 2008;**30**:565–82.
13. Lundberg O, Lahelma E. Nordic health inequalities in the European context. In: Kautto M, Fritzell J, Hvinden B, *et al*. eds. *Nordic Welfare States in the European Context*. London: Routledge, 2001:**42**–65.
14. Dahl E, Fritzell J, Lahelma E, *et al*. Welfare state regimes and health inequalities. In: Siegrist J, Marmot M, eds. *Social Inequalities in Health*. Oxford: Oxford University Press, 2006:193–222.
15. Huijts T, Eikemo TA. Causality, selectivity or artefacts? Why socioeconomic inequalities in health are not smallest in the Nordic countries. *Eur J Public Health* 2009;**19**:452–3.
16. Bartley M. *Health Inequality: An Introduction to Theories, Concepts and Methods*. Cambridge: Polity Press, 2004.
17. Macintyre S. The Black Report and Beyond: what are the issues? *Soc Sci Med* 1997;**44**:723–45.

18. Bambra C. *Work, Worklessness and the Political Economy of Health*. Oxford: Oxford University Press, 2011.
19. Skalická V, Lenthe F, Bambra C, *et al.* Material, psychosocial, behavioural and biomedical factors in the explanation of socio-economic inequalities in mortality: evidence from the HUNT study. *Int J Epidemiol* 2009;**38**:1272–84.
20. Bambra C. Social inequalities in health: interrogating the Nordic welfare state "puzzle". In: Kvist J, Fritzell J, Hvinden B, *et al*, eds. *Changing Equality: The Nordic Welfare Model in the 21st Century*. Bristol: Policy Press, 2011.
21. Borrell C, Espelt A, Rodríguez-Sanz M, *et al. Explaining Variations Between Political Traditions in the Magnitude of Socio-economic Inequalities in Self-perceived Health. Tackling Health Inequalities in Europe: Eurothine*. Rotterdam: Erasmus Medical Centre, 2007:213–29.
22. Dibben C, Popham F. Are socio-economic groupings the most appropriate method for judging health equity between countries? *J Epidemiol Community Health* 2010;**65**:4–5.
23. Mackenbach J, Bakker M, Kunst A, *et al.* Socio-economic inequalities in health in Europe: an overview. In: Mackenbach J, Bakker M, eds. *Reducing inequalities in health: A European perspective*. London: Routledge, 2002:3–24.
24. Cavelaars A, Kunst A, Geurts J, *et al.* Educational differences in smoking: international comparison. *BMJ* 2000;**320**:1102–7.
25. White M, Adams J, Heywood P. How and why do interventions that increase health overall widen inequalities within populations? In: Babones S, ed. *Social Inequality and Public Health*. Bristol: Policy Press, 2009.
26. Fritzell J, Ritakallio V. Societal shifts and changed patterns of poverty, International Journal of Social Welfare, 2010. doi:10.1111/j.1468-2397.2010.00728.x.
27. Diderichsen F. Impact of income maintenance policies. In: Mackenbach J, Bakker M, eds. *Reducing Inequalities in Health: A European Perspective*. London: Routledge, 2002:53–66.
28. Tudor-Hart J. The inverse care law. *Lancet* 1971;**297**:405–12.
29. Watt G. The inverse care law today. *Lancet* 2002;**360**:252–4.
30. Stirbu I. *Inequalities in Health: Does Health Care Matter?* Rotterdam: Erasmus MC, 2008.
31. Wiking E, Johansson SE, Sundquist J. Ethnicity, acculturation, and self reported health. A population based study among immigrants from Poland, Turkey, and Iran in Sweden. *J Epidemiol Community Health* 2004;**58**:574–82.
32. Yngwe M, Fritzell J, Lundberg O, *et al.* Exploring relative deprivation: is social comparison a mechanism in the relation between income and health? *Soc Sci Med* 2003;**57**:1463–73.
33. Whitehead M, Burström B, Diderichsen F. Social policies and the pathways to inequalities in health: a comparative analysis of lone mothers in Britain and Sweden. *Soc Sci Med* 2000;**50**:255–70.
34. Lundberg O. Commentary: politics and public health – some conceptual considerations concerning welfare state characteristics and public health outcomes. *Int J Epidemiol* 2008;**37**:1105–8.
35. Bredgarrd T, Larsen F, Kongshoj Madsen P. *The Flexible Danish Labour Market: A Review*. Aalborg: Aalborg Univeristy, 2005.
36. Lundberg O, Yngwe M, Bjork L, *et al. The Nordic Experience: Welfare states and Public Health (NEWS)*. Stockholm: Centre for Heath Equity Studies, 2008.
37. Dragano N, Siegrist J, Wahrendorf M. Welfare regimes, labour policies and unhealthy psychosocial working conditions: a comparative study with 9917 older employees from 12 European countries. *J Epidemiol Community Health*. 2011;**65**:793–9.
38. Lundberg O, Yngwe M, Kölegård Stjärne M, *et al.* The role of welfare state principles and generosity in social policy programmes for public health: an international comparative study. *Lancet* 2008;**372**:1633–40.

39. Fritzell J, Lundberg O. Fighting inequalities in health and income: one important road to welfare and social development. In: Kangas O, Palme J, eds. *Social Policy and Economic Development in the Nordic Countries*. Basingstoke: Palgrave Macmillan, 2005.
40. Avendano M, Jürges H, Mackenbach JP. Educational level and changes in health across Europe: longitudinal results from SHARE. *J Eur Soc Policy* 2009;**19**:301–16.
41. Dahl E, Thielens K, van der Wel K. Health inequalities and work in a comparative perspective: a multilevel analysis of EU SILC. *13th Biennial Congress of the European Society for Health and Medical Sociology*; 26–28 August 2010, 2010 Ghent, Belgium.
42. Zambon A, Boyce W, Cois E, *et al*. Do welfare regimes medicate the effect of socio-economic position on health in adolescence? A cross national comparison in Europe, North America and Israel. *Int J Health Serv* 2006;**36**:309–29.
43. Borrell C, Espelt A, Rodríguez-Sanz M, *et al*. Analysing differences in the magnitude of socio-economic inequalities in self-perceived health by countries of different political traditions in Europe. *Int J Health Serv* 2009;**39**:321–41.
44. Eikemo T, Skalicka V, Avendano M. Variations in relative health inequalities: are they a mathematical artefact? *Int J Equity Health* 2009;**8**:32.
45. Krieger N, Rehkopf D, Chen JT, *et al*. The fall and rise of US inequities in premature mortality: 1960–2002. *PLoS Med* 2008;**5**:227–41.
46. Rose G. *The Strategy of Preventive Medicine*. Oxford: Oxford University Press, 1992.
47. Stanistreet D, Bambra C, Scott-Samuel A. Is patriarchy the source of male mortality? *J Epidemiol Community Health* 2005;**59**:873–6.
48. Bambra C, Pope D, Swami V, *et al*. Gender, health inequality and welfare state regimes: a cross-national study of thirteen European countries. *J Epidemiol Community Health* 2009;**63**:38–44.
49. Bartley M. Health inequalities and societal institutions. *Soc Theory Health* 2003;**1**:108–29.
50. Raphael D, Bryant T. The welfare state as a determinant of women's health: support for women's quality of life in Canada and four comparison nations. *Health Policy* 2004;**68**:63–79.
51. Weber L. Reconstructing the landscape of health disparities research: Promoting dialogue and collaboration between feminist, intersectional and biomedical paradigms. In: Schulz AJ, Mullings L, eds. *Gender, Race, Class and Health: Intersectional Approaches*. San Francisco: Jossey Bass, 2006:21–59.
52. Bambra C. Changing the world? Reflections on the interface between social science, epidemiology and public health. *J Epidemiol Community Health* 2009;**63**:867–8.

18

Is Wealthier Always Healthier in Poor Countries? The Health Implications of Income, Inequality, Poverty, and Literacy in India

Keertichandra Rajan, Jonathan Kennedy
and Lawrence King

Introduction

A large body of research has linked higher average income levels in less developed countries (LDCs) to improved public health through materialist mechanisms (Preston, 1975; Pritchett & Summers, 1996). Other factors that affect social well-being such as inequality, especially through non-materialist pathways, are assumed to be insignificant in LDCs. The policy prescription is simple: social well-being in poor countries is best improved by increasing GDP per capita (Anand & Ravallion, 1993; Dollar & Kraay, 2002). This paper uses state-, district-, and individual-level data to test the associations between public health and average income, poverty, income inequality, and literacy in India. It demonstrates that this simple policy prescription must be qualified.

The policy debate arises between three main positions: pro-market liberalizers, the psycho-social school, and a pro-poor position. Pro-market liberalizers – who are dominant in the policy debate – argue that raising average incomes through economic liberalization is the most effective way

Source: *Social Science & Medicine*, 88 (2013): 98–107.

to improve public health. They point to seminal work by Preston (1975) and Pritchett and Summers (1996) that shows the relationship between average income and health is curvilinear and concave, and that the causal direction is from wealth to health. Their argument is based on reducing material deprivation: higher average incomes allow public investment in health infrastructure at the societal-level and sufficient expenditure on diet and medicine at the individual-level to protect health (see also Anand & Ravallion, 1993; Dollar & Kraay, 2002).

The psycho-social school, focussing on developed countries, accepts these materialist pathways and the important role of average income levels but also introduces non-materialist pathways and income inequality. For individuals with relatively low incomes, inequality generates stress that damages health directly through 'psycho-neuro-endocrine' mechanisms and indirectly through unhealthy behaviours associated with stress, like smoking and alcohol abuse. Socially, these feelings manifest as reduced civic participation and anti-social behaviour, affecting the health of others, including those higher up the income range (Lynch, Smith, Kaplan, & House, 2000:1201; Marmot, 2002; Murali & Oyebode, 2004; Wilkinson, 1996, 1997). This view is closely related to the 'social capital' paradigm, in which inequality reduces 'civic engagement' and 'levels of mutual trust' (Kawachi & Kennedy, 1999; Kawachi, Kennedy, Lochner, & Prothrow-Stith, 1997:1492). In this paradigm it is this fraying of social bonds that gives rise to both the individual and social effects that, in turn, manifest as poorer public health. These effects are often captured in objective measures of public health like infant or under-five mortality or life expectancy. But more subjective measures of well-being such as 'life satisfaction' and self-reported health have received increasing attention following work by Stiglitz, Sen, and Fitoussi (2008) advocating more holistic measures of development, including public health.

Wilkinson (1994) locates materialist and non-materialist pathways on either side of the inflection point in the Preston curve – the 'epidemiological transition' (Fig. 1). Before this transition, the leading cause of mortality is material deprivation; after it the effects of inequality predominate. Frey and Stutzer (2002) and Inglehart (2002) make analogous policy prescriptions for subjective measures like life satisfaction: poor countries must prioritise raising average incomes; only policy in rich countries can afford to be broader.

The pro-poor position extends the psycho-social school's paradigm beyond developed countries and posits that both materialist and non-materialist mechanisms operate in LDCs too. It shows that the effects of economic growth are strongly mediated by inequality and poverty. Biggs, King, Basu, and Stuckler's (2010) study of 22 Latin America countries over 47 years suggests that although average income is the key determinant of public health, its positive effects are almost absent when growth is accompanied by rising inequality and poverty. Here one effect of inequality may be political: "the greater the income gap, the greater the disparity in interests.

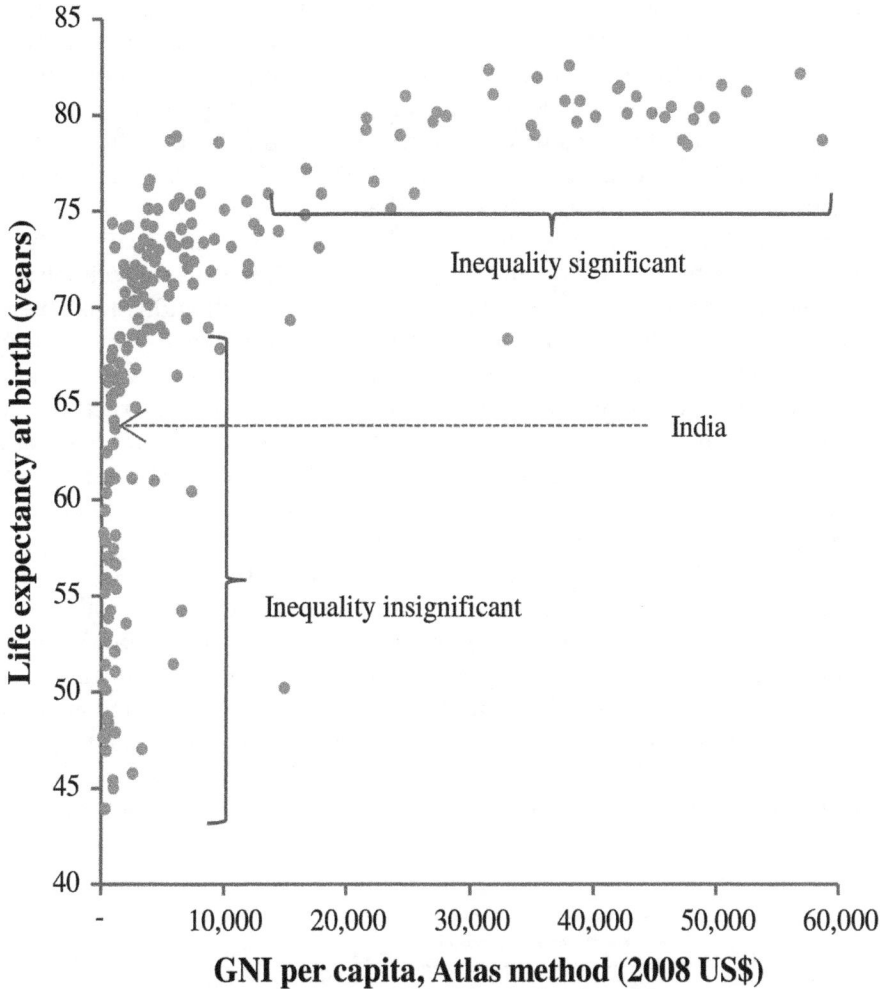

ᵃIn 2008 the average life expectancy in India was 64 years and gross national income per capita US$1080.
ᵇThe curve would be more linear if in purchasing price parity (ppp) terms. But it would still slope upwards: ppp would attenuate but not completely undermine either societal- or individual-level operators.
Source: World Development Bank Indicators.

Figure 1: The preston curve and the epidemiological shift[a,b]

This translates, because of the clout of the elite, into constant pressure for lower taxes and reduced public spending [on public health]" (Krugman, cited in Kawachi & Kennedy, 1999:221). (Bertola (1993) and Perotti (1993) have constructed models that connect income inequality to support for a tax to fund a public good such as public healthcare.) The pro-poor position echoes the 'Easterlin paradox', which juxtaposes substantial increases in per capita incomes with paltry rises or even falls in subjective measures of well-being, especially in transitional economies. Materialist variables like average income and poverty may be the chief determinants of objective measures of public health like infant mortality rates but this work suggests that even in developing countries inequality, among other factors, undermines more subjective measures, including life satisfaction and self-reported health, and thereby undercuts the gains made by increasing income levels (Brockmann, Delhey, Welzel, & Yuan, 2009; Easterlin, 2010, 2003; Easterlin, Morgan, Switek, & Wang, 2012; Knight & Gunatilaka, 2011).

In summary, these theories implicate three main income-variables: average income, poverty, and income inequality; and four causal mechanisms: investment in infrastructure; personal protection of health; individual stress; and social capital. Investment in infrastructure and personal protection of health are materialist, whereas individual stress and social capital are non-materialist. By level of operation, however, investment in infrastructure and social capital are at societal-level whereas personal protection of health and individual stress are at individual-level (Fig. 2). (In reality these mechanisms are interdependent and not easily isolated – see Pickett & Wilkinson, 2009, on 'compositional' and 'contextual' factors.)

Although this study's central aim is to compare the effects of average income levels with those of income distribution on public health, the analysis below also introduces literacy rate as an alternative predictor to income-measures. The predictive power of literacy has been well-established in both developed and developing countries and can be located in both materialist and non-materialist mechanisms (see literature surveys in DeWalt, Berkman, Sheridan, Lohr, & Pignone, 2004; Kabir, 2008:186187; WHO, 2007). Literacy mediates the investment in infrastructure pathway by

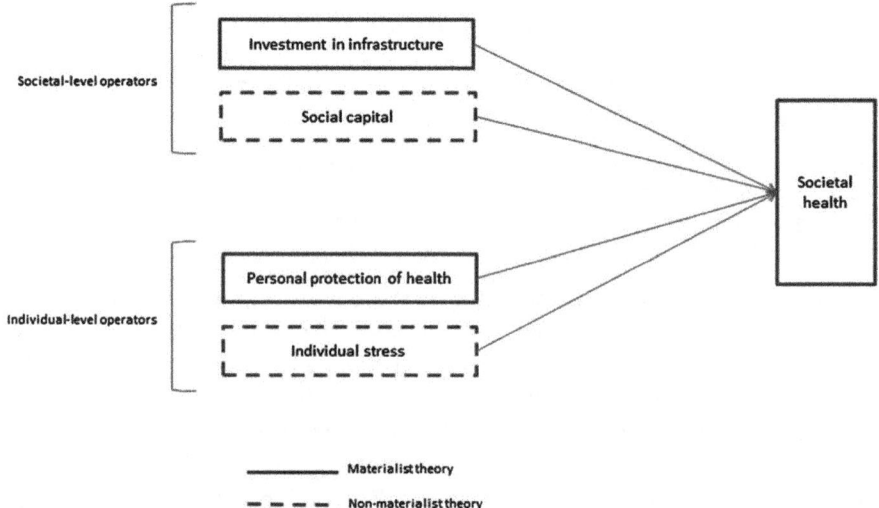

Figure 2: Pathways to public health – levels of operation and theoretical bases

enabling a population to engage with the healthcare infrastructure available and respond to public health campaigns (DeWalt et al., 2004:1232). In poor countries female illiteracy in particular is associated with child mortality (Caldwell, 1986:184–187; Sen, 1999:195–198). At the individual-level it is associated with better personal protection of health, including healthier behaviours such as not smoking and improved diets (Kabir, 2008:186). And, to the extent that it is a marker of an individual's socio-economic status, it may also be implicated in the non-materialist individual stress pathway (DeWalt et al., 2004:1237).

Case Selection

We focus on India, home to over one sixth of the world's population and one third of the world's poor, in which the effects of liberalizing reforms since the mid-1980s are hotly contested. The World Bank (undated) estimates 37% of India's population live on less than US$1.25 per day. Oxford University's Multidimensional Poverty Index (2010) gives an even higher figure of 55% – over 600 million people. Several analysts have noted that India's public health indicators have failed to keep pace with its GDP (Horton & Das, 2011; Subramanyam, Kawachi, Berkman, & Subramanian, 2010). World Bank (2013) indicators show life expectancy at birth in China is 73 compared with 65 in India and infant mortality rate per 1000 live births is 13 compared with 47 (2011 data). At the extreme, these failures have galvanized anti-state movements including the Maoist insurgents or Naxalites (Kennedy & King, 2011). Others have noted that India's health crisis tends to be explained in terms of access to healthcare (Narayan, 2011). While this is clearly a very important factor, these accounts tend to neglect socio-economic factors such as poverty and inequality, which have proven to be powerful determinants of public health in Western Europe.

Analyses of the role of inequality in public health tend to bypass the poorest countries (exceptions include Biggs et al., 2010; Ram, 2006). In a meta-analysis of literature in the area, however, Subramanian and Kawachi (2004) note more studies of poor societies are needed – and at a sub-national level. This study, then, supplements the existing literature in two main ways. First, it focuses on a poor country. In per capita terms, India is considerably poorer than China or the Eastern European transition economies that are at the heart of studies of the 'Easterlin paradox'. Second, it complements those framed at country-level (Biggs et al., 2010, on 22 Latin American countries; Ram, 2006, on a broader sample of countries) or state-level (Kawachi et al., 1997, studying 39 US states; James & Syamala, 2010, studying 15 Indian states) by testing the relationship between inequality and public health at the lower level of the Indian district, the almost 600 administrative units below Indian states. As well as providing ample statistical power, framing analysis at district-level may also render non-materialist pathways more plausible, since district-level inequality is likely to be more immediate to the individual than national- or state-level inequality. And since the Indian district is an administrative unit, some political mechanisms may also be implied in our results.

Methodology and Hypotheses

Data

Data for state-level analysis are drawn from various official sources – see Appendix A. For district- and individual-level analyses we use two sources: the 2001 Indian census and the 60th survey round of the Indian National Statistical Survey Office (NSSO), conducted in 2004. Under-five and infant mortality rates for 479 districts across 17 major states, accounting for roughly 95% of India's total population, have been calculated by Rajan, Nair, Sheela, Jagatdeb, and Mishra (2008) using the 2001 census. These are used alongside self-reported ailment by individuals and district-wise measures of average income, poverty gap, income inequality, and literacy based on data from the NSSO's 60th survey round on healthcare and morbidity conducted between January and June 2004. The round surveyed 383,346 individuals across 583 districts in all 35 states and Union Territories (a small number of remote areas were not surveyed – see NSSO, 2006:2). NSSO provide stratus weights that allow aggregate district-level and state-level population estimates from survey data. These weights, reflecting NSSO's stratified sample selection process and relating the estimated proportion an individual respondent represents in the total Indian population, are applied throughout so that, for example, a total population of 958,927,836 individuals is estimated across the same 583 districts. Aggregated estimates were checked wherever possible against NSSO's own reported state-level and national-level estimates (NSSO, 2006). Rounded weights were used to obviate the absurdity of fractions of individuals. This created a rounding error that reduced the total estimated population by just 0.01%; or a maximum of 0.06% for any one state. The two data sources are assumed to be sufficiently commensurate: under-five and infant mortality rates are not expected to have changed rapidly between 2001 and 2004.

Dependent Variables

Two dependent variables are used. Rajan et al.'s (2008) calculations using census data provide the district-wise number of deaths per 1000 live births for infants less than one year of age and for children less than five years of age. Unsurprisingly, these alternative measures are closely correlated ($r = 0.967$, see Appendix B). Under-five mortality rate is selected for presentation below it provides a smoother data series. (Given their close correlation, the two series yield very similar results – available on request – with coefficient estimates in the same directions and the same pattern of statistical significance.)

To test the importance of non-materialist pathways in multilevel logistic models, self-reporting of 'ailment' by individuals is selected as the binary dependent variable. 'Ailment' here covers all health complaints, including accidents, in the 15 days before being surveyed whether or not they are diagnosed and/or treated. Such measures have been widely used and, importantly, capture the *perception* of malady which, even if not real, is considered crucial in non-materialist theories. In a review of the literature on self-reported measures of subjective wellbeing, including health, Stiglitz, Sen, and Fitoussi commend them despite their need for further development (2008:150–51; see also Krueger & Schkade, 2008). We operationalize the measure in three ways: for all individuals, for men only, and for women only, since gender may be an important factor in self-reported health. One salient weakness of self-reported measures is their dependency on culture. We use state-level fixed effects to control for state-level variations in culture (see below).

Other Variables

All income-variables are derived from NSSO data on monthly consumption expenditure per capita over the month before survey date (used interchange-ably with 'income' henceforth). For each district, average income is the esti-mated mean of individual incomes. Poverty gap is the income-weighted proportion of the district population with incomes below a threshold of 60% of the estimated national mean, with the largest weights attaching to individ-uals with the greatest shortfall below this threshold. (For example, an indi-vidual with income Rs. 10 less than the 60% threshold is given a weight of 10, one with a shortfall of Rs. 20 a weight of 20, etc.) Poverty gap expresses the depth of material deprivation, reflecting the higher likelihood of ill health effects attaching to the lowest incomes in materialist theory. A 60% threshold is somewhat arbitrary but captures a reasonable portion of the estimated population (26.7%, compared with 5.9% with a 40% threshold, and 0.1% with a 20% threshold). Poverty gap has the added benefit of being much less strongly correlated with average income levels than headcount ratios and therefore obviates issues related to multicollinearity (Appendix B). Income inequality is the estimated Gini coefficient of all incomes in a district. (Across the 17 major states it ranged from 7.2 in Nainital district, Uttarakhand, to 46.7 in Sundargarh district, Odisha.) It is highly correlated with percentile ratios but has the advantage of capturing the entire (estimated) income spec-trum in a district.

NSSO data also provide a variable for general education in eleven ascending categories ranging from illiteracy to graduate-level and above. District-wise literacy, the rate of individuals in category 2 (literate) and above, is introduced as a control to highlight the prevalence of under-five mortality among the illiterate as well as the poor. Models that substitute literacy with average education – the mean of individual education scores across a district – are also estimated. These models are then re-estimated using specifically female/male literacy and average education.

Other control variables, such as employment rates or public hygiene and sanitation, are eschewed for two reasons, following Pritchett and Summers (1996) and Biggs et al. (2010:268–269). First, since there are multiple mechanisms linking income-variables to health, we exclude other controls to capture the whole effect. Second, to obviate multicollinearity: variables such as average income and public health are likely to be correlated with several other societal variables.

Appendix C presents descriptive statistics for the main district-level variables used.

Data Treatment

Rajan et al.'s (2008) district-wise under-five and infant mortality rates were used untreated and little cleaning of NSSO data was necessary. A small number of missing values and some outlying income outturns (0.47% of the surveyed sample of 383,346) were removed to reduce spurious variation. There were 180 households that reported zero consumption expenditure but given India's poverty these are likely to be genuine and were not removed. Since incomes were heavily skewed towards the lowest (skewness 13.8), outliers were removed only from the top of the range. High incomes were removed from districts whose ratio of standard deviation to mean was more than 2.0 (covering 97.7% of assumed normal distributions). Incomes higher than Rs. 80,000 (the 99.99th percentile) were also removed. These two criteria removed just 64 outliers. All cleaning reduced the total sample size by less than 0.49%, lowered the sample mean income by 0.45%, but reduced the standard deviation of incomes by 12.01%. Trial histograms showed all calculated district-level variables to be mildly skewed. Data were not transformed, however, because taking logs increased the skew of inequality, the variable of chief interest (although it reduces skews of other independent variables). (Log–log models were also estimated and produced similar results but are not presented – see note in Table 2.)

State-Level Fixed Effects

State-level fixed effects are used in district- and individual-level models to control for substantial cultural, physical, and public policy variations across India's states (see Zimmerman, 2008, on 'unobserved confounders' – although these arguments are most germane to longitudinal analyses). India's states

Table 1: Coefficient estimates of state-level infant mortality rates linearly regressed on state-level net state domestic product per capita, poverty rate, income inequality, and literacy rate (with standard errors and p-scores); all states and union territories[a,b]

| | Single variables | | | | Pairwise comparison with NSDP/cap | | | All income-variables/all | |
	Model 1	Model 2	Model 3	Model 4	Model 5	Model 6	Model 7	Model 8	Model 9
NSDP/cap	**-0.008****	–	–	–	**-0.009*****	**-0.006***	-0.002	**-0.007***	-0.002
	(0.002)				**(0.002)**	**(0.002)**	(0.003)	**(0.003)**	(0.003)
Inequality	–	0.584	–	–	1.085	–	–	0.814	0.944
		(0.692)			(0.673)			(0.701)	(0.595)
Poverty	–	–	**0.835****	–	–	0.509	–	0.395	0.243
			(0.260)			(0.303)		(0.317)	(0.272)
Literacy	–	–	–	**-1.197*****	–	–	**-1.116****	–	**-1.079****
				(0.218)			**(0.321)**		**(0.311)**
Adjusted R^2	0.259	0.000	0.215	0.462	0.296	0.301	0.459	0.309	0.478
N	32	35	35	35	32	32	32	32	32

[a]Standard errors in parentheses. ***p-score < 0.001, **p < 0.010, *p < 0.050. All results with p < 0.05 are considered statistically significant and are in bold.
[b]All linear terms, for regressand and regressors.

are organised mainly along linguistic-cultural lines. Diet and lifestyle vary widely (compare rice-based, low fat diets in Tamil Nadu with wheat-based, high-fat diets in Punjab; alcohol is prohibited in five states, including Gujarat), as do climate and geography (compare warm winters in Kerala with freezing ones in Uttar Pradesh; Odisha's forest cover with Rajasthan's desert). Also, the positive association between 'health-consciousness' at state-level and the perception, and therefore reporting, of ailment in India is well-established (NSSO, 2006:18–20). With high education levels and good healthcare infrastructure, Kerala, for example, stands out as highly health-conscious and reports markedly higher rates of ailment. State-level fixed effects, then, also control for differences in health consciousness, state-level provision of public goods, including access to and quality of healthcare, and state social spending.

As well as being theoretically sound, state-level fixed effects are pragmatic. A Hausman test of consistent and efficient estimators under fixed and random effects for Model 18 (Table 2) returns a large score of 165.00 ($p = 0.000$), reflecting the substantial differences in coefficient magnitudes under fixed and random effects. Using state-level fixed effects in a cross-sectional study of districts such as this is equivalent to estimating a multi-level model in which states are assigned dummy variables. The several statistically significant coefficients estimated for these state-dummies – only two were statistically insignificant ($p > 0.050$) – corroborate the Hausman test and underscore the importance of controlling for state-level effects. Random effects models, then, generate biased estimates since state-level variations powerfully influence the relationship under study. Note, however, that under random effects the direction of coefficients and their statistical significance at 95% confidence does not change. The difference in the size of coefficient estimates may be partly because fixed effects models block the (materialist) investment in infrastructure pathway at state-level. But since this is only one of several possible state-level mechanisms, and given the large difference in coefficients estimates (reflected in the large Hausman test score), we opt to use fixed effects models and thereby focus this analysis on district-level mechanisms.

Hypotheses and Models

The central policy debate addressed by this paper is expressed in the following hypotheses:

H1. Controlling for average incomes, under-five mortality rates across Indian states/districts are positively associated with income inequality levels.

H2. Controlling for average incomes, under-five mortality rates across Indian states/districts are positively associated with poverty gaps.

H3. Controlling for average incomes, under-five mortality rates across Indian states/districts are negatively associated with literacy rates.

H1, H2 and H3 are tested with linear regression models first at state-level, then at district-level with state-level fixed effects. These models do not assume any specific causal mechanism. To study the importance of inequality on measures of well-being through specifically non-materialist pathways a multi-level hypothesis is generated:

> **H4.** Controlling for district average income, individual income, and individual education, the higher the level of inequality in the district in which an individual lives, the more likely s/he is to report an ailment.

A multi-level logistic regression model is developed to test H4. State-level fixed effects again account for unspecified cross-state variances. There is no poverty term: personal protection of health depends on personal income and is theoretically independent of exposure to district-level poverty. District- rather than state-level income inequality is used since individuals can be expected to be more sensitive to inequality across their district than across their state (the mean population across the 583 districts in the cleaned NSSO sample is 1.6 million, compared with 27.3 million for states). But these are applied to the surveyed sample – not estimated population – of over 380,000 individuals across 583 districts.

Results and Interpretation

Inequality and Public Health – State-Level

Regressing state-level infant mortality rates on net state domestic product per capita across India's 35 states and Union Territories returns a negative coefficient (Model 1, Table 1). Inequality is neither significant by itself nor once income is controlled (Models 2, 5). Average income, poverty rate, and literacy by themselves are all statistically significant, and literacy has the largest effect (Models 1, 3, 4). Poverty, a positive associate, loses its significance once income is controlled (Model 6), and income loses its significance once literacy is controlled (Model 7). When all four variables are included only literacy remains statistically significant (Model 9 – estimating Model 9 with only the 17 major states reduces literacy's statistical significance, $b_1 = -1.016, p = 0.080$).

Inequality and Public Health – District-Level

Regressing district-level under-five mortality rates on average income and income inequality produces results in line with standard materialist theory (Models 10,11,14, Table 2). Average income is negatively and significantly associated with under-five mortality. Income inequality's apparent negative association, however, is due to its correlation with average income ($r = 0.279$). Once average income is controlled for, the inequality coefficient becomes statistically insignificant ($p = 0.560$). H1 cannot be accepted:

Table 2: Coefficient estimates of district-level under-five mortality rates linearly regressed on district-level average income, income inequality, poverty gap, and literacy rate (with standard errors and p-scores); state-level fixed effects[a,b,c].

	Single variables				Pairwise comparison with average income			All income-variables/all	
	Model 10	Model 11	Model 12	Model 13	Model 14	Model 15	Model 16	Model 17	Model 18
Average income	**-0.028*** (0.004)**	–	–	–	**-0.027*** (0.004)**	**-0.026*** (0.004)**	**-0.011** (0.004)**	**-0.022*** (0.005)**	-0.005 (0.005)
Income inequality	–	**-0.366*** (0.103)**	–	–	-0.062 (0.107)	–	–	-0.190 (0.120)	-0.145 (0.113)
Poverty gap	–	–	**0.368*** (0.082)**	–	–	0.151+ (0.085)	–	**0.221* (0.096)**	**0.202* (0.090)**
Literacy rate	–	–	–	**-0.572*** (0.052)**	–	–	**-0.480*** (0.062)**	–	**-0.475*** (0.062)**
Adjusted R^2	0.739	0.711	0.715	0.765	0.739	0.740	0.768	0.741	0.770
N	479	479	479	479	479	479	479	479	479

+Coefficient estimate close to statistical significance: $p = 0.077$.
[a]Mortality rates calculated by Rajan et al. (2008) using 2001 census data; all other variables calculated by authors using survey data from NSSO 60th round in 2004.
[b]Standard errors in parentheses. ***p-score < 0.001, **$p < 0.010$, *$p < 0.050$. All results with $p < 0.05$ are considered statistically significant and are in bold.
[c]All linear terms, for regressand and regressors. Log–log models returned similar results, with statistically significant coefficient estimates in the same directions and slightly higher R^2 values of ~ 0.800. The splitting of the poverty term into a log-term and a dummy term (= 1 when poverty is zero, to deal with ln[0] in 42 districts) makes the coefficients less straightforward to interpret. (The dummy poverty term effectively becomes a coefficient of wealth.)

controlling for average income, income inequality is not a predictor of public health. Poverty gap is strongly and positively associated with under-five mortality (Model 12). (Note that poverty's larger coefficient estimate is mainly due to arbitrary differences in measurement units: using daily rather than monthly income would increase the average income coefficient roughly 30-fold.) Again, however, controlling for average income renders the poverty estimate statistically insignificant (Model 15), albeit at a strict 95% confidence level ($p = 0.077$) – H2 cannot be accepted. Controlling for both average income and inequality, however, a strong and significant coefficient is again estimated for poverty (Model 17).

Once literacy rate is controlled for in Model 16, the effect of average income, with which it is strongly correlated ($r = 0.654$), becomes much weaker – H3 is accepted. Moreover, literacy is a stronger predictor of under-five mortality than even poverty (both are measured as percentages). Including all variables, average income is rendered statistically insignificant and only poverty and literacy remain as strong predictors of under-five mortality (Model 18). Substituting literacy rate alternatively with female or male literacy rates ($r = 0.868$) does not substantially affect the non-income coefficient in Model 18 (female: $b_1 = -0.431, p = 0.000$; male: $b_1 = -0.406, p = 0.000$). Replacing literacy rate with average education, female average education, or male average education, however, produces much smaller non-income coefficients ($b_1 = -0.103, p = 0.000$; female: $b_1 = -0.093, p = 0.000$; male: $b_1 = -0.084, p = 0.000$; female/male average education are again highly correlated, $r = 0.892$).

Inequality and Self-Reported Health – Individual-Level

Logistic regression of individual-level self-reported health on district-level inequality with state-level fixed effects returns large and statistically significant odds ratios greater than 1 (Model 19, Table 3). Larger odds ratios are returned for women than for men in gender-specific estimations. Introducing district average income, individual income, and individual education as controls substantially attenuates the odds ratios but they remain high and above 1 (Model 20) – H4 is accepted. Again, larger estimates are returned for women than for men.

Discussion

Pro-market liberalizers emphasise the need for further increases in Indian average incomes and can point to differences across states. Our state-level results support their position: inequality is not a significant predictor of public health at this level once average income is controlled. We do not replicate James and Syamala's (2010) finding, based on 1990s Indian state-level data, that controlling for per capita income, inequality becomes a positive

Table 3: Individual reports of ailment in 15 days before survey logistically regressed on district-level inequality: Odds ratios (with standard errors and p-scores); state-level fixed effects[a,b]

	Model 19			Model 20: Controlling for district average income, individual income and individual education		
	All	*Men*	*Women*	*All*	*Men*	*Women*
District inequality	**5.752*** (0.563)**	**5.167*** (0.704)**	**6.465*** (0.908)**	**1.834*** (0.201)**	**1.610** (0.247)**	**2.116*** (0.334)**
N[c]	381,475	194,755	186,720	381,114	194,565	186,579

[a]Standard errors in parentheses. ***p-score < 0.001, **p < 0.010, *p < 0.050. All results with p < 0.05 are considered statistically significant and are in bold.
[b]All linear terms, for regressand and regressors.
[c]Sample sizes are smaller in Model 20 due to missing education data.

predictor of child mortality. Controlling for literacy, however, we find the effect of average income becomes statistically insignificant. Nevertheless, the state is a highly aggregated analytical unit. India's largest state, Uttar Pradesh, had a population of over 166 million in 2001; its smallest 'major state', Chhattisgarh, had a population of almost 21 million. In addition, their number provides little statistical power. This study's findings at district-level may be more illuminating.

Model 17, using only income-variables, provides strong evidence for the materialist explanation of variances in public health across Indian districts. Average income and poverty – key materialist variables – have statistically significant effects on public health in the expected direction and income inequality does not. The model predicts that *ceteris paribus* it would take roughly Rs. 45 more per month in average consumption expenditure or a five percentage point narrowing of the poverty gap for one less underfive death per thousand live births. For a 'typical' district with average consumption expenditure of Rs. 557 per month and poverty gap of 18% (the respective averages across the 479 districts), this translates to an 8% rise in average incomes or a 25% reduction in poverty gap.

Even if Model 17 predicted perfectly (an adjusted R^2 value of 0.741 is high but far from perfect), reducing poverty by 25% may be as difficult as raising average incomes by 8% – which should policy focus on? Introducing another broad – and non-income – measure of development complicates the policy debate. Including literacy in Model 18 suggests the average income variable is capturing the effects of poverty and literacy. But whereas it would still take a five percentage point decrease in poverty to save one under-five per thousand, it would take only a two percentage point increase in literacy. For the typical district with a literacy rate of 55%, this translates to a 25% reduction in poverty gap or just a 4% increase in literacy.

Our district-level analysis highlights two caveats to the straightforward 'wealthier is healthier' policy prescription. First, public health is indeed undermined by material deprivation but literacy, another tractable (non-income) form of disadvantage plays an important role too. Although wealthier is healthier, wealth cannot be understood simply in terms of average income. Second, the effect of average income levels is indirect: higher average incomes may improve public health but only through reduced poverty and improved literacy. To this extent the distribution of income and other development goods does matter. Note also that although we find no statistically significant association between inequality and public health, inequality may affect public health by effectively sustaining poverty, not only through purely distributional effects but over time by reducing the growth elasticity of poverty – see Kapoor (2013) whose longitudinal study was at Indian state-, not district-level.

The small differences between female and male literacy variables' predictive powers suggest that in the early 2000s child mortality across these

479 districts was slightly more strongly associated with women's literacy and education than with men's. More germane to this study's central concern, however, is the much larger difference between literacy and average education variables, which underscores the efficacy of focussing on the most disadvantaged rather than on improving the average. Substituting literacy with average education calculated from NSSO data in Model 9 repeats the pattern at state-level: the non-income coefficient is attenuated ($b_1 = -0.246$, $p = -0.000$). Literacy may act as a floor, capturing the minimum skill – or 'functional health literacy' – required to understand medicine labelling, access healthcare, and engage with public health programmes (Nutbeam, 2000:263–265).

Although income inequality may not predict under-five or infant mortality, it may still affect social well-being via non-materialist pathways. Individual-level logistic models estimate exposure to district-level income inequality is associated with a much greater likelihood of an individual reporting an ailment in the 15 days before being surveyed. Controlling only for cross-state variations in unspecified variables, a unit-increase in district inequality – or a 4% increase in inequality for the 'typical district' with Gini coefficient 0.23 (the average across the 583 districts) – is associated with an odds ratio of 5.8 for an individual reporting ailment. With district average income and individual income controlled to capture the materialist pathways of level of district development (investment in infrastructure) and individual ability to protect health, and individual education controlled to capture individual health consciousness as well as the protective effects of education, these odds are substantially attenuated. But they remain high: the same 4% increase in inequality is still associated with an 83% increase in the odds of an individual reporting ailment. Since the model's controls include the chief variables of materialist theory, this is tentative evidence that increased inequality has a negative effect on individual well-being via non-materialist pathways. Re-estimating these models by gender returns higher odds ratios for women than for men. Women in high inequality districts are almost twice as likely as men to report an ailment. This corroborates similar findings elsewhere of poorer self-reported health among women (Case & Paxson, 2005).

Conclusions

Standard policy prescriptions for improving public health in less developed countries focus on raising average income levels since it is widely accepted that 'wealthier is healthier'. Only after the 'epidemiological transition' is inequality hypothesized to become a significant predictor of health. In the case of India in 2004 wealthier is indeed healthier. But our analysis suggests it is low poverty and high literacy rather than wealth *per se* that improves public health. Infant mortality rates are negatively associated with average income levels and positively associated with poverty at both state- and district-level. Inequality, however, is not associated with public health at state- or district-level, where linear regression models controlling for average income and variations in unspecified state level variables show income inequality is not a statistically significant predictor of infant or under-five mortality rates. But controlling for poverty gap and literacy rate renders the average income coefficient statistically insignificant too. This implies that expanding economic output improves public health not by raising average income but by reducing poverty and increasing literacy – undermining the dominant pro-market liberalization position and supporting the pro-poor position. Moreover, of the two predictors, literacy has a markedly stronger effect than poverty. These models are not designed to isolate any particular causal mechanism but their estimates are in line with materialist theory, operating at both societal- and individual-level (investment in health infrastructure and personal protection of healthcare).

The analysis, however, also finds evidence for the negative affect of inequality operating through non-materialist pathways, even in this LDC before the 'epidemiological transition.' Multi-level logistic models that control for district average income and individual income, the chief materialist variables, and for individual education show both men and women are more likely to self-report an ailment if exposed to higher district-level inequality. This evidence is, however, only tentative: other pathways, such as elite capture of public health resources in high inequality districts, cannot be dismissed.

These findings have important policy implications. First, although wealthier is indeed healthier, policymakers should focus on alleviating poverty rather than simply raising average incomes to improve public health. Second, addressing other (non-income) development issues such as illiteracy may be more effective than raising incomes. Policy must also be more subtle. While inequality cannot predict infant or under-five mortality rates, it is strongly connected to self-reported health, even in a major LDC. Economic policies narrowly focused on growth, therefore, may be insufficient. They must be coupled with a broader understanding of societal well-being and the factors that promote it.

Limitations

This study complements existing work by focussing on a less developed country at district-level, and using a multi-level model, as recommended by Subramanian and Kawachi (2004). It is, however, clearly limited and much further work is needed. An obvious lacuna is an understanding of how the associations analysed here have changed over time, especially important in a fast-changing LDC like India. Time-series data could also reveal how long average income and poverty take to manifest in health outcomes ('incubation' periods may be longer via societal-level pathways than via individual-level pathways); and whether these effects are mediated by changes in, rather than levels of, inequality (as reported for Latin America by Biggs et al., 2010). Second, focussing on 17 major states neglects several interesting cases, including Delhi, the rich and unequal capital, and the sparsely populated and less developed districts of India's north east. Third, income-variables derived from consumption expenditure data are likely to underestimate the true extent of inequality (as well as average income and poverty). Fourth, the causal connections underlying these results must be crystallised by connecting inequality directly to the intermediary outcomes implicated by theory, for example, levels of investment in public healthcare for a materialist, societal-level theory or individuals' levels of stress-related hormones for a non-materialist, individual-level theory. Ethnographic studies, especially those built on social capital theory, may more fully address the changes in health-affecting and/or health-reporting behaviours that inequality brings about, how these changes vary across close-knit and loosely associated groups, how they vary across groups of different sizes (in 2001 the populations of districts in India's major states ranged from 21 thousand to 9.4 million), and what they are contingent on (including public health programmes). Such studies could in turn develop the underlying theory, locating points of interdependence between individual- and societal-level, and materialist and non-materialist mechanisms.

The expanding literature on the sociology of health, nonetheless, should help turn policymakers' attention away from simple metrics of success like GDP growth and towards a more qualified understanding of social priorities.

Appendix A

Infant mortality rate, net state domestic product per capita, and poverty pate across Indian states and union territories

	State	Population[a]	% of India pop	IMR[b]	NSDP/capita[c]	Poverty rate[d]	Inequality[e]	Literacy rate[f]
	India	1,028,737	100	58	24,143	28	32.3	64.8
1	**Uttar Pradesh**	**166,198**	**16**	**72**	**12,840**	**33**	**28.1**	**56.3**
2	**Maharashtra**	**96,879**	**9**	**36**	**35,915**	**31**	**34.8**	**76.9**
3	**Bihar**	**82,999**	**8**	**36**	**7759**	**41**	**22.0**	**47.0**
4	**West Bengal**	**80,176**	**8**	**40**	**22,654**	**25**	**32.4**	**68.6**
5	**Andhra Pradesh**	**76,210**	**7**	**59**	**25,321**	**16**	**32.9**	**60.5**
6	**Tamil Nadu**	**62,406**	**6**	**41**	**30,105**	**23**	**33.1**	**73.5**
7	**Madhya Pradesh**	**60,348**	**6**	**79**	**15,442**	**38**	**27.4**	**63.7**
8	**Rajasthan**	**56,507**	**5**	**67**	**18,565**	**22**	**26.8**	**60.4**
9	**Karnataka**	**52,851**	**5**	**49**	**26,745**	**25**	**30.8**	**66.6**
10	**Gujarat**	**50,671**	**5**	**53**	**32,021**	**17**	**30.1**	**69.1**
11	**Odisha**	**36,805**	**4**	**77**	**17,380**	**46**	**30.7**	**63.1**
12	**Kerala**	**31,841**	**3**	**12**	**31,871**	**15**	**30.1**	**90.9**
13	**Jharkhand**	**26,946**	**3**	**49**	**18,512**	**40**	**27.4**	**53.6**
14	**Assam**	**26,656**	**3**	**66**	**16,782**	**20**	**23.8**	**63.3**
15	**Punjab**	**24,359**	**2**	**45**	**32,948**	**8**	**27.2**	**69.7**
16	**Haryana**	**21,145**	**2**	**61**	**37,842**	**14**	**25.3**	**67.9**
17	**Chhattisgarh**	**20,834**	**2**	**60**	**18,559**	**41**	**27.5**	**64.7**
18	Delhi	13,851	1	32	61,560	15	28.8	81.7
19	Jammu & Kashmir	10,144	1	49	21,314	5	23.9	55.5
20	Uttarakhand	8489	1	42	24,740	40	29.8	71.6
21	Himachal Pradesh	6078	1	51	32,564	10	27.4	76.5
22	Tripura	3199	0	32	24,394	19	32.6	73.2
23	Meghalaya	2319	0	54	23,793	19	21.6	62.6
24	Manipur	2294	0	14	18,527	17	16.0	70.5
25	Nagaland	1990	0	17	20,234	19	19.1	66.6
26	Goa	1348	0	17	76,426	14	27.6	82.0
27	Arunachal Pradesh	1098	0	38	27,271	18	32.0	54.3
28	Pondicherry	974	0	24	48,573	22	34.7	81.2
29	Chandigarh	901	0	21	74,442	7	28.3	81.9
30	Mizoram	889	0	19	24,662	13	23.0	88.8
31	Sikkim	541	0	32	26,693	20	24.8	68.8

(Continued)

Appendix A: (Continued)

	State	Population[a]	% of India pop	IMR[b]	NSDP/capita[c]	Poverty rate[d]	Inequality[e]	Literacy rate[f]
32	Andaman & Nicobar Islands	356	0	19	40,921	23	24.2	81.3
33	Dadra & Nagar Haveli	220	0	48		33	27.4	57.6
34	Daman & Diu	158	0	37		11	19.2	78.2
35	Lakshadweep	61	0	30		16	20.5	86.7
	ALL STATES – Correlation with IMR[g]				**-0.532****	**0.488****	0.145	**-0.692*****
	N				**32**	**35**	35	35
	17 MAJOR STATES – Correlation with IMR[g]				**-0.611****	0.442	-0.435	**-0.727*****
	N				**17**	**17**	**17**	17

Major states are in bold.
[a]Population figures from 2001 census.
[b]Infant mortality rates from Sample Registration System, Registrar General (India). 2004 data.
[c]Net state domestic product per capita from Central Statistical Organisation (India). As at 2004, current prices.
[d]Poverty rate from Databook For Deputy Chairman, Planning Commission of India. 2004 data.
[e]Income inequality calculated by authors as Gini co-efficient of monthly consumption expenditure per capita data from 60th round of National Sample Survey Office survey. 2004 data.
[f]Literacy rate from Office of the Registrar General. 2001 data.
[g]***p-score < 0.001, **p < 0.010, *p < 0.050.

Marmot, M. (2002). The influence of income on health: views of an epidemiologist. *Health Affairs, 21*(2), 31–46.

Murali, V., & Oyebode, F. (2004). Poverty, social inequality and mental health. *Advances in Psychiatric Treatment, 10*, 216–224.

Narayan, R. (2011). Universal health care in India: missing core determinants. *The Lancet, 377*(9769), 883–885.

National Sample Survey Organisation. (2006). *Morbidity, health care and the condition of the aged: NSS 60th round (January–June 2004). Ministry of Statistics and Programme Implementation: Government of India*. Report 507 (60/25.0/1).

Nutbeam, D. (2000). Health literacy as a public health goal: a challenge for contemporary health education and communications strategies into the 21st century. *Health Promotion International, 15*, 259–267.

Oxford Poverty and Human Development Initiative. (2010). *Multidimensional poverty index*. Oxford: Oxford University/UNDP Press.

Perotti, R. (1993). Political equilibrium, income distribution, and growth. *Review of Economic Studies, 60*(4), 7557–7576.

Pickett, K. E., & Wilkinson, R. G. (2009). Greater equality and better health. *British Medical Journal, 339*, 1154–1155.

Preston, S. H. (1975). The changing relation between mortality and level of economic development. *Population Studies, 29*(2), 231–248.

Pritchett, L., & Summers, L. H. (1996). Wealthier is healthier. *The Journal of Human Resources, 31*(4), 841–868.

Rajan, S. I., Nair, P. M., Sheela, K. L., Jagatdeb, L., & Mishra, N. R. (2008). *Infant and child mortality in India: District level estimates*. New Delhi: Population Foundation of India.

Ram, R. (2006). Further examination of the cross-country association between income inequality and population health. *Social Science & Medicine, 62*(2006), 779–791.

Sen, A. (1999). *Development as freedom*. Oxford: Oxford University Press.

Stiglitz, J.,E., Sen, A., & Fitoussi, J.-P. (2008). *Report by the commission on the measurement of economic performance and social progress*. Paris: Commission on the Measurement of Economic Performance and Social Progress.

Subramanian, S. V., & Kawachi, I. (2004). Income inequality and health: what have we learned so far? *Epidemiological Reviews, 26*(2004), 78–91.

Subramanyam, M. A., Kawachi, I., Berkman, L. F., & Subramanian, S. V. (2010). Is economic growth associated with reduction in child undernutrition in India? *PLoS Medicine, 8*(3), e1000424 http://dx.doi.org/10.1371/journal.pmed.1000424.

WHO Commission on the Social Determinants of Health. (2007). *Achieving health equity: From root causes to fair outcomes*. Geneva: World Health Organisation.

Wilkinson, R. G. (1994). The epidemiological transition: from material scarcity to social disadvantage? *Daedalus, 123*(4), 61–77.

Wilkinson, R. G. (1996). *Unhealthy societies: The afflictions of inequality*. New York: Routledge.

Wilkinson, R. G. (1997). Socioeconomic determinants of health: health inequalities: relative or absolute material standards? *British Medical Journal, 314*, 591–595.

World Bank. (undated). *India: Country results profile*. At: http://web.worldbank.org/ WBSITE/EXTERNAL/NEWS/0,contentMDK:22888405~menuPK:41310~pagePK:343 70~piPK:34424~theSitePK:4607,00.html Accessed online on 12.04.13.

World Bank. (2013). *World development indicators*. At: http://data.worldbank.org/data-catalog/ world-development-indicators. Accessed online on 12.04.13.

Zimmerman, F. J. (2008). A commentary on "Neo-materialist theory and the temporal relationship between income inequality and longevity change". *Social Science & Medicine, 66*(2008), 1882–1894.